AF581675

Partial Differential Equations in Ecology

Partial Differential Equations in Ecology: 80 Years and Counting

Editor

Sergei Petrovski

MDPI • Basel • Beijing • Wuhan • Barcelona • Belgrade • Manchester • Tokyo • Cluj • Tianjin

Editor
Sergei Petrovski
University of Leicester
UK

Editorial Office
MDPI
St. Alban-Anlage 66
4052 Basel, Switzerland

This is a reprint of articles from the Special Issue published online in the open access journal *Mathematics* (ISSN 2227-7390) (available at: https://www.mdpi.com/journal/mathematics/special_issues/pdee).

For citation purposes, cite each article independently as indicated on the article page online and as indicated below:

LastName, A.A.; LastName, B.B.; LastName, C.C. Article Title. *Journal Name* **Year**, *Volume Number*, Page Range.

ISBN 978-3-0365-0296-0 (Hbk)
ISBN 978-3-0365-0297-7 (PDF)

Contents

About the Editor

Sergei Petrovski is an applied mathematician with thirty years of research experience in mathematical ecology and ecological modelling. His research spans across a broad variety of problems in ecology and population dynamics, with a particular emphasis on modelling complex multiscale ecological, agro-ecological and socio-ecological systems. Some of his older results on ecological pattern formation and biological invasion modelling have become textbook material. His recent research on the effect of global warming on atmospheric oxygen, where he discovered a new type of ecological catastrophe, was highlighted by the media around the world. He published four books and more than 130 papers in peer-reviewed journals. He currently holds the position of Chair in Applied Mathematics at the University of Leicester (UK). He is the Editor-in-Chief of *Ecological Complexity* (Elsevier) and the Section Editor-in-Chief for the "Theoretical and Mathematical Ecology" section of *Mathematics* (MDPI). He is also the founder and the scientific coordinator of the Models in Population Dynamics and Ecology (MPDE) conference series.

Preface to "Partial Differential Equations in Ecology: 80 Years and Counting"

Application of partial differential equations (PDEs) in ecology has a long history dating back to 1937. It was at this time that Ronald Fisher and Andrey Kolmogorov et al., through their research on the spread of an advantageous gene, discovered the travelling wave solution of a scalar diffusion-reaction equation. Fifteen years later, Alan Turing's work on chemical morphogenesis demonstrated that, due to diffusive instability, a system of two coupled PDEs gives rise to pattern formation: an interesting result that was later shown to have a variety of ecological applications. These seminal papers led to an outbreak of research on all aspects of the population dynamics in space and time using PDEs of the diffusion-reaction type. Nowadays, on appropriate spatial and temporal scales, PDEs remain a fully relevant and powerful modelling framework; they are widely used both to bring new light to old problems and to gain insight into new ones. This volume, originally published as a Special Issue of *Mathematics*, presents a small collection of specially selected papers and aims to highlight the current role of PDE-based models in ecology and population dynamics. A variety of models is used, including traditional reaction-diffusion equations, cross-diffusion, the Cahn–Hilliard equation, among others, and a broad range of problems is addressed.

Sergei Petrovski
Editor

Article

Individual Variability in Dispersal and Invasion Speed

Aled Morris [1,2], Luca Börger [1,3] and Elaine Crooks [1,2,*]

1 Centre for Biomathematics, College of Science, Swansea University, Swansea SA2 8PP, UK
2 Department of Mathematics, College of Science, Swansea University, Swansea SA2 8PP, UK
3 Department of Biosciences, College of Science, Swansea University, Swansea SA2 8PP, UK
* Correspondence: e.c.m.crooks@swansea.ac.uk

Received: 31 July 2019; Accepted: 22 August 2019; Published: 1 September 2019

Abstract: We model the growth, dispersal and mutation of two phenotypes of a species using reaction–diffusion equations, focusing on the biologically realistic case of small mutation rates. Having verified that the addition of a small linear mutation term to a Lotka–Volterra system limits it to only two steady states in the case of weak competition, an unstable extinction state and a stable coexistence state, we exploit the fact that the spreading speed of the system is known to be linearly determinate to show that the spreading speed is a nonincreasing function of the mutation rate, so that greater mixing between phenotypes leads to slower propagation. We also find the ratio at which the phenotypes occur at the leading edge in the limit of vanishing mutation.

Keywords: invasive species; linear determinacy; population growth; mutation; spreading speeds; travelling waves

---

1. Introduction

The speed at which a species expands its range is a fundamental parameter in ecology, evolution and conservation biology. Knowledge of this speed enables us to predict the ability of a species to keep up with the rate at which the climate changes or the rate at which an exotic species invades, representing two prominent ecological challenges [1,2]. It is known that traits such as dispersal and population growth affect the rate at which a species expands its range, and there has been a suggestion in recent work that polymorphism in traits could cause a species invasion to occur at a faster rate than a single morph would in isolation [3,4]. Understanding the effect that each trait of a species has, and could potentially have, on its rate of spread is therefore important to understanding how the spread of a species can evolve.

Most common models of invasions in population dynamics incorporate aspects of dispersal and growth, e.g., works by the authors of [5–8], however the mutation of one phenotype to another has been a less common inclusion. Even the addition of a simple mutation term can dramatically affect the behaviour of a model. A review of Cosner [9] singles out two models that involve mutation and multiple dispersal strategies in a population of a species: the model introduced in Elliott and Cornell [3] to investigate dispersal polymorphism for two morphs, in which a simple linear mutation is used, and the model of Bouin et al. [10], motivated by the destructive invasion of cane toads across northern Australia, in which mutations are considered to act as a diffusion process in the phenotype space.

Elliott and Cornell assume that the spread rate of the two phenotypes in their system, usually referred to as the spreading speed, is determined by the linearisation of their system at the extinction state zero. This assumption can be rigorously proved to hold under reasonable conditions on the parameters; see the framework of Girardin [11], which applies in particular to this model, and also, for an alternative approach, Morris [12], where the assumption is proved using earlier results of Wang [13] in the case where the mutation rate is small, which is generally the case for all organisms since natural selection typically acts to minimize mutation rate [14]. Moreover, in addition to being linearly determined, the spreading speed equals the minimal speed of a class of travelling waves [11,12], mimicking well-known results on travelling waves and spreading speeds for the Fisher–KPP equation [15–17].

Knowing that the rate of spread is linearly determinate and linked to travelling wave speeds provides a powerful tool that we exploit here to deduce ecologically-important information about the invasion of trait-structured species using the model introduced in [3]. In particular, we establish results on the dependence of spreading speeds on the mutation rate, and on the composition of the leading edge of minimal speed travelling waves in the limit of vanishing mutation. Some of our results focus, as in [3], on the case when different morphs have varying dispersal abilities or strategies, and in addition, there is a trade-off between dispersal and growth. Such trade-offs are exhibited by many species, including certain plants, insects and terrestrial arthropods; see, for instance, the review of Bonte et al. [18] on the costs of dispersal.

Elliott and Cornell's model examines the interaction between an establisher phenotype with population density n_e, and a disperser phenotype with population density n_d, using a Lotka–Volterra competition system:

$$\begin{aligned}\frac{\partial n_e}{\partial t} &= D_e\frac{\partial^2 n_e}{\partial x^2} + r_e n_e(1 - m_{ee}n_e - m_{ed}n_d) - \mu e n_e + \mu d n_d \\ \frac{\partial n_d}{\partial t} &= D_d\frac{\partial^2 n_d}{\partial x^2} + r_d n_d(1 - m_{de}n_e - m_{dd}n_d) - \mu d n_d + \mu e n_e.\end{aligned} \tag{1}$$

The first term on the right hand side of each equation represents the dispersal of the phenotype through diffusion, where D_e and D_d are the dispersal rates of each morph. The second term describes the growth rate of the phenotype using a logistic term, this is similar to what is used in Fisher's model [16]. We use r_e and r_d to represent the growth rate of each morph, m_{ee} and m_{dd} represent the intramorph competition, while m_{ed} and m_{de} represent the intermorph competition. The third and fourth terms represent a linear mutation between the phenotypes at mutation rates of μe and μd, where μ, e and d are constants. Note that we slightly modify the model in the work by the authors of [3] here by replacing the parameters μ_e, μ_d in the work by the authors of [3] with μe and μd to enable dependence on mutation to be investigated by variation of the single parameter μ. It is assumed that all parameters of the system are positive real numbers.

As in the work by the authors of [3], we suppose a basic trade-off between dispersal and growth, namely, that the establisher phenotype has the larger growth rate, while the disperser phenotype has the larger dispersal rate,

$$r_e > r_d, \quad D_d > D_e. \tag{2}$$

While trade-off (2) is not needed either in the proof of linear determinacy or in some of our results on the dependence of spreading speed on mutation rate, we will make use of (2) to discuss parameter-dependent options for the vanishing-mutation limit of spreading speeds in Section 3, and in Section 4, where we characterize the composition of the leading edge of solutions of (1). Further discussion on interesting possible implications of dispersal–growth trade-offs for this model is presented in [12]. Following classical competition theory [19], we suppose throughout that the intramorph competition is greater than the intermorph competition,

$$m_{dd} > m_{ed}, \quad m_{ee} > m_{de}. \tag{3}$$

We also have in mind throughout that the mutation rate μ is relatively small in comparison to the other parameters, to remain biologically realistic [14].

Kolmogorov, Petrovskii and Piskunov [15] studied the existence of monotonic travelling wave solutions of the scalar form of the equation

$$\frac{\partial u}{\partial t} = A\frac{\partial^2 u}{\partial x^2} + f(u). \tag{4}$$

Throughout this work, we will consider travelling wave solutions to be solutions of the Equation (4) of the form $u(x,t) = w(x-ct)$, where $w : \mathbb{R}^n \to \mathbb{R}^n$ is called the wave profile and $c \in \mathbb{R}$ is the speed of the wave. Kolmogorov, Petrovskii and Piskunov studied the case when $n = 1$, $A = d$ and $f(u) = ru(1-u)$, proposed by Fisher [16], and proved there is a continuum of values of c for which a monotonic travelling wave solution exists, specifically if $c \geq c^*$, where $c^* = 2\sqrt{rd}$ is the minimal travelling wave speed, as well as establishing stability properties of the minimal-speed front. Aronson and Weinberger [17] further studied this system and characterised c^* as a spreading speed. These results were extended to cooperative systems of equations for a suitable class of nonlinearities f by Volpert, Volpert and Volpert [20].

The system (1) is of the form (4) if we let $u = (n_e, n_d)^T \in \mathbb{R}^2$, A be a diagonal matrix containing the dispersal rates,

$$A = \begin{pmatrix} D_e & 0 \\ 0 & D_d \end{pmatrix}, \tag{5}$$

and f be a nonlinear function containing the growth, competition and mutation terms,

$$f(n_e, n_d) = \begin{pmatrix} r_e n_e(1 - m_{ee}n_e - m_{ed}n_d) - \mu e n_e + \mu d n_d \\ r_d n_d(1 - m_{de}n_e - m_{dd}n_d) + \mu e n_e - \mu d n_d \end{pmatrix}. \tag{6}$$

In the following, we will use the notation $u > v$ to denote that the ith component of each vector satisfies $u_i > v_i$, for each i, similarly for $u < v$, $u \geq v$ and $u \leq v$. We say that $u \in \mathbb{R}^n$ is positive if $u > 0$. The notation $u \in (a,b]$ denotes that the ith component of each vector satisfies the inequality $a_i < u_i \leq b_i$ for each i.

Elliott and Cornell investigated numerically the effect of varying the parameters on the spreading speed of the system and interestingly found that, for certain values of growth and dispersal rate, the system would spread faster in the presence of both phenotypes than just one phenotype would spread in the absence of mutation [3]. They predict the spreading speed obtained for each set of parameters in the limit of small mutation, using the front propagation method of van Saarloos [21], making the assumption that the spreading speed of system (1) is linearly determinate in order to do so. As $\mu \to 0$ the three possible limiting speeds are

$$v_e = 2\sqrt{r_e D_e}, \quad v_d = 2\sqrt{r_d D_d}, \quad v_f = \frac{|r_e D_d - r_d D_e|}{\sqrt{(r_e - r_d)(D_d - D_e)}}. \tag{7}$$

Condition (2) is enough to ensure that v_f exists and is faster than v_e and v_d, which are the spreading speeds of the two Fisher–KPP equations that would be satisfied by each phenotype in isolation. The faster speed v_f is predicted for parameters in the region of the positive quadrant of $(r_d/r_e, D_e/D_d)$-space, which satisfies the inequalities

$$\frac{D_d}{D_e} + \frac{r_d}{r_e} > 2, \quad \frac{D_e}{D_d} + \frac{r_e}{r_d} > 2, \tag{8}$$

represented by the shaded area in Figure 1.

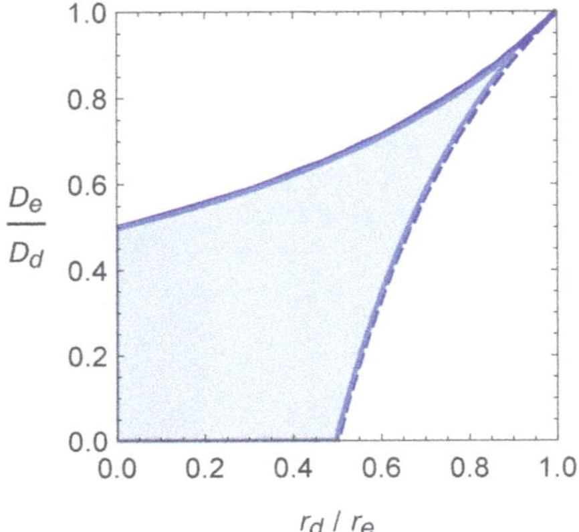

Figure 1. Parameter regions showing when the faster invasion speed observed by the authors of [3] occurs. In the upper left unshaded region the solution travels at the speed at which the establisher travels without competition. In the lower right unshaded region the solution travels at the speed at which the disperser travels without competition. In the shaded region the faster spreading speed is observed.

The spreading speed of the system is said to be linearly determinate if it is the same as the spreading speed of the system obtained when (1) is linearised about the $(0, 0)$ equilibrium, namely,

$$\begin{aligned}\frac{\partial n_e}{\partial t} &= D_e \frac{\partial^2 n_e}{\partial x^2} + (r_e - \mu e) n_e + \mu d n_d \\ \frac{\partial n_d}{\partial t} &= D_d \frac{\partial^2 n_d}{\partial x^2} + (r_d - \mu d) n_d + \mu e n_e.\end{aligned} \tag{9}$$

This assumption is one that is suggested by the numerical studies in [3], but is not always true even in the scalar case [21–24]. Stokes [25] calls the minimal wave speed c^* "pulled" if it is equal to the linearised spreading speed, that is, the speed of the front is determined by the individuals at the leading edge. Similarly the minimal wave speed is said to be "pushed" if its speed is greater than the linearised spreading speed, in this case the speed is determined by individuals behind the leading edge. Typically there are qualitative differences in wave behaviour depending on whether the wave is pushed or pulled, e.g., stability in the scalar case is discussed in the work by the authors of [26].

In the case of systems, most sufficient conditions for linear determinacy require a cooperative assumption on the system. A system is cooperative when the off-diagonal elements of the Jacobian matrix of f are always non-negative, i.e.,

$$\frac{\partial f_i(u)}{\partial u_j} \geq 0, \quad \text{if } i \neq j. \tag{10}$$

In biological terms this would mean that each phenotype benefits from the presence of others. A cooperative system is useful mainly due to the existence of a comparison principle for such systems [27,28], which is useful in particular in the proof of linear determinacy.

Theorem 1 (Comparison Principle [13], Theorem 3.1). *Let A be a positive definite diagonal matrix. Assume that f is a vector-valued function in $\mathbb{R}^n$ that is continuous and piecewise continuously differentiable in $\mathbb{R}$,* *and that the underlying system (4) is cooperative. Suppose that $u(x,t)$ and $v(x,t)$ are bounded on $\mathbb{R} \times [0,\infty)$ and satisfy*

$$\frac{\partial u}{\partial t} - A\frac{\partial^2 u}{\partial x^2} - f(u) \le \frac{\partial v}{\partial t} - A\frac{\partial^2 v}{\partial x^2} - f(v)$$

If $u(x,t_0) \le v(x,t_0)$ for $x \in \mathbb{R}$, then

$$u(x,t) \le v(x,t), \quad \text{for } x \in \mathbb{R},\ t \ge t_0.$$

Linear determinacy was shown to hold for some cooperative systems by Lui [27], and the result was extended to more general cooperative systems by Weinberger, Lewis and Li [28], who assumed, in particular, that for any positive eigenvector q of $f'(0)$,

$$f(\alpha q) \le \alpha f'(0)q, \quad \text{for } \alpha > 0. \tag{11}$$

This reduces to a condition imposed by Hadeler and Rothe [23] in the scalar case.

Unfortunately, we see from the Jacobian matrix (12) that our system (1) is only partially cooperative,

$$J_f(n_e,n_d) = \begin{pmatrix} r_e(1-2m_{ee}n_e - m_{ed}n_d) - \mu e & \mu d - r_e m_{ed} n_e \\ \mu e - r_d m_{de} n_d & r_d(1 - m_{de}n_e - 2m_{dd}n_d) - \mu d \end{pmatrix}. \tag{12}$$

In fact it is typically only cooperative at small population densities due to the relative smallness of the mutation term (see Figure 2).

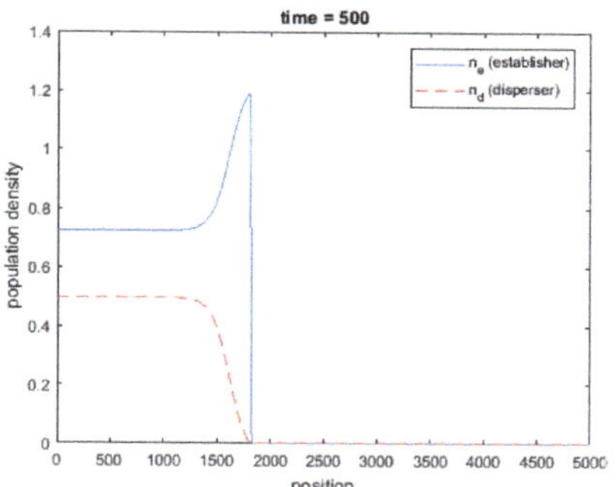

Figure 2. Solution of the system (1) with parameter values: $D_e = 0.3$, $D_d = 1.5$, $r_e = 1.1$, $r_d = 0.2$, $m_{ee} = 1.0/1.2$, $m_{dd} = 1.0$, $m_{ed} = 0.8$, $m_{de} = 0.7$, $\mu e = 0.001$, $\mu d = 0.00025$. A Heaviside step function was used as initial condition for each component.

However, Girardin [11] has recently established a linear determinacy result for a class of non-cooperative reaction–diffusion systems that includes the case considered here; see Theorem 1.7, together with Theorems 1.5 and 1.6, of Girardin [11], which apply here because when $(n_e,n_d) = (0,0)$, the Jacobian (12) always has at least one positive eigenvalue, which can be seen from arguments similar to those used in the proof of Proposition 3.1 below. An alternative proof of linear determinacy for the particular system (1) is given by Morris [12], Theorems 4.5 and 2.15, using the linear determinacy framework outlined by Wang [13]. This latter result is proved under the assumption of sufficiently small mutation and intermorph competition, and uses the fact that (1) is cooperative at low population densities to trap the nonlinearity f between two cooperative nonlinearities, f^- and f^+.

Exploiting this linear determinacy, we answer ecologically important questions pertaining to our system in the case of small mutation rate. We first take advantage of a Perron–Frobenius structure to investigate the effect of mutation on spreading speed, and show that an increase in mutation between morphs results in a decrease in the value of the spreading speed; see Theorem 3. This slowing of the speed of propagation as the mutation rate increases is mathematically related, in fact, to the so-called 'reduction phenomenon', that greater mixing lowers growth, discussed in Altenberg [29]. Secondly we investigate the composition of the leading edge of invasion in the limit of small mutation rate, and demonstrate the effects of dispersal, growth rate and mutation on this composition. As a by-product, we also characterise the vanishing-mutation limit of the spreading speed for three different regimes of diffusion and growth parameters in Theorem 2, which yields, in particular, a rigorous explanation of the parameter-dependent selection criteria for the three possible limiting speeds (7) that were discussed in the work by the authors of [3].

We draw the reader's attention to two further interesting references that tackle questions for systems related to (1). Griette and Raoul [30], motivated by an epidemiological model, studied the existence and properties of travelling waves for a special case of system (1). They assume, in particular, that $D_e = D_d$, of which advantage can be taken to prove an explicit formula for the spreading speed and to characterise the shape of travelling wave solutions, including proving non-monotone behaviour in one phenotype and asymptotic behaviour at $\pm\infty$. Clearly this assumption of equal dispersal rates, though realistic for the modelling of wild type and mutant types of a virus in the work by the authors of [30] and extremely useful mathematically, is not reasonable for the dispersal polymorphism that is our focus here. Cantrell, Cosner and Yu [31] study (1) from the perspective not of propagation phenomena but of dynamics on a bounded domain. They provide a detailed study of equilibria, the phase plane and dynamics for a range of different parameters, including various regimes for the competition parameters $m_{ee}, m_{ed}, m_{ed}, m_{de}$.

The rest of the paper is organised as follows. Section 2 presents preliminary material on equilibria of the system (1) and their relationship to equilibria for the related competition–diffusion system when $\mu = 0$. The effect of mutation on spreading speeds is discussed in Section 3, using the characterisation of the spreading speed as the linearly-determined minimal speed of a family of travelling waves, which can be expressed and analysed in terms of Perron–Frobenius matrix theory. Section 4 focusses on the parameter regime, in which the dispersal and growth of both morphs play a role in the vanishing-mutation limit of the spreading speed and derives an expression for the ratio of the morphs in the leading edge of the invasion in this case. Some conclusions and remarks are given in Section 5.

2. Equilibria of the System

We begin with a brief discussion of the equilibria of the system (1) under our assumption (3) on the competition parameters; see also the work by the authors of [31] for further investigation of equilibria of (1). A much studied competition–diffusion system, similar to (1) but where there is no mutation between phenotypes and both intramorph competition values equal one, is the Lotka–Volterra system of equations [32–35],

$$\begin{aligned} \frac{\partial n_e}{\partial t} &= D_e \frac{\partial^2 n_e}{\partial x^2} + r_e n_e (1 - n_e - m_{ed} n_d), \\ \frac{\partial n_d}{\partial t} &= D_d \frac{\partial^2 n_d}{\partial x^2} + r_d n_d (1 - n_d - m_{de} n_e). \end{aligned} \tag{13}$$

Lewis, Li and Weinberger [5] note that a coexistence equilibrium for this system (13) exists if, and only if, $(1 - m_{ed})(1 - m_{de}) > 0$; that is, either when both $m_{ed} < 1$ and $m_{de} < 1$, or both $m_{ed} > 1$ and $m_{de} > 1$. Note that the case where $m_{ed} < 1$ and $m_{de} < 1$ corresponds to the condition (3) in our system (1). A stability analysis shows that this coexistence equilibrium is stable when $m_{ed} < 1$ and $m_{de} < 1$, and unstable when $m_{ed} > 1$ and $m_{de} > 1$, where here stability is understood in the sense of stability of the ODE system given by (13) with $D_e = D_d = 0$. It should also be noted that the works by the authors of [5,34] also study the case where one species invades the territory of another, while we, and also the authors of [3,30], are concerned with two morphs of a species invading a previously unoccupied territory.

For our system (1) we can easily see that there only exist two constant equilibria by plotting the nullclines of (6),

$$r_e n_e(1 - m_{ee}n_e - m_{ed}n_d) - \mu e n_e + \mu d n_d = 0, \tag{14}$$
$$r_d n_d(1 - m_{de}n_e - m_{dd}n_d) - \mu d n_d + \mu e n_e = 0, \tag{15}$$

and observe where they intersect. The nullclines confirm for a specific choice of parameters that we only have two non-negative equilibria and that they are the extinction equilibrium and a single coexistence equilibrium (Figure 3a). The nullclines appear as they do in Figure 3a if the parameters satisfy the conditions

$$\frac{\mu d}{r_e m_{ed}} < \frac{r_e - \mu e}{r_e m_{ee}}, \qquad \frac{\mu e}{r_d m_{de}} < \frac{r_d - \mu d}{r_d m_{dd}}, \tag{16}$$

which aligns with our assumption that the mutation is relatively small. We note that it is clearly also possible to deal with cases in which the mutation does not satisfy assumptions (16), however we are only interested in the case of small mutation here.

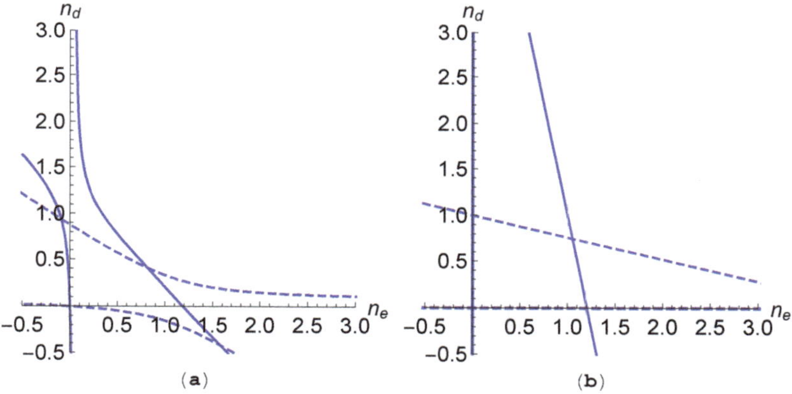

Figure 3. (**a**) Nullclines of Equation (6) and (**b**) Nullclines of Equation (17). Parameter values $D_e = 0.3$, $D_d = 1.5$, $r_e = 1.1$, $r_d = 0.2$, $m_{ee} = 1.0/1.2$, $m_{dd} = 1.0$, $m_{ed} = 0.8$, $m_{de} = 0.7$, $\mu e = 0.01$, $\mu d = 0.025$. Each point at which the nullclines intersect represents an equilibrium. We can see that for this choice of parameters Equation (6) has only two non-negative equilibria, while (17) has four.

However, simply plotting the nullclines of our system does not tell us the stability of each equilibrium, we therefore consider a modified version of (6) without the mutation terms, which we call g,

$$g(n_e, n_d) = \begin{pmatrix} r_e n_e(1 - m_{ee}n_e - m_{ed}n_d) \\ r_d n_d(1 - m_{de}n_e - m_{dd}n_d) \end{pmatrix}. \tag{17}$$

Due to the relative smallness of the mutation terms, we can then introduce them as a perturbation before using the implicit function theorem.

First we evaluate the equilibria of g. We can easily see that there are four equilibria by plotting the nullclines,

$$r_e n_e(1 - m_{ee}n_e - m_{ed}n_d) = 0, \tag{18}$$
$$r_d n_d(1 - m_{de}n_e - m_{dd}n_d) = 0, \tag{19}$$

which we do in Figure 3b for certain parameters. The equilibria of (17) consist of an extinction equilibrium $(0,0)$, two equilibria on the axes where one phenotype is present while the other is extinct, $(1/m_{ee},0)$, $(0,1/m_{dd})$, and a coexistence equilibrium

$$\left(\frac{m_{dd}-m_{ed}}{m_{ee}m_{dd}-m_{ed}m_{de}}, \frac{m_{ee}-m_{de}}{m_{ee}m_{dd}-m_{ed}m_{de}}\right)$$

which we refer to as (n_e^*, n_d^*) for simplicity. Note that (n_e^*, n_d^*) is a coexistence equilibria due to the condition (3) specified earlier.

The Jacobian of (17) is

$$J_g(n_e,n_d) = \begin{pmatrix} r_e(1-2m_{ee}n_e-m_{ed}n_d) & -r_e m_{ed} n_e \\ -r_d m_{de} n_d & r_d(1-m_{de}n_e-2m_{dd}n_d) \end{pmatrix}. \tag{20}$$

Substituting in values of n_e and n_d at each of the equilibria to the trace and determinant of (20) we see that the equilibrium (n_e^*, n_d^*) is stable, while the other three are unstable. Note also that the determinant of (20) is non-zero when evaluated at each of the equilibria of (17).

We now use the implicit function theorem [36] to determine how each equilibrium moves when mutation is introduced to the system (17) as a perturbation. To do so, we suppose that there exists $\mu > 0$ such that

$$f(n_e,n_d) = g(n_e,n_d) + \mu M \begin{pmatrix} n_e \\ n_d \end{pmatrix} \tag{21}$$

where g is defined in (17) above, μ is a non-negative scalar parameter which we use to vary the mutation and M is the matrix of mutation coefficients

$$M = \begin{pmatrix} -e & d \\ e & -d \end{pmatrix}. \tag{22}$$

The equilibria for our original system (1) satisfy $f(n_e,n_d)=0$, where f is the nonlinearity (6), so that

$$g(n_e,n_d) + \mu M \begin{pmatrix} n_e \\ n_d \end{pmatrix} = 0. \tag{23}$$

As a consequence of the implicit function theorem, in a neighbourhood of $\mu = 0$ and an equilibrium $(\bar{n}_e, \bar{n}_d)$ of g, there is a unique solution of (23) which is a continuously differentiable function of μ, say $h_{(\bar{n}_e,\bar{n}_d)}(\mu)$, provided g is invertible at $(\bar{n}_e, \bar{n}_d)$. This ensures we can differentiate (23) in order to obtain an expression describing how an equilibrium $(\bar{n}_e, \bar{n}_d)^T$ is perturbed upon the introduction of mutation μ. Since the determinant of the Jacobian matrix J_g is not equal to zero at any of the equilibria, we may invert J_g and obtain the expression

$$\Theta(\bar{n}_e,\bar{n}_d) := \left.\frac{\mathrm{d}}{\mathrm{d}\mu} h_{(\bar{n}_e,\bar{n}_d)}(\mu)\right|_{\mu=0} = -J_g(\bar{n}_e,\bar{n}_d)^{-1} M \begin{pmatrix} \bar{n}_e \\ \bar{n}_d \end{pmatrix}. \tag{24}$$

Clearly the extinction equilibrium $(0,0)$ remains at $(0,0)$, and the implicit function theorem ensures the local uniqueness of this equilibrium for small $\mu > 0$. Evaluating (24) at each of the other equilibria of g, we see that the equilibrium $(1/m_{ee},0)$ is perturbed into the lower right quadrant, because

$$\Theta\left(\frac{1}{m_{ee}},0\right) = \frac{\mu e}{r_e r_d(m_{ee}-m_{de})}\left(\frac{r_e m_{ed}}{m_{ee}} - \frac{r_d}{m_{ee}}(m_{ee}-m_{de}), -r_e\right)^T. \tag{25}$$

Note that the term $(m_{ee} - m_{de})$ is positive due to the condition (3). Similarly, the equilibrium $(0, 1/m_{dd})$ is perturbed into the upper left quadrant, since

$$\Theta\left(0, \frac{1}{m_{dd}}\right) = \frac{\mu d}{r_e r_d (m_{dd} - m_{ed})}\left(-r_d, \frac{r_d m_{de}}{m_{dd}} - \frac{r_e}{m_{dd}}(m_{dd} - m_{ed})\right)^T. \tag{26}$$

Finally, the coexistence equilibrium is perturbed a small amount in a direction which is dependant on the parameters of the system,

$$\Theta\left(n_e^*, n_d^*\right) = \begin{pmatrix} r_e n_e^* n_d^* \left[\mu e(m_{dd} - m_{ed}) - \mu d(m_{ee} - m_{de})\right] \\ r_d n_e^* n_d^* \left[\mu d(m_{ee} - m_{dd}) - \mu e(m_{dd} - m_{ed})\right] \end{pmatrix}. \tag{27}$$

Moreover, since the Jacobian is a continuous function of μ, we know that for small $\mu \neq 0$, the stability of each of the equilibria remains the same as when $\mu = 0$. Therefore, by introducing a small amount of mutation to our system, we are left with two non-negative equilibria: an unstable extinction state $(0,0)$ and a stable coexistence state (n_e^*, n_d^*).

3. The Role of Mutation in Spreading Speeds

In this and the following section, we derive predictions about the spreading of species modelled by (1) by exploiting the linear determinacy of the system together with the fact that the spreading speed can be characterised using travelling waves.

We being by deriving a μ-dependent expression for the minimal travelling wave speed of the linearisation of (1) about the origin. If the general reaction–diffusion system (4) admits a travelling wave solution $u(x,t) = w(x - ct)$, then by substituting $w(x - ct)$ in to the general form (4) we may write the system in the form of a travelling wave equation:

$$Aw''(\xi) + cw'(\xi) + g(w(\xi)) + \mu M w(\xi) = 0. \tag{28}$$

We now have an ordinary differential equation in the single variable $\xi = x - ct$, the linearisation of which about the origin is

$$Aw''(\xi) + cw'(\xi) + g'(0)w(\xi) + \mu M w(\xi) = 0. \tag{29}$$

Further, if we substitute the ansatz solution $w(\xi) = e^{-\beta\xi}q$ into (29), we obtain the eigenvalue problem

$$\left(\beta A + \beta^{-1}(g'(0) + \mu M)\right) q = cq, \tag{30}$$

where $\beta > 0$ is the spatial decay and $q > 0$ denotes the phenotypic distribution at the leading edge. We define the matrix on the left hand side to be

$$H_{\beta,\mu} := \beta A + \beta^{-1}(g'(0) + \mu M) = \beta A + \beta^{-1} f'(0). \tag{31}$$

For $\mu > 0$ the matrix $H_{\beta,\mu}$ has strictly positive off-diagonal elements and therefore by the Perron–Frobenius theorem has a Perron–Frobenius eigenvalue, which is plotted in Figure 4 as a function of β. This Perron–Frobenius eigenvalue, which we denote $\eta_{\mathrm{PF}}\left(H_{\beta,\mu}\right)$, is the larger of the two real eigenvalues of $H_{\beta,\mu}$ and has a one-dimensional eigenspace spanned by a positive eigenvector. Further, this Perron–Frobenius eigenvalue is positive for every $\beta, \mu > 0$.

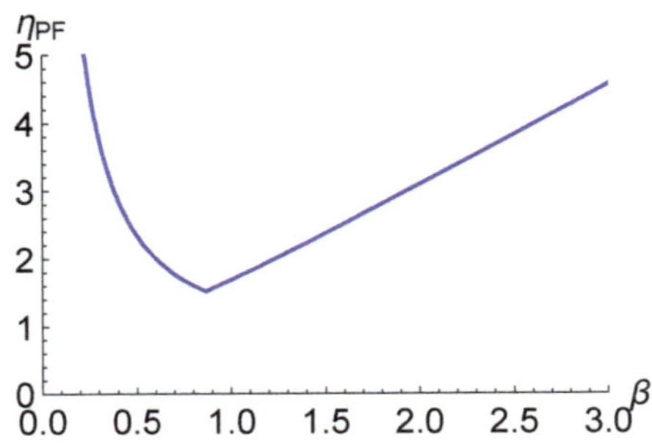

Figure 4. Perron–Frobenius eigenvalue of H_β, which exists when $\mu > 0$. Parameters used are $D_e = 0.3$, $D_d = 1.5$, $r_e = 1.1$, $r_d = 0.2$, $e = 0.001$, $d = 0.00025$, $\mu = 1$.

Proposition 1. *The Perron–Frobenius eigenvalue of the matrix $\beta A + \beta^{-1} f'(0)$ is strictly positive for all $\mu > 0$, $\beta > 0$*

Proof. Multiplying the matrix $\beta A + \beta^{-1} f'(0)$ on the right by $(d, e)^T$ we obtain

$$(\beta A + \beta^{-1} f'(0)) \begin{pmatrix} d \\ e \end{pmatrix} = \begin{pmatrix} \beta D_e d + \beta^{-1} r_e d \\ \beta D_d e + \beta^{-1} r_d e \end{pmatrix} > 0. \tag{32}$$

By Corollary 1.6 of Crooks [37], this implies

$$\eta_{\mathrm{PF}}(\beta A + \beta^{-1} f'(0)) > 0. \tag{33}$$

□

Definition 1. *The minimal speed of the travelling wave solution $w(\xi) = e^{-\beta\xi} q$ of (28) for a given $\mu > 0$ is given by*

$$\eta(\mu) := \inf_{\beta > 0} \eta_{\mathrm{PF}} \left(H_{\beta,\mu} \right). \tag{34}$$

We define $\beta(\mu)$ to be the value of β at which $\eta(\mu)$ is attained.

Note that it follows from Gershgorin's Circle Theorem that $\eta_{\mathrm{PF}}(H_{\beta,\mu}) \to \infty$ as $\beta \to 0$ and $\beta \to \infty$, which, together with Lemma 1.1 (5)–(6) of Wang [13] and the fact that $\beta \mapsto \eta_{PF}(\beta^2 A + f'(0))$ is a strictly convex function of β by Lemma 3.7 of Crooks [37], implies that there exists a unique value $\beta = \beta(\mu)$ at which $\inf_{\beta>0} \eta_{\mathrm{PF}}(H_{\beta,\mu})$ is attained.

In the case $\mu = 0$, the matrix $H_{\beta,0}$ is diagonal, and therefore does not have a Perron–Frobenius eigenvalue. Instead the matrix $H_{\beta,0}$ has the two eigenvalues

$$\beta D_d + \beta^{-1} r_d, \quad \text{and} \quad \beta D_e + \beta^{-1} r_e, \tag{35}$$

which we plot in Figure 5a as functions of β (for certain parameters). The minimum of the first curve in (35) is $2\sqrt{r_d D_d}$, which we note is the Fisher speed of the disperser v_d, and occurs at the β-value

$$\beta_d = \sqrt{r_d / D_d}. \tag{36}$$

Similarly, the minimum of the second curve in (35) is the Fisher speed of the disperser $v_e = 2\sqrt{r_e D_e}$, and occurs at the β-value

$$\beta_e = \sqrt{r_e / D_e}. \tag{37}$$

The third point of interest is the point at which the the two curves in (35) cross, which is v_f, and occurs at the β-value

$$\beta_f = \frac{\sqrt{r_e - r_d}}{\sqrt{D_d - D_e}}. \tag{38}$$

By analogy with the Perron–Frobenius eigenvalue we consider the maximum of these two eigenvalues for each β, which we plot in Figure 5b. Note here the similarity to the plot of the Perron–Frobenius eigenvalue of $H_{\beta,\mu}$, seen in Figure 4.

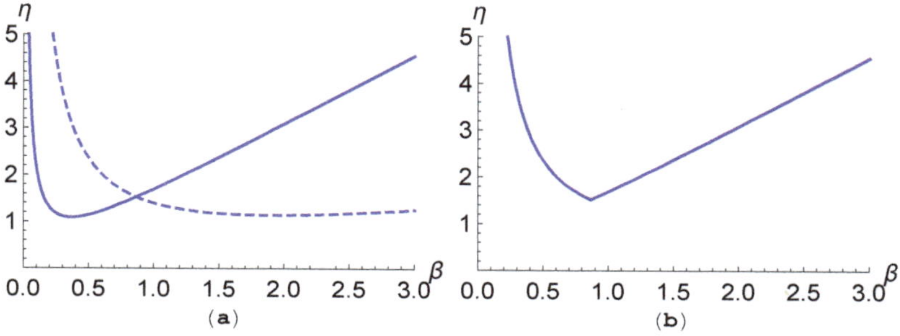

Figure 5. (**a**) Eigenvalues of H_β in the case of $\mu = 0$ when $D_e = 0.3$, $D_d = 1.5$, $r_e = 1.1$, $r_d = 0.2$. (**b**) η corresponding to each β for $\mu = 0$.

Definition 2. *We define the minimum over β of the maximum function of eigenvalues defined in (35) to be η_0. Likewise, we denote the value of β at which this minimum is obtained by β^*.*

Note that, in Figure 5, the minimum over β of the maximum of the eigenvalues (35) is the point at which they both meet, and therefore in this case $\eta_0 = v_f$ and $\beta^* = \beta_f$.

However there are also regions of parameters in which the eigenvalues (35) meet in such a way that this is not the case. The condition that the minima of the two eigenvalues lie on either side of the point at which they meet, which is equivalent to the minimum value of β of the maximum of the eigenvalues being at the crossing point, as in Figure 5, is

$$\sqrt{\frac{r_d}{D_d}} < \sqrt{\frac{r_e - r_d}{D_d - D_e}} < \sqrt{\frac{r_e}{D_e}}, \tag{39}$$

which implies the condition

$$\frac{r_e}{r_d} + \frac{D_e}{D_d} > 2 \quad \text{and} \quad \frac{r_d}{r_e} + \frac{D_d}{D_e} > 2, \tag{40}$$

imposed by Elliott and Cornell [3] ensuring that the faster spreading speed v_f (7) is obtained in the limit as $\mu \to 0$. Further, when condition (40) is satisfied, $H_{\beta^*,0}$ is a multiple of the identity and has a repeated eigenvalue with a two-dimensional eigenspace. We consider this case, where v_f is obtained as the limiting speed as $\mu \to 0$, to be of most biological interest, as unlike v_e and v_d, v_f is then dependent on the traits of both morphs and occurs as a result of polymorphism.

In the case that the condition (40) is not satisfied we would expect one of the individual speeds be selected. For example if the dispersal and growth rates of each phenotype satisfy

$$\frac{r_e}{r_d} + \frac{D_e}{D_d} < 2 \quad \text{and} \quad \frac{r_d}{r_e} + \frac{D_d}{D_e} > 2, \tag{41}$$

then the individual speed of phenotype d is selected. In this case the minimum over β of the maximum of the eigenvalues (35) is no longer the crossing point of the eigenvalues, and is instead the minimum of the eigenvalue curve $\beta D_d + \beta^{-1} r_d$, therefore $\eta_0 = v_d$ and $\beta^* = \beta_d$. Similarly, in the case

$$\frac{r_e}{r_d} + \frac{D_e}{D_d} > 2 \quad \text{and} \quad \frac{r_d}{r_e} + \frac{D_d}{D_e} < 2, \tag{42}$$

where the individual speed of phenotype e is selected, we see that $\eta_0 = v_e$ and $\beta^* = \beta_e$. It is not possible for both inequalities in (40) to be reversed while the condition (2) holds.

Proposition 2. *If (40) holds, then $\eta_0 = v_f$ and $\beta^* = \beta_f$. If (41) holds, then $\eta_0 = v_d$ and $\beta^* = \beta_d$. If (42) holds, then $\eta_0 = v_e$ and $\beta^* = \beta_e$.*

The values η_0 and β^*, defined in Definition 2 using the eigenvalues of the matrices $H_{\beta,0}$, are shown in Morris [12] Theorems 5.6–5.11 to be the limits of $\eta(\mu)$ and $\beta(\mu)$ as the mutation rate $\mu \to 0$. In light of the linear determinancy of the spreading speed established in the work by the authors of [11,12], this yields a rigorous characterisation of the limit of the spreading speed as $\mu \to 0$, which shows, in particular, that the limiting speeds (7) predicted by Elliott and Cornell [3] using the front propagation method of van Saarloos [21] are indeed what is obtained in the limit of small mutation. Here we summarise the results and refer to Theorems 5.6–5.11 of Morris [12] for details of the proofs.

Theorem 2. *In all three of the parameter regions (40)–(42),*

$$\lim_{\mu \to 0} \eta(\mu) = \eta_0, \quad \lim_{\mu \to 0} \beta(\mu) = \beta^*,$$

where η_0, β^* are as defined in Definition 2, that is,

(i) if (40) holds, then

$$\lim_{\mu \to 0} \eta(\mu) = \frac{|r_e D_d - r_d D_e|}{\sqrt{(r_e - r_d)(D_d - D_e)}}, \qquad \lim_{\mu \to 0} \beta(\mu) = \frac{\sqrt{r_e - r_d}}{\sqrt{D_d - D_e}},$$

(ii) if (41) holds, then

$$\lim_{\mu \to 0} \eta(\mu) = 2\sqrt{r_d D_d}, \qquad \lim_{\mu \to 0} \beta(\mu) = \sqrt{r_d / D_d},$$

(iii) if (42) holds, then

$$\lim_{\mu \to 0} \eta(\mu) = 2\sqrt{r_e D_e}, \qquad \lim_{\mu \to 0} \beta(\mu) = \sqrt{r_e / D_e}.$$

Note that Tang and Fife [38] established the existence of travelling fronts for (1) for all speeds greater than or equal to $\max\{2\sqrt{r_e D_e}, 2\sqrt{r_d D_d}\}$. Therefore, interestingly, in the case when (2) and (40) are satisfied, the limit as $\mu \to 0$ of the minimal front speed $\eta(\mu)$ is strictly larger than the minimal front speed for (1) in the absence of mutation.

Since calculating the spreading speed $\eta(\mu)$ involves minimizing in β and the diffusion matrix A is not a multiple of the identity, finding an explicit expression for $\eta(\mu)$ is not very tractable. However, by adapting an argument from the work by the author of [37], we show that $\eta(\mu)$ is a nonincreasing function of the mutation rate μ. A key ingredient of the proof is the following classical result of Cohen [39] on convexity properties of Perron–Frobenius eigenvalues.

Lemma 1 (Cohen [39]). *Let $P, Q \in \mathbb{R}^{n\times n}$ such that P is diagonal and Q has positive off-diagonal elements. Then the Perron–Frobenius eigenvalue of $P+Q$ is a convex function of P; that is, given diagonal matrices P_1 and P_2 and $0 < \alpha < 1$,*

$$\eta_{\mathrm{PF}}(\alpha P_1 + (1-\alpha)P_2 + Q) \leq \alpha\eta_{\mathrm{PF}}(P_1+Q) + (1-\alpha)\eta_{\mathrm{PF}}(P_2+Q). \tag{43}$$

This convexity, together with the fact that $\eta_{PF}(M) = 0$, can now be exploited to obtain the following.

Theorem 3. *$\eta(\mu)$ is a nonincreasing function of μ.*

Proof. Take $\mu_0 > 0$, define Z to be the 2×2 zero matrix, and define

$$P := \beta(\mu_0)^2 A - \beta(\mu_0)\eta(\mu_0)I + g'(0) \tag{44}$$

Note that $\beta(\mu_0) > 0$ and

$$\eta_{\mathrm{PF}}(\beta^2 A - \beta\eta(\mu_0)I + g'(0) + \mu_0 M) \geq 0, \quad \forall\, \beta > 0, \tag{45}$$

$$\eta_{\mathrm{PF}}(\beta(\mu_0)^2 - \beta(\mu_0)\eta(\mu_0)I + g'(0) + \mu_0 M) = 0. \tag{46}$$

Now for any $\mu > 0$, we know that

$$\eta_{\mathrm{PF}}\left(P + \mu M\right) = \mu\,\eta_{\mathrm{PF}}\left(\frac{1}{\mu}P + M\right), \tag{47}$$

and in particular, $\eta_{\mathrm{PF}}(P+\mu M)$ and $\eta_{\mathrm{PF}}\left(\frac{1}{\mu}P + M\right)$ have the same sign. Moreover,

$$\eta_{\mathrm{PF}}\left(\frac{1}{\mu_0}P + M\right) = 0. \tag{48}$$

For any $\mu > \mu_0$, we can write,

$$\frac{1}{\mu}P = \frac{\mu_0}{\mu}\left(\frac{1}{\mu_0}P\right) + \left(1 - \frac{\mu_0}{\mu}\right)Z. \tag{49}$$

Now in Lemma 1, let P/μ_0 and Z be P_1 and P_2, respectively, and let $\alpha := \mu_0/\mu$. Then by the convex dependence of η_{PF},

$$\begin{aligned} \eta_{\mathrm{PF}}\left(\frac{1}{\mu}P + M\right) &\leq \frac{\mu_0}{\mu}\eta_{\mathrm{PF}}\left(\frac{1}{\mu_0}P + M\right) + \left(1 - \frac{\mu_0}{\mu}\right)\eta_{\mathrm{PF}}\left(Z + M\right) \\ &= \frac{\mu_0}{u}\eta_{\mathrm{PF}}\left(\frac{1}{\mu_0}P + M\right) + \left(1 - \frac{\mu_0}{\mu}\right)\eta_{\mathrm{PF}}\left(M\right). \end{aligned} \tag{50}$$

We know from (48) that the first term is equal to zero, and $\eta_{\mathrm{PF}}(M) = 0$ implies the second term is also zero. The inequality (50) therefore becomes

$$\eta_{\mathrm{PF}}\left(\frac{1}{\mu}P + M\right) \leq 0, \qquad \mu > \mu_0. \tag{51}$$

Using (47), this implies

$$\eta_{\mathrm{PF}}\left(P + \mu M\right) \leq 0, \qquad \mu > \mu_0, \tag{52}$$

and substituting in the full expression for P and dividing by $\beta(\mu_0)$ then yields

$$\eta_{\mathrm{PF}}(\beta(\mu_0)A + \beta(\mu_0)^{-1}(g'(0) + \mu M)) \leq \eta(\mu_0), \quad \mu > \mu_0. \tag{53}$$

Hence

$$\inf_{\beta>0} \eta_{\mathrm{PF}}(\beta A + \beta^{-1}(g'(0) + \mu M)) \leq \eta(\mu_0). \tag{54}$$

Since the expression on the left hand side of (54) is the definition of $\eta(\mu)$, we have

$$\eta(\mu) \leq \eta(\mu_0). \tag{55}$$

Since this holds for any $\mu > \mu_0$, we have therefore shown that $\eta(\mu)$ is a nonincreasing function of μ. □

In light of the linear determinancy of the spreading speed [11,12], Theorem 3 establishes that the spreading speed for the nonlinear system (1) is a nonincreasing function of the mutation rate μ. Interestingly, this is related mathematically to the so-called "reduction phenomenon" discussed by Altenberg [29], which roughly says that, under certain conditions, greater mixing results in lowered growth. On the other hand, our result can be interpreted as showing that greater mixing results in a slower speed of propagation. It is possible, in fact, to make use of Theorem 6(iii) [29] to give a slightly different proof of Proposition 3.

4. Behaviour at the Leading Edge in the Limit $\mu \to 0$

Here we study the Perron–Frobenius eigenvector $q_\beta(\mu)$ corresponding to the eigenvalue $\eta(\mu)$ in the limit as the mutation rate $\mu \to 0$, with the aim of determining the ratio of the phenotypes in the leading edge of an invasion. We do this under conditions (2) and (8) on the dispersal and growth parameters, which, by Theorem (2)(i), ensures that the faster speed v_f is obtained as $\mu \to 0$ and, as we will see, results in both phenotypes being present in the leading edge. Throughout this section we therefore assume that

$$\frac{r_e}{r_d} + \frac{D_e}{D_d} > 2 \quad \text{and} \quad \frac{r_d}{r_e} + \frac{D_d}{D_e} > 2, \tag{56}$$

as well as

$$r_e > r_d, \; D_d > D_e, \tag{57}$$

and for convenience in the following, define the quantities

$$a := {\beta^*}^2 D_d - r_d, \quad b := -{\beta^*}^2 D_e + r_e. \tag{58}$$

Note that a straightforward calculation shows that the condition

$$a > 0 \quad \text{and} \quad b > 0 \tag{59}$$

is equivalent to the condition (56). We consider this parameter regime to be of greatest biological interest, since unlike v_e and v_d, the value v_f is dependent on the traits of both morphs and occurs as a result of polymorphism. Note that it is shown by (Girardin [40], Theorem 1.1) that the eigenvector $q_\beta(\mu)$, which arises in the explicit solution of the linearised problem (29), does indeed also characterise the asymptotic behaviour of travelling wave solutions of the nonlinear system (1) close to the extinction state.

In the absence of mutation ($\mu = 0$), Figure 5 illustrates the β-dependence of eigenvalues of the matrix $H_{\beta,0}$ for a choice of dispersal and growth parameters for which conditions (56) and (57) are satisfied. Under these conditions, the minimal travelling wave speed η_0 is obtained at the point at which the two eigenvalues of $H_{\beta,0}$ meet, so that

$$\eta_0 = \beta^* D_d + {\beta^*}^{-1} r_d = \beta^* D_e + {\beta^*}^{-1} r_e, \qquad \beta^* = \sqrt{\frac{r_e - r_d}{D_d - D_e}}. \tag{60}$$

In this case, $H_{\beta^*,0}$ is a multiple of the identity, so has a repeated eigenvalue with a two-dimensional eigenspace, making it not obvious a priori what ratio between the two components one would expect in the limit of vanishing mutation.

We therefore investigate the Perron–Frobenius eigenvector q of $H_{\beta(\mu),\mu}$ in the limit $\mu \to 0$, restricting attention to the region of parameters in which (56) and (57) hold. Recall from Theorem 2 that $\lim_{\mu\to 0} \eta(\mu) = \eta_0$ and $\lim_{\mu\to 0} \beta(\mu) = \beta^*$, and assume that the following limits exist.

$$\eta'(0) = \lim_{\mu\to 0} \frac{\eta(\mu) - \eta_0}{\mu}, \quad \beta'(0) = \lim_{\mu\to 0} \frac{\beta(\mu) - \beta^*}{\mu}. \tag{61}$$

Note that we provide further numerical justification of these assumptions in Figure 6. Rewriting (30) as a system of scalar equations and taking $\beta = \beta(\mu)$, we obtain

$$\left(\beta(\mu) D_e + \frac{(r_e - \mu e)}{\beta(\mu)} - \eta(\mu)\right) + \frac{\mu d}{\beta(\mu)} \frac{q_2}{q_1} = 0, \tag{62}$$

$$\left(\beta(\mu) D_d + \frac{(r_d - \mu d)}{\beta(\mu)} - \eta(\mu)\right) \frac{q_2}{q_1} + \frac{\mu e}{\beta(\mu)} = 0, \tag{63}$$

for $\mu > 0$. Then adding and subtracting η_0 from (62) and (63), and dividing by $\mu\beta(\mu)^{-1}$ yields

$$\beta^* \frac{(\beta D_e + \beta^{-1} r_e) - (\beta^* D_e + {\beta^*}^{-1} r_e)}{\mu} - \beta^* \frac{\eta - \eta_0}{\mu} - e + d\frac{q_2}{q_1} = 0 \tag{64}$$

$$\left(\beta^* \frac{(\beta D_d + \beta^{-1} r_d) - (\beta^* D_d + {\beta^*}^{-1} r_d)}{\mu} - \beta^* \frac{\eta - \eta_0}{\mu} - d\right) \frac{q_2}{q_1} + e = 0 \tag{65}$$

Now bearing in mind (61) and the fact that $\beta^* \neq 0$, it follows from (64) that

$$\frac{q_{0_2}}{q_{0_1}} := \lim_{\mu\to 0} \frac{q_2}{q_1}$$

exists, so that taking the limit $\mu \to 0$ in (64), (65), we obtain a system of two equations in the three unknowns q_{0_2}/q_{0_1}, $\eta'(0)$ and $\beta'(0)$,

$$\beta^* \left[D_e \beta'(0) - \frac{r_e \beta'(0)}{\beta^{*2}} - \eta'(0) \right] - e + d\left(\frac{q_{0_2}}{q_{0_1}}\right) = 0, \tag{66}$$

$$\beta^* \left[D_d \beta'(0) - \frac{r_d \beta'(0)}{\beta^{*2}} - \eta'(0) \right] \left(\frac{q_{0_2}}{q_{0_1}}\right) - d\left(\frac{q_{0_2}}{q_{0_1}}\right) + d = 0. \tag{67}$$

To obtain a third equation, recall that the eigenvalues λ of $H_{\beta,\mu}$ satisfy $\det(H_{\beta,\mu} - \lambda I) = 0$, so that

$$(\lambda - \beta D_e - \beta^{-1}(r_e - \mu e))(\lambda - \beta D_d - \beta^{-1}(r_d - \mu d)) - \beta^{-2}\mu^2 ed = 0, \tag{68}$$

and further, at $\beta = \beta(\mu)$, we have

$$\frac{\partial \lambda}{\partial \beta} = 0, \quad \text{when } \lambda = \eta(\mu). \tag{69}$$

Differentiating (68) with respect to β and setting $\beta = \beta(\mu)$, $\lambda = \eta(\mu)$, we then multiply the result by β^3 to obtain

$$(\beta^2(D_e + D_d) - (r_e + r_d - \mu e - \mu d))\beta\,\eta(\mu) - 2\mu^2 ed = \\ (\beta^2 D_e - (r_e - \mu e))(\beta^2 D_d + (r_d - \mu d)) + (\beta^2 D_e + (r_e - \mu e)(\beta^2 D_d - (r_d - \mu d)). \tag{70}$$

Then differentiating with respect to μ and letting $\mu \to 0$ yields a third equation relating $\beta'(0)$ and $\eta'(0)$, namely

$$2\beta^*(bD_d - aD_e)\beta'(0) + \eta_0(a-b)\beta'(0) + \beta^*(a-b)\eta'(0) = (bd - ae), \tag{71}$$

where a and b are as defined in (58). Recall from (59) that a, b are both positive due to the assumption (56) on the parameters D_e, D_d, r_e and r_d.

We thus have a system of three equation (66), (67) and (71), which we solve. Eliminating $\eta'(0)$ and $\beta'(0)$ gives the explicit expression for q_{0_2}/q_{0_1}:

$$q_{\text{ratio}} := \frac{q_{0_2}}{q_{0_1}} = \sqrt{\frac{be}{ad}} = \sqrt{\frac{(r_e - \beta^{*2} D_e)e}{(\beta^{*2} D_d - r_d)d}} > 0. \tag{72}$$

Recall that this is the ratio of the two components of the eigenvector of (30) in the limit $\mu \to 0$, where q_2 represents the disperser and q_1 the establisher. This quantity therefore predicts the ratio of the two morphs present in the leading edge of the minimal-speed travelling wave solution to our system [40], which gives insight into the expected long-time behaviour of the solution of (1) with Heaviside initial conditions that models the invasion of the two morphs into an empty region. Note that since $q_{\text{ratio}} > 0$, both morphs play a role in the leading edge.

The presence of the two positive quantities a and b in (72) emphasizes that this expression holds only in the zone where (56), which is equivalent to (59), is satisfied, in which case the faster speed v_f is obtained in the limit $\mu \to 0$. If instead either condition (41) or condition (42) holds, in which case the $\mu \to 0$ limits of $\beta(\mu)$ and $\eta(\mu)$ are given by Theorem 3.5 (ii) or (iii) and depend either only on D_e, r_e or only on D_d, r_d, then up to normalisation, the $\mu \to 0$ limit q_0 of the eigenvector $q_\beta(\mu)$ is either $(1,0)^T$ or $(0,1)^T$, corresponding to $q_{\text{ratio}} = 0$ or ∞. This is because the diagonal matrix $H_{\beta^*,0} = \text{diag}(\beta^* D_e + \frac{r_e}{\beta^*}, \beta^* D_d + \frac{r_d}{\beta^*})$ is then no longer a multiple of the identity and the eigenvector q_0 satisfies $H_{\beta^*,0}\, q_0 = \eta_0\, q_0$. So in these parameter zones, one of the two phenotypes will dominate the leading edge of the front in the limit $\mu \to 0$, in contrast to the zone when (56) holds, when (72) is valid and both components play a role.

4.1. Discussion of Leading Edge Behaviour

In order to investigate the effects of parameters on the proportion of each morph in the leading edge, note that using the substitutions $r = r_d/r_e$, $D = D_e/D_d$, $m = e/d$, we can rewrite (72) as a function of only three variables, the ratio of the growth rates r, the ratio of the dispersal rates D, and the ratio of the mutation rates m, namely,

$$q_{\text{ratio}} = \sqrt{\left(\frac{2D - rD - 1}{2r - rD - 1}\right) m}. \tag{73}$$

This is both biologically interesting and mathematically elegant.

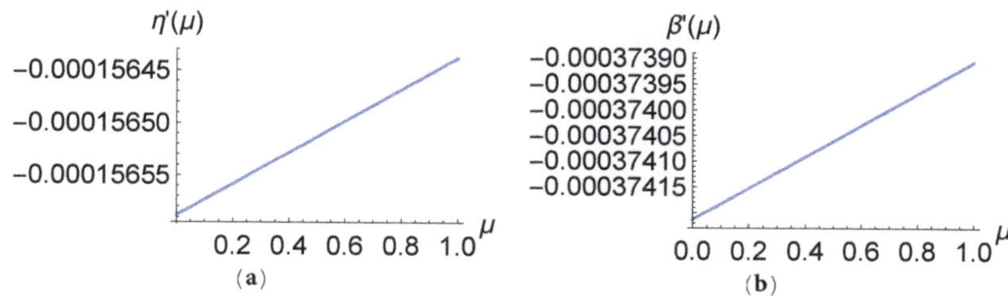

Figure 6. Numerical solutions of **(a)** $\eta'(\mu)$, **(b)** $\beta'(\mu)$ demonstrating convergence as $\mu \to 0$, obtained using Mathematica. Parameter values are $r_e = 1.1$, $r_d = 0.2$, $D_e = 0.3$, $D_d = 1.5$, $e = 0.001$ and $d = 0.00025$.

It is instructive and biologically relevant to comment on the effect on q_{ratio} of varying parameters r, D, and m. Figure 7 illustrates how the ratio of growth, dispersal and mutation rates affects the proportion of each morph in the leading edge of the invasion wave.

We see in Figure 7a that as r_e increases (or r_d decreases) there is an increase in the establisher morph in the leading edge. However, as q_{ratio} does not fall below 1, there is always a larger amount of the disperser morph in the leading edge. As we increase r_e to infinity (or decrease r_d to 0) q_{ratio} tends to $\sqrt{(1-2/D)m}$. Observe also that in (i), q_{ratio} blows up at the value of $r = r_d/r_e$ at which, for the fixed value of $D = D_e/D_d$ in the plot, our assumption that $a > 0$ and $b > 0$ stops holding. This corresponds to the (normalised) eigenvector q_0 approaching the eigenvector $(0,1)^T$ and the parameters D, r entering the zone where (41) holds and $\lim_{\mu\to 0}\eta(\mu)$ is the Fisher speed of the disperser morph v_d instead of v_f.

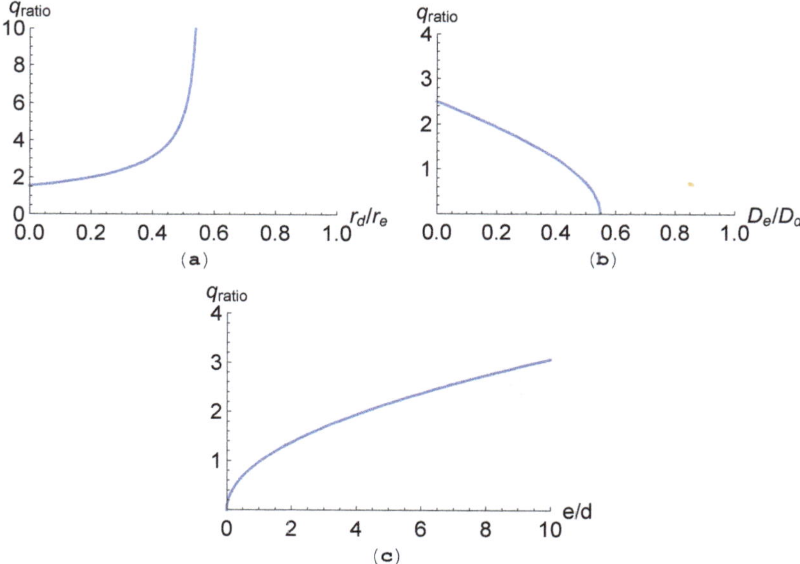

Figure 7. Parameter sweeps of Equation (73). We fix D and m in **(a)**, r and m in **(b)**, and r and D in **(c)**. When fixed, parameters take the values $r = 0.2/1.1$, $D = 0.3/1.5$, $m = 0.001/0.00025$.

Figure 7b tells us that as we increase D_d (or decrease D_e), there is an increase in the disperser morph in the leading edge. If D_e and D_d are close enough in value, it is possible that the establisher morph outnumbers the disperser in the leading edge, which we see in the region in which q_{ratio} drops below one. As we increase D_d to infinity (or decrease D_e to 0) q_{ratio} tends to $\sqrt{m/(1-2r)}$. In the case of (ii), the ratio q_{ratio} becomes zero at the value of $D = D_e/D_d$ at which, for the fixed value of $r = r_d/r_e$ in the plot, our assumption that $a > 0$ and $b > 0$ no longer holds. This corresponds to the (normalised) eigenvector q_0 approaching the eigenvector $(1,0)^T$ and the parameters D, r entering the zone where (42) holds and $\lim_{\mu\to 0} \eta(\mu)$ is the Fisher speed of the establisher morph v_e instead of v_f.

In Figure 7c we see that if the mutation rate of a morph is increased it results in a decrease in the density of that morph in the leading edge. Since the ratio m is not restricted to a particular range, like r and D due to (2), we see that the mutation rate is also able to change which morph is more prevalent in the leading edge. If we fix all parameters but the mutation rates in (73) we obtain $q_{\text{ratio}} = \sqrt{km}$, where $k > 0$ is a constant. Thus q_{ratio} goes to zero as m grows smaller and blows up as m tends to infinity. These provide us with interesting predictions which could be tested experimentally with real biological systems.

Note that, of course, the results of the works by the authors of [11,12] on linear determinacy and existence of travelling waves are valid for all parameter choices D_e, D_d, r_e, r_d. The fact that the expression q_{ratio} only holds in some parameter zones is just an expression of the fact that only for certain parameters does the eigenvector that describes the leading edge of the front, which is strictly positive whenever $\mu > 0$, tend to a strictly positive eigenvector as $\mu \to 0$.

Recall that we made assumptions (61) on the existence of $\eta'(0)$ and $\beta'(0)$, for which we now provide some justification by numerically solving Equations (62), (68) and (70) as a system of three equations in q_2/q_1, $\eta(\mu)$ and $\beta(\mu)$. Differentiating the latter two with respect to μ we plot $\eta'(\mu)$, $\beta'(\mu)$ and q_2/q_1 as functions of μ in Figure 6 demonstrating the existence of $\eta'(0)$, $\beta'(0)$ and q_{ratio} for this set of parameters. Note that as μ increases we see an increase in the proportion of the phenotype n_d, which is expected as for this set of parameters $e > d$.

Finally, we also plot $\beta(\mu)$ and $\eta(\mu)$ as functions of μ in Figure 8. As $\mu \to 0$ we see $\beta(\mu)$ and $\eta(\mu)$ converge to the values β^* and η_0, respectively. Further, as we increase μ we observe a decrease in the minimal spreading speed $\eta(\mu)$, this is consistent with our result in Section 4.1.

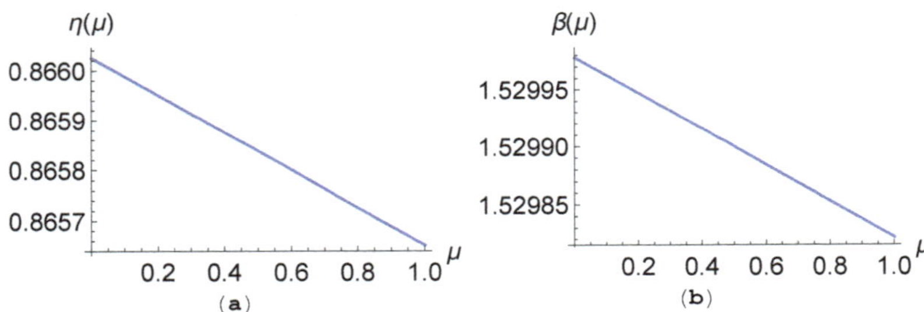

Figure 8. Numerical solutions of (**a**) $\beta(\mu)$ and (**b**) $\eta(\mu)$ demonstrating convergence to β^* and η_0, respectively, as $\mu \to 0$, obtained using Mathematica. The behaviour of $\eta(\mu)$ is consistent with our result that $\eta(\mu)$ is a nonincreasing function of μ. Parameter values: $r_e = 1.1$, $r_d = 0.2$, $D_e = 0.3$, $D_d = 1.5$, $e = 0.001$ and $d = 0.00025$.

5. Conclusions and Remarks

This article exploits linear determinacy of the competition–diffusion-mutation system (1) to establish ecologically-motivated results about the invasion properties of a species with two morphs. These results constitute theoretical predictions of the model, of which the most significant are

(i) the spreading speed decreases as the mutation rate increases;
(ii) there are three possible spreading speeds in the limit of vanishing mutation rate, the choice of which depends on the dispersal and growth parameters; and
(iii) the ratio of the phenotypes in the leading edge of the invasion.

It would, of course, be valuable to compare these predictions with experimental or empirical data. Note that on the one hand, for the parameter regime (40), so-called 'anomalous spreading' in [3], the spreading speed in the vanishing mutation limit is faster than either phenotype would spread in isolation. On the other hand, once mutation is present, the spreading speed is a nonincreasing function of mutation. Although these predictions are rigorous consequences of the model, their juxtaposition is slightly counter-intuitive from an ecological point of view.

Activity on propagation phenomena in models incorporating mutation has increased markedly in recent years (see the works by the authors of [10,11,30,40] and many others) and is producing a growing body of interesting results and open questions. A natural progression of our work would be a detailed analysis of systems for N morphs, in particular, dependence of spreading speeds and the composition of the morphs in the leading edge on parameters. The methods developed in Sections 3 and 4 have the clear potential to yield results in the N morph case. Also interesting is the role of functional trade-offs in determining spreading speeds in competition–diffusion mutation models, some first results on which are established in [12]. More wide-ranging questions include how best to model mutation (for instance, should the rate μ be replaced by a density-dependent rate), whether continuous or discrete trait models are most appropriate in a given setting, and how to pass rigorously from a discrete setting with a large number of traits to the continuous-trait setting.

Author Contributions: All authors contributed equally and significantly in writing this article.

Funding: This research received no external funding.

Acknowledgments: We would like to acknowledge Natalia Petrovskaya at the University of Birmingham and Sergei Petrovskii at the University of Leicester for their feedback and for organising the META Mathematical Ecology: Theory and Applications series of events. We would also like to thank Léo Girardin at Université Paris-Sud, Stephen Cornell and Vincent Keenan at the University of Liverpool, Jason Matthiopoulos at the University of Glasgow, and Jonathan Potts at the University of Sheffield for their insight and several interesting discussions. The first author would like to acknowledge Swansea University and the College of Science at Swansea University for financial support in the form of a PhD studentship held in the Swansea Centre for Biomathematics. We are also grateful to the two reviewers for useful remarks that have helped to improve the manuscript.

Conflicts of Interest: The authors declare no conflict of interest.

References

1. Parmesan, C.; Yohe, G. A globally coherent fingerprint of climate change impacts across natural systems. *Nature* **2003**, *421*, 37–42. [CrossRef] [PubMed]
2. Sakai, A.; Allendorf, F.; Holt, J.; Lodge, D.; Molofsky, J.; With, K.; Baughman, S.; Cabin, R.; Cohen, J.; Ellstrand, N.; et al. The population biology of invasive species. *Annu. Rev. Ecol. Syst.* **2001**, *32*, 305–332. [CrossRef]
3. Elliott, E.; Cornell, S. Dispersal polymorphism and the speed of biological invasions. *PLoS ONE* **2012**, *7*, e40496. [CrossRef] [PubMed]
4. Elliott, E.; Cornell, S. Are anomalous invasion speeds robust to demographic stochasticity? *PLoS ONE* **2013**, *8*, e67871. [CrossRef]
5. Lewis, M.; Li, B.; Weinberger, H. Spreading speed and linear determinacy for two-species competition Models. *J. Math. Biol.* **2002**, *45*, 219–233. [CrossRef] [PubMed]
6. Lewis, M.; Petrovskii, S.; Potts, J. *The Mathematics Behind Biological Invasions*; Springer International Publishing: Cham, Switzerland, 2016.
7. Murray, J. *Mathematical Biology*, 2nd ed.; Springer: Cham, Switzerland, 1994.
8. Turchin, P. *Quantitative Analysis of Movement: Measuring and Modeling Population Redistribution in Animals and Plants*; Sinauer Associates: Sunderland, MA, USA, 1998.
9. Cosner, C. Challenges in modeling biological invasions and population distributions in a changing climate. *Ecol. Complex.* **2014**, *20*, 258–263. [CrossRef]
10. Bouin, E.; Calvez, V.; Meunier, N.; Mirrahimi, S.; Perthame, B.; Raoul, G.; Voituriez, R. Invasion fronts with variable motility: Phoenotype selection, spatial sorting and wave acceleration. *C. R. Math.* **2012**, *350*, 761–766. [CrossRef]

11. Girardin, L. Non-cooperative Fisher–KPP systems: Traveling Waves and Long-Time Behavior. *Nonlinearity* **2017**, *31*, 108–164. [CrossRef]
12. Morris, A. The Role of Polymorphism, Mutation and Trade-Offs in Spreading Speeds of Biological Invasion. Ph.D. Thesis, Swansea University, Swansea, UK, 2019.
13. Wang, H. Spreading speeds and traveling waves for non-cooperative reaction-diffusion systems. *J. Nonlinear Sci.* **2011**, *21*, 747–783. [CrossRef]
14. Lynch, M.; Ackerman, M.; Gout, J.F.; Long, H.; Sung, W.; Thomas, W.; Foster, P. Genetic drift, selection and the evolution of the mutation rate. *Nat. Rev. Genet.* **2016**, *17*, 704–714. [CrossRef]
15. Kolmogorov, A.; Petrovsky, I.; Piskunov, N. Study of the diffusion equation with growth of the quantity of matter and its application to a biological problem. *Appl. Math. Non-Phys. Phenom.* **1937**, *1*, 1–26.
16. Fisher, R. The wave of advance of advantageous genes. *Ann. Eugen.* **1937**, *7*, 355–369. [CrossRef]
17. Aronson, D.; Weinberger, H. Multidimensional Nonlinear Diffusion Arising in Population Genetics. *Adv. Math.* **1978**, *30*, 33–76. [CrossRef]
18. Bonte, D.; Van Dyck, H.; Bullock, J.; Coulon, A.; Delgado, M.; Gibbs, M.; Lehouck, V.; Matthysen, E.; Mustin, K.; Saastamoinen, M.; et al. Costs of Dispersal. *Biol. Rev.* **2012**, *87*, 290–312. [CrossRef] [PubMed]
19. Tilman, D. *Resource Competition and Community Structure*; Princeton University Press: Princeton, NJ, USA, 1982.
20. Volpert, A.; Volpert, V.; Volpert, V. *Travelling Wave Solutions of Parabolic Systems*; American Mathematical Society: Providence, RI, USA, 1994; Volume 140.
21. van Saarloos, W. Front propagation into unstable states. *Phys. Rep.* **2003**, *386*, 29–222. [CrossRef]
22. Benguria, R.; Depassier, M. Validity of the linear speed selection mechanism for fronts of the nonlinear diffusion equation. *Phys. Rev. Lett.* **1994**, *73*, 2272–2274. [CrossRef] [PubMed]
23. Hadeler, K.; Rothe, F. Travelling fronts in nonlinear diffusion equations. *J. Math. Biol.* **1975**, *2*, 251–263. [CrossRef]
24. Lucia, M.; Muratov, C.; Novaga, M. Linear vs. nonlinear selection for the propagation speed of the solutions of scalar reaction-diffusion equations invading an unstable equilibrium. Commun. *Pure Appl. Math.* **2004**, *57*, 616–636. [CrossRef]
25. Stokes, A. On two types of moving front in quasilinear diffusion. *Math. Biosci.* **1976**, *31*, 307–315. [CrossRef]
26. Rothe, F. Convergence to pushed fronts. *Rocky Mt. J. Math.* **1981**, *11*, 617–634. [CrossRef]
27. Lui, R. Biological growth and spread modeled by systems of recursions I. Mathematical theory. *Math. Biosci.* **1989**, *93*, 269–295. [CrossRef]
28. Weinberger, H.; Lewis, M.; Li, B. Analysis of linear determinacy for spread in cooperative models. *J. Math. Biol.* **2002**, *45*, 183–218. [CrossRef] [PubMed]
29. Altenberg, L. Resolvent positive linear operators exhibit the reduction phenomenon. *Proc. Natl. Acad. Sci. USA* **2012**, *109*, 3705–3710. [CrossRef] [PubMed]
30. Griette, Q.; Raoul, G. Existence and qualitative properties of travelling waves for an epidemiological model with mutations. *J. Differ. Equ.* **2016**, *260*, 7115–7151. [CrossRef]
31. Cantrell, R.; Cosner, C.; Yu, X. Dynamics of populations with individual variation in dispersal on bounded domains. *J. Biol. Dyn.* **2018**, *12*, 288–317. [CrossRef] [PubMed]
32. Dancer, E.; Hilhorst, D.; Mimura, M.; Peletier, L. Spatial segregation limit of a competition–diffusion system. *Eur. J. Appl. Math.* **1999**, *10*, 97–115. [CrossRef]
33. Ni, W.M. *The Mathematics of Diffusion*; Society for Industrial and Applied Mathematics: Philadelphia, PA, USA, 2011.
34. Roques, L.; Hosono, Y.; Bonnefon, O.; Bolvin, T. The effect of competition on the neutral intraspecific diversity of invasive species. *J. Math. Biol.* **2015**, *71*, 465–489. [CrossRef] [PubMed]
35. Shigesada, N.; Kawasaki, K.; Teramoto, E. Spatial segregation of integrating species. *J. Math. Biol.* **1979** *79*, 83–99.
36. Apostol, T. *Mathematical Analysis*, 2nd ed.; Addison-Wesley: Boston, MA, USA, 1974.
37. Crooks, E. On the Volpert theory of travelling wave solutions for parabolic Systems. *Nonlinear Anal. Theory Methods Appl.* **1996**, *26*, 1621–1642. [CrossRef]
38. Tang, M.; Fife, P. Propagating fronts for competing species equations with diffusion. *Arch. Ration. Mech. Anal.* **1980**, *73*, 69–77. [CrossRef]

39. Cohen, D.; Motro, U. More on Optimal Rates of Dispersal: Taking into Account the Cost of the Dispersal Mechanism. *Am. Nat.* **1989**, *134*, 659–663. [CrossRef]
40. Girardin, L. Non-cooperative Fisher–KPP systems: Asymptotic Behavior of Traveling Waves. *Math. Models Methods Appl. Sci.* **2018**, *28*, 1067–1104. [CrossRef]

Article

Optimal Control of a PDE Model of an Invasive Species in a River

Rebecca Pettit and Suzanne Lenhart *

Department of Mathematics, University of Tennessee, Knoxville, TN 37996-1320, USA; beccapettit08@gmail.com
* Correspondence: slenhart@utk.edu

Received: 6 September 2019; Accepted: 10 October 2019; Published: 15 October 2019

Abstract: Managing invasive species in rivers can be assisted by appropriate adjustment of flow rates. Using a partial differential equation (PDE) model representing an invasive population in a river, we investigate controlling the water discharge rate as a management strategy. Our goal is to see how controlling the water discharge rate will affect the invasive population, and more specifically how water discharges may force the invasive population downstream. We complete the analysis of a flow control problem, which seeks to minimize the invasive population upstream while minimizing the cost of this management. Using an optimality system, consisting of our population PDE, an adjoint PDE, and corresponding optimal control characterization, we illustrate some numerical simulations in which parameters are varied to determine how far upstream the invasive population reaches. We also change the river's cross-sectional area to investigate its impact on the optimal control.

Keywords: optimal control; partial differential equation; invasive species in a river

---

1. Introduction

The need to develop methods for managing invasive species in rivers is growing [1–3], and modelling can give insight into management strategies. Jacobsen et al. [4], Jin and Lewis [5,6], and Gagnon et al. [7] used integrodifference models to represent such populations in rivers. With a more realistic model, Jacobsen et al. concluded that longer stream length and lower flow rates increase the chance of a species persisting while higher flow streams must be even longer to maintain persistence [4]. They also showed that flow velocity variability can produce persistence similar to the mean velocity [4]. Jin and Lewis [5,6] also conducted a theoretical analysis using integrodifference equations assuming periodicity with respect to time. In 2011, they first explored how the critical domain size would affect the species persistence and determined that a periodic dispersal kernel and a weighted time-averaged dispersal kernel have the same effect. However, this only holds for the estimation of the critical domain size [5]. In 2012, Jin and Lewis [6] studied spreading speeds for the same integrodifference equations in [5]. Jin and Lewis again found that the periodic dispersal kernel and the weighted time-averaged dispersal kernel both yield the same spreading speeds. Once again, this only applied for the spreading speeds, and there is no proof that the dynamics of the systems are the same [6]. In [7], Gagnon et al. took an applied approach having a specific species, *Codium fragile*. Gagnon et al. focused on wind-driven spread of floating fragments of this algae. They found that prevalent winds were vital in determining the spreading speeds as well as the fragmentation into non-buoyant and buoyant pieces was the most important feature in determining spread [7].

In a recent paper, Jin et al. built a reaction–diffusion partial differential equation (PDE) model to better represent a population in a river by using not just the population growth, but how the river itself changes over time [8]. The idea was that this new model would better represent a population trying to move up a river. Clifford et al. [9] discussed how water flow affects the ecology of the river and the shift from focusing purely on river flow dynamics to how these flow dynamics create various habitats [9].

Huang et al. [10] used a PDE system to examine how a species that usually lives in the benthic layer of a river may move in the stream using drift. They used three values to measure the persistence of a species in the river and found that the net reproductive rate determined if a population could persist in the river or not [10]. In a later paper in 2017, Jin et al. [11] explored how the cross-sectional area of a river would affect whether the population persisted. Assuming the meandering rivers are periodic, the cross-sectional area changes over time [11]. They found that increasing growth rates increases the chance of persistence by the population and allows for spreading both up and downstream while increasing flow spread decreases the chance of persistence and upstream spread but increases downstream spread.

Jin et al. [8] explored simpler forms of the models by setting the cross-sectional area and water discharge to constant values to see how a population would move in the stream. They began with traveling waves in constant environments and temporally or spatially heterogeneous environments. To see the effect on the population's movement upstream, Jin et al. then varied the water discharge and cross-sectional area between two time periods to illustrate a seasonal period of summer versus winter. The water discharge and cross-sectional area in the summer are two values, and in the winter are other values. To investigate time varying flows and corresponding management, we use optimal control theory on the water flow discharge to control the movement of an invasive species upstream.

We are interested in modeling an invasive species (like carp) which moves upstream. To manage an invasive species in a downstream position in a river, one needs an action to keep that species from moving upstream. Current methods to prevent further expansion include electric fences or nets in the river [1]. In addition, adjusting the flow rate via dams or water release mechanisms may be a reasonable containment mechanism. Poff and Zimmerman [12] mention that extreme changes in river flow can affect a populations survival success in a habitat. They discuss flows changing due to climate and controlling runoff by catchment [12]. Using the Jin et al. [8] model in our work, we choose the water discharge "flow" function to affect the invasive species movement upstream. Taking the advection coefficient as a time varying control function, we will be utilizing optimal control theory to analyze this model and this management mechanism to contain the species downstream.

In the following sections, we first formulate the problem and describe the terms of the PDE model. We define the weak solution and the solution space. We then describe the control set and the objective functional to be minimized. Using a priori bounds on the state in the weak solution space, we prove the existence of the optimal control. Then, we prove differentiability of the control-to-state solution map while giving the sensitivity system leading to the adjoint system. We differentiate the objective functional and derive an optimal control characterization using the sensitivity and adjoint systems. Finally, we present some numerical simulations to show the effects of our optimal control on the movement of the population. Specifically, we illustrate how the population moves upstream with and without control.

2. Optimal Control Problem Formulation

We use the model adapted from Jin et al. [8]. To formulate our state PDE and corresponding optimal control problem for the weak solutions, our solution spaces are $V = L^2(0,T; H^1_{\{0\}}(0,L))$ and $V^* = L^2(0,T; (H^1_{\{0\}}(0,L))^*)$, where the subscript $\{0\}$ indicates that the functions vanish at $x = 0$.

Definition 1. *A function $N \in V \cap L^\infty((0,T)\times(0,L))$ with $N_t \in V^*$ is a weak solution of the state PDE:*

$$\begin{aligned}
&N_t = -A_t(x,t)\frac{N}{A(x,t)} + \frac{1}{A(x,t)}\left(D(x,t)A(x,t)N_x\right)_x - \frac{Q(t)}{A(x,t)}N_x + rN\left(1-\frac{N}{K}\right) && \\
&N(0,t) = 0 && \text{on } (0,T), x = 0, \\
&N_x(L,t) = 0 && \text{on } (0,T), x = L, \\
&N(x,0) = N_0(x) && \text{on } (0,L), t = 0
\end{aligned} \tag{1}$$

in $\Lambda = (0, L) \times (0, T)$, *if*

$$\int_0^T \langle N_t, \phi \rangle dt + \int_0^T \int_0^L \left(A_t \frac{N\phi}{A} + DAN_x \left(\frac{\phi}{A} \right)_x + \frac{Q}{A} N_x \phi - rN\phi \left(1 - \frac{N}{K} \right) \right) dxdt = 0$$

for all $\phi \in V$*, where* $\langle \, , \rangle$ *in the integrand denotes the duality* $H^1_{\{0\}}(0, L), H^1_{\{0\}}(0, L)^*$.

The state $N(x, t)$ represents the population density in the river, $A(x, t)$ is the cross-sectional area of the river at location x at time t, $D(x, t)$ is the diffusion coefficient, r is the population intrinsic growth rate, and K is the carrying capacity. The control function $Q(t)$ represents the water discharge rate. An important note is that the location $x = 0$ is upstream while $x = L$ is downstream. For our boundary conditions, at upstream where $x = 0$, we assume no population is initially there, as they are invading by moving from downstream to upstream. Likewise, at downstream, $x = L$, we assume no flux, meaning the population is not increased from an outside source. To simplify the calculations, the model uses a one-dimensional, spatial domain corresponding to the length of the river with the assumption of homogeneous mixing in any cross-sectional area.

Our goal is to control the water discharge rate $Q(t)$ to prevent an invasive species from moving upstream. Our control set for $0 < m \leq M$ is

$$U = \{Q \in L^\infty(0, T) \,|\, m \leq Q(t) \leq M\}.$$

Because of the damming of rivers, $Q(t)$ is bounded above by some value M and bounded below by m positive since water discharge by definition must be positive. We assume we can fully control the water discharge rate by controlling how much water moves through a dam. Note that $Q(t)$ is not a function of space, but just time. The idea explored in Jin et al. is that the water discharge rate being higher during certain seasons and lower in others affects the population level at various locations [8].

Our objective functional which we intend to minimize is:

$$J(Q) = \int_0^T \int_0^L W(x) N(x, t) dxdt + \int_0^T \epsilon Q^2(t) dt,$$

where $\epsilon > 0$ is small. The coefficient $W(x)$ is a weight function that is large at $x = 0$ and small at $x = L$ since our goal is to keep the population downstream (at $x = L$). The first integration term seeks to minimize the size of the invasive species with an emphasis on keeping the invader downstream (due to the weight function) by decreasing N near $x = L$. The $Q^2(t)$ term seeks to minimize the cost of effort needed to stop the species from moving upstream with an ϵ coefficient to keep it small. Thus, the term with $W(x)N(x, t)$ should dominate the objective functional. We seek $Q^* \in U$ such that

$$J(Q^*) = \inf_{Q \in U} J(Q).$$

Assuming A is a C^1 function and D is a continuous function on our domain, we make the following additional assumptions:

$$r > 0, K > 0,$$

$$0 < m \leq Q(t) \leq M,$$

$$0 < \epsilon_A \leq A(x, t) \leq M_A,$$

$$|A_x(x, t)| \leq M_B, \; |A_t(x, t)| \leq M_C,$$

$$0 < \epsilon_D \leq D(x, t) \leq M_D,$$

$$0 \leq N_0(x, t) \leq M_1.$$

To solve the optimal control problem for a given $Q \in U$, we need the existence of a unique weak state solution to (1), and this solution must be non-negative and bounded above. The corresponding two theorems follow from results as in [13,14].

Theorem 1 (Positivity and Existence of State Solution). *Given N_0 non-negative, $L^\infty(0,L)$ bounded and in $H^1_{\{0\}}(0,L)$, then, for each $Q \in U$, there is a unique non-negative weak solution $N = N(Q)$ of tjhe state PDE (1) with $0 \le N(x,t) \le C$, where C holds for all $Q \in U$.*

Theorem 2 (A priori estimates). *There exists positive constants α_1 and α_2 such that, for any $Q \in U$, and $N \in V \cap L^\infty(\Lambda)$ where $N_t \in L^2(0,T;(H^1_{\{0\}}(0,L))^*)$, the weak solution of the PDE corresponding to Q with $N \ge 0$ a.e. satisfies*

$$\left(\iint_\Lambda N_x^2 dxdt + \iint_\Lambda N^2 dxdt\right) \le \alpha_1 \tag{2}$$

and $\|N_t\|_{V^*} \le \alpha_2$.

We denote the weak solution of the state PDE (1) corresponding to control Q as $N(Q)$.

Theorem 3 (Existence of Optimal Control). *There exists an optimal control, $Q^* \in U$, satisfying*

$$J(Q^*) = \inf_{Q \in U} J(Q).$$

Proof. We choose a minimizing sequence $\{Q^n\} \subset U$ with $n = 1,2,3,\ldots$ such that

$$\lim_{n\to\infty} J(Q^n) = \inf_{Q \in U} J(Q)$$

and $N^n = N(Q^n)$ the corresponding solution to tje state PDE (1). By the a priori estimates in Theorem 2, there exists $N^* \in V \bigcap L^\infty((0,T)\times(0,L))$, with $N_t^* \in V^*$, and $Q^* \in U$ such that, on a subsequence, these weak convergences hold:

$$\begin{aligned} &N^n \rightharpoonup N^* \text{ in } V, \\ &N_t^n \rightharpoonup N_t^* \text{ in } V^*, \\ &Q^n \rightharpoonup Q^* \text{ in } L^2((0,T)\times(0,L)). \end{aligned} \tag{3}$$

We must show that $N^* = N(Q^*)$, where N^* is the population solution associated with Q^*. Using the compactness result in [15] coupled with the convergences in (3), we obtain this strong L^2 convergence,

$$N^n \to N^* \text{ in } L^2((0,T)\times(0,L)).$$

Now, we must show that Q^* is an optimal control and that $N^* = N(Q^*)$. Using the convergences in (3) and the L^∞ bounds on Q, the convergences above give limits for these terms:

$$\begin{aligned} &\int_0^T \langle N_t^n, \phi\rangle dt \rightharpoonup \int_0^T \langle N^*, \phi\rangle dt \\ &\int_0^T\int_0^L \frac{A_t}{A} N^n \phi dxdt \to \int_0^T\int_0^L \frac{A_t}{A} N^* \phi dxdt \\ &\int_0^T\int_0^L DAN_x^n \left(\frac{\phi}{A}\right)_x dxdt \rightharpoonup \int_0^T\int_0^L DAN_x^* \left(\frac{\phi}{A}\right)_x dxdt \\ &\int_0^T\int_0^L -rN^n \phi dxdt \to \int_0^T\int_0^L -rN^* \phi dxdt \\ &\int_0^T\int_0^L \frac{r\phi}{K}(N^n)^2 dxdt \to \int_0^T\int_0^L \frac{r\phi}{K}(N^*)^2 dxdt. \end{aligned}$$

To handle a term with Q^n and N_x^n, integrating by parts and rearranging gives

$$\begin{aligned}
&\int_0^T \int_0^L \frac{1}{A} \left(Q^n N_x^n - Q^* N_x^*\right) \phi dx dt \\
&\quad = -\int_0^T \int_0^L Q^n \left(\frac{\phi}{A}\right)_x (N^n - N^*)\, dx dt - \int_0^T \int_0^L N^* \left(\frac{\phi}{A}\right)_x (Q^n - Q^*)\, dx dt \\
&\qquad + \int_0^T \frac{Q^n \phi}{A} (N^n - N^*)(L,t)\, dt + \int_0^T \frac{N^* \phi}{A} (Q^n - Q^*)(L,t)\, dt. \qquad (4)
\end{aligned}$$

By the Cauchy–Schwartz inequality, the strong convergence of $\{N^n\}$, ϕ being in L^2 and the L^∞ bounds on $\{Q^n\}$, the first and third terms of Equation (4) give

$$\begin{aligned}
&\lim_{n\to\infty} \left| -\int_0^T \int_0^L Q^n \left(\frac{\phi}{A}\right)_x (N^n - N^*)\, dx dt + \int_0^T \frac{Q^n \phi}{A} (N^n - N^*)(L,t)\, dt \right| \\
&\quad \le \lim_{n\to\infty} M \left[\left(\int_0^T \int_0^L |N^n - N^*|^2 dx dt \right)^{1/2} \left(\int_0^T \int_0^L \left| \frac{\phi}{A} \right|_x^2 dx dt \right)^{1/2} \right. \\
&\qquad \left. + \left(\int_0^T \left| \frac{\phi}{A}(L,t) \right|^2 dt \right)^{1/2} \left(\int_0^T |N^n - N^*|^2 (L,t)\, dt \right)^{1/2} \right] \\
&\quad = 0
\end{aligned}$$

using the bounds in (2) and the convergences in (3) and a result from [16] for

$$N^n \to N^* \text{ in } L^2\left(\{L\} \times (0,T)\right).$$

By the weak convergence of $\{Q^n\}$, ϕ being fixed in L^2 and N^* being fixed with L^∞ bounds, the second and fourth terms of Equation (4) converge, which gives

$$\lim_{n\to\infty} \int_0^T \int_0^L \frac{1}{A} \left(N_x^n Q^n - N_x^* Q^*\right) \phi dx dt = 0.$$

Hence, as $n \to \infty$, N^* satisfies the state PDE with the control Q^* in (1), i.e., $N^* = N(Q^*)$. Again using the weak convergences (3) in the objective functional on the minimizing sequence $\{Q^n\}$ with L^∞ bounds and with lower semicontinuity with respect to L^2 weak convergence [17],

$$\int_0^T (Q^*)^2 dt \le \underline{\lim}_n \int_0^T (Q^n)^2 dt,$$

we obtain $\inf_{Q\in U} J(Q) \ge J(Q^*)$ and then Q^* is an optimal control. □

We now differentiate the control-to-state map $Q \to N(Q)$ as a directional derivative to find the sensitivity function.

Theorem 4 (Differentiability of solution map). *The solution map*

$$Q \in U \to N = N(Q) \in V$$

is differentiable in the following sense: There exists $\psi \in V$, *such that*

$$\frac{N(Q + \epsilon l) - N(Q)}{\epsilon} \rightharpoonup \psi$$

as $\epsilon \to 0$, *where* $Q + \epsilon l \in U$ *and* $l \in L^\infty(0,T)$. *Moreover,* $\psi = \psi(Q,l)$ *with* $\psi_t \in V^*$ *satisfies*

$$\begin{array}{lll} \psi_t = \frac{-A_t}{A}\psi + \frac{1}{A}(DA\psi_x)_x - \frac{Q}{A}\psi_x - \frac{l}{A}N_x + r\psi - \frac{2r}{K}\psi N & in & \Lambda = (0,L)\times(0,T), \\ \psi(0,t) = 0 & on & (0,T), x = 0, \\ \psi_x(L,t) = 0 & on & (0,T), x = L, \\ \psi(x,0) = 0 & on & (0,T), t = 0. \end{array}$$

Proof. To justify the convergence of the quotients, we use a priori estimates. Thus, we change the variables such that $w(x,t) = N(x,t)e^{-\alpha t}$ and $w^\epsilon(x,t) = N^\epsilon(x,t)e^{-\alpha t}$, where $N = N(Q)$ and $N^\epsilon = N(Q+\epsilon l)$. With some standard PDE estimates, we obtain the desired estimate for α large

$$\frac{1}{2}\int_0^L \left(\frac{w^\epsilon - w}{\epsilon}\right)^2 (T,x)dx + (\alpha - C_1)\iint_\Lambda \left(\frac{w^\epsilon - w}{\epsilon}\right)^2 dxdt + \frac{\epsilon_D}{2}\iint_\Lambda \left(\frac{w^\epsilon - w}{\epsilon}\right)_x^2 dxdt \le C_0.$$

The bounds on $\frac{w^\epsilon - w}{\epsilon}$ give corresponding bounds on $\frac{N^\epsilon - N}{\epsilon}$. From its PDE, we obtain a uniform estimate on $\frac{N_t^\epsilon - N_t}{\epsilon}$ in V^*. Thus, there exists $\psi \in V$ and $\psi_t \in V^*$ such that $\frac{N(Q+\epsilon l) - N(Q)}{\epsilon} \rightharpoonup \psi$ in V and $\frac{N_t(Q+\epsilon l) - N_t(Q)}{\epsilon} \rightharpoonup \psi_t$ in V^* as $\epsilon \to 0$.

Using weak convergence of the quotients, ψ satisfies the sensitivity PDE, by passing to the limit in

$$\begin{aligned} \left(\frac{N^\epsilon - N}{\epsilon}\right)_t =& \frac{-A_t}{A}\left(\frac{N^\epsilon - N}{\epsilon}\right) + \frac{1}{A}\left(DA\left(\frac{N^\epsilon - N}{\epsilon}\right)_x\right)_x \\ & - \frac{Q}{A}\left(\frac{N^\epsilon - N}{\epsilon}\right)_x - \frac{\epsilon l}{A\epsilon}N_x^\epsilon + r\left(\frac{N^\epsilon - N}{\epsilon}\right) - \frac{r}{K}\left(\frac{N^\epsilon - N}{\epsilon}(N^\epsilon + N)\right). \end{aligned}$$

Since $N(Q+\epsilon l) \to N(Q)$ strongly in V, we obtain

$$\psi_t = \frac{-A_t}{A}\psi + \frac{1}{A}(DA\psi_x)_x - \frac{Q}{A}\psi_x - \frac{l}{A}N_x + r\psi - \frac{2r}{K}\psi N.$$

Our initial and boundary conditions become $\psi(0,t) = 0$, $\psi_x(L,t) = 0$, and $\psi(x,0) = 0$. □

Next, we build our adjoint PDE using the sensitivity PDE. We arrange the sensitivity equation corresponding to an optimal state $N^* = N(Q^*)$ with direction l above as

$$\psi_t + \frac{A_t}{A}\psi - \frac{1}{A}(DA\psi_x)_x + \frac{Q^*}{A}\psi_x - r\psi + \frac{2r}{K}\psi N^* = -\frac{l}{A}N_x^*$$

to obtain the linear operator on ψ,

$$L\psi = \psi_t + \frac{A_t}{A}\psi - \frac{1}{A}(DA\psi_x)_x + \frac{Q^*}{A}\psi_x - r\psi + \frac{2r}{K}\psi N^*,$$

and thus the sensitivity PDE can be written as $L\psi = -\frac{l}{A}N_x^*$.

For our adjoint function, we wish to find an operator L^* such that $\langle \lambda, L\psi\rangle = \langle L^*\lambda, \psi\rangle$ in V, V^* duality. Starting with the weak form of the LHS of ψ PDE and test function λ with $\lambda(0,t) = 0$ and using integration by parts, to obtain $\langle \lambda, L\psi\rangle = \langle L^*\lambda, \psi\rangle$, our adjoint boundary condition for $x = L$ becomes

$$D(L,t)A(L,t)\left(\frac{\lambda(L,t)}{A(L,t)}\right)_x + Q(t)\frac{\lambda(L,t)}{A(L,t)} = 0.$$

Thus, we found that

$$\begin{array}{lll} L^*\lambda = -\lambda_t - \left(DA\left(\frac{\lambda}{A}\right)_x\right)_x - Q\left(\frac{\lambda}{A}\right)_x + \frac{A_t\lambda}{A} - r\lambda + 2\frac{r\lambda}{K}N^* & \text{in} & \Lambda, \\ L^*\lambda = W(x) & \text{in} & \Lambda, \\ \lambda(x,T) = 0 & \text{on} & (0,L), t = T, \\ \lambda(0,t) = 0 & \text{on} & (0,T), x = 0, \\ D(L,t)A(L,t)\left(\frac{\lambda(L,t)}{A(L,t)}\right)_x + Q(t)\frac{\lambda(L,t)}{A(L,t)} = 0 & \text{on} & (0,T), x = L. \end{array} \tag{5}$$

We choose the RHS of the adjoint PDE such that $L^*\lambda = W(x)$, where $W(x)$ is the derivative of the integrand of the objective functional with respect to $N(x,t)$. We use the sensitivity and adjoint PDEs in the differentiation of map $Q \to J(Q)$ to obtain our optimal control characterization.

Theorem 5. *Given an optimal control Q^* in U, there exists a solution λ in $V \cap L^\infty(\Lambda)$ with $\lambda_t \in V^*$ to the adjoint problem (5). Moreover, we obtain a characterization of our optimal control:*

$$Q^*(t) = min\left(M, max\left(\frac{1}{2\epsilon}\int_0^L \frac{\lambda}{A}N_x^*(x,t)dx, m\right)\right), \tag{6}$$

where $\epsilon > 0$.

Proof. Given the existence of an optimal control, the corresponding solution of the adjoint problem exists since the adjoint problem is linear in the adjoint function [18].

Suppose $Q^*(t)$ is an optimal control. Let $l(t) \in L^\infty(0,T)$ such that $Q + \delta l \in U$ for $\delta > 0$. We denote $N(Q+\delta l)$ by N^δ. The derivative of $J(Q)$ with respect to Q at Q^* in the direction l satisfies

$$\begin{aligned} 0 &\le \lim_{\delta\to 0}\frac{J(Q^*+\delta l) - J(Q^*)}{\delta} \\ &= \lim_{\delta\to 0}\frac{1}{\delta}\left[\int_0^T\int_0^L W(x)N^\delta(x,t)dxdt + \int_0^T \epsilon\left(Q^*+\delta l\right)^2(t)dt \right. \\ &\quad \left. - \left(\int_0^T\int_0^L W(x)N^*(x,t)dxdt + \int_0^T \epsilon\left(Q^*\right)^2(t)dt\right)\right] \\ &= \int_0^T\int_0^L W(x)\psi dxdt + \int_0^T 2l\epsilon Q^* dt \\ &= \int_0^T\int_0^L \left(-\lambda_t\psi - \left(DA\left(\frac{\lambda}{A}\right)_x\psi_x\right) - Q\psi\left(\frac{\lambda}{A}\right)_x + \frac{A_t}{A}\lambda\psi - r\lambda\psi + 2\frac{r}{K}N^*\lambda\psi\right)dxdt \\ &\quad + \int_0^T 2l\epsilon Q^* dt \\ &= \int_0^T\int_0^L \left(\psi_t\lambda + \frac{A_t}{A}\lambda\psi - \left(DA\left(\frac{\lambda}{A}\right)_x\psi_x\right) + \frac{Q}{A}\psi_x\lambda - r\psi\lambda + \frac{2r}{K}\psi\lambda N^*\right)dxdt \\ &\quad + \int_0^T 2l\epsilon Q^* dt \\ &= \int_0^T l\left(2\epsilon Q^* - \int_0^L \frac{\lambda}{A}N_x^* dx\right)dt. \end{aligned}$$

Choosing appropriate variations for l gives the desired characterization of our optimal control. □

The optimality system consists of the state and adjoint systems (1), (5), and the optimal control characterization (6). The uniqueness of the optimal control follows from the uniqueness of the solutions to the optimality system. Techniques that were used in [13,19] and this extra estimate $\int_0^L \bar{p}_x^2 dx \le K_5 e^{2\alpha T}$ [20] gives the desired uniqueness result.

Theorem 6 (Uniqueness of Optimal Control). *For sufficiently small T, the solution to the optimality system (1), (5), and (6) is unique.*

If one has a valid generalization of this model to two of three spatial dimensions, these theoretical results for the optimal control analysis and the corresponding optimality system could be generalized, with appropriate a priori estimates . The assumptions on the diffusion coefficient and the cross-sectional area can be generalized as long as the needed a priori estimates and corresponding compactness results can be obtained. For some background on such optimal control problems, see the book by Li and Yong [21].

3. Numerical Results

In the previous sections, we have that a unique optimal control exists for our optimality system. Now, we illustrate numerical solutions for the optimality system for several scenarios. To solve for an optimal control numerically, we must solve the optimality system consisting of the state and adjoint PDEs with the optimal control characterization. For this, we use the Forward-Backward Sweep method [22,23]. We make an initial guess for the control, Q. Then, using the initial condition, we solve the state PDE, N, forward in time. Next, we solve the adjoint PDE, λ, backward in time. The PDEs are solved using finite differences as discussed below. We update Q using our calculated N and λ values in the optimal control characterization, and, finally, we check convergence of control values at successive iterations. A discussion about convergence and stability of this method can be found in [22]. In the numerical approximations, we use an upwind or downwind difference as appropriate. For an example, in our state PDE for the advection term, N_x, we use a downwind difference [24] since its coefficient, $-\frac{Q(t)}{A(x,t)}$, is negative. This difference goes backwards in space. Similarly, we use a downwind difference for the A_t term as well. We use an upwind difference in the adjoint advection term since $Q(t) > 0$, which goes forward in space [24].

3.1. Cross-Sectional Area Is Constant

We begin these numerical simulations assuming our cross-sectional area, A, is constant. Later, we will vary it with respect to space and time. To illustrate the spread of a population in a river with control, we use the following Base Case parameters:

$$\begin{aligned} &T = 10 \text{ (final time)}, L = 10 \text{ (length of river)}, \\ &r = 0.6, K = 200, D = 0.1, A = 20, \epsilon = 0.05,\ 0 \le Q(t) \le 10. \end{aligned} \tag{7}$$

Note that the downstream is represented by $x = L$. Figure 1 shows the initial condition and the weight function used in our simulations. In Figure 1a, we place the initial population downstream to see their movement upstream. In Figure 1b, we use a weight function that has values 100 upstream (at $x = 0$) and nearly 0 upstream (at $x = 10$). Recall that we are minimizing the objective functional, so this weight function helps push the population downstream.

We first illustrate how far the population moves upstream with no control, constant control, and optimal control in Figure 2. To generate this graph, we use a lower bound of 0.5 for the population, so, for each time, we find the lowest location x (more upstream) with $N(x,t) > 0.5$ and record that river location. The constant control has the value of the upper bound of the optimal control. The population with no control moves further upstream than the population with constant control, which is to be expected. The population with no control moves from 8 to about 2.8, which is about two-thirds upstream. The population with constant control starts at 8 and ends about 7.1. Having a constant control at the upper bound ($Q = 10$) resulted in the least movement upstream. With the optimal control, the population moves further upstream than the constant control case. Note that the optimal control is not at its upper bound for the whole time period due to costs in J. The objective functional balances keeping the population downstream with costs. The population in the optimal control case starts at 8 and ends around 5.5.

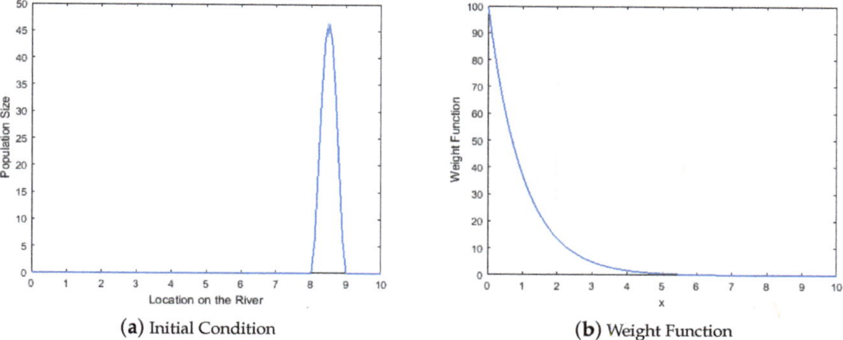

(**a**) Initial Condition (**b**) Weight Function

Figure 1. (**a**) The initial condition for the population; (**b**) the weight function in the objective functional.

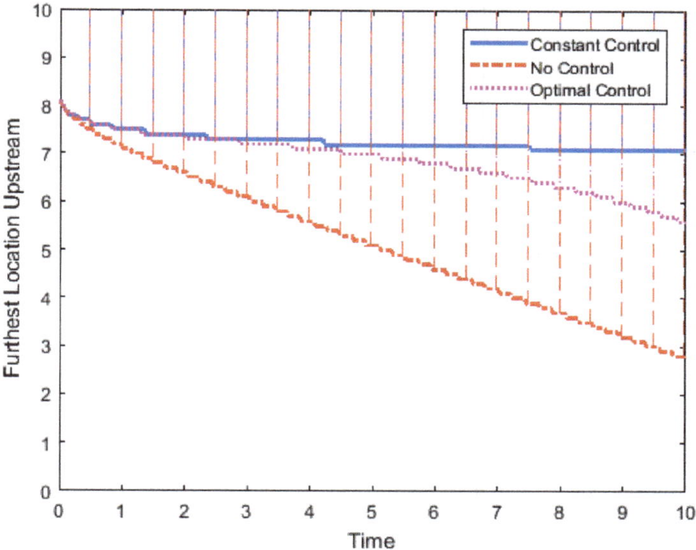

Figure 2. Location of the population in the river for the no control case (red), for the constant control case with $Q = 10$ (blue), and the optimal control case (magenta) over the time, with parameter values fromtje Base Case list (7).

Figure 3 shows the population plots for no control (Figure 3a) and for optimal control (Figure 3b). The population in the no control case spreads farther upstream than in the optimal control case. The population also grows larger in the no control case. In Figure 3c, the populations at the final time are plotted for the length of the river. The no control population is further upstream and is approaching the carrying capacity while the optimal control population only reached a value of about 140 at any given river location. The population size was decreased, and the population was kept more downstream in the optimal control case. Our objective functional value for the no control case is 239.57, while, for the optimal control case, is 40.12, which is an 83% improvement.

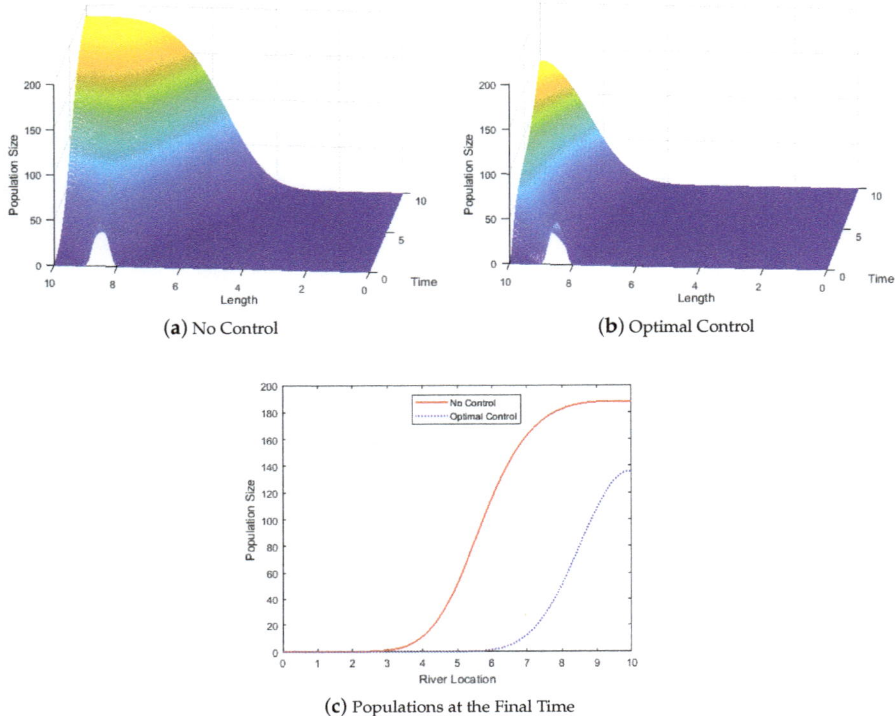

(**a**) No Control

(**b**) Optimal Control

(**c**) Populations at the Final Time

Figure 3. Population plots for (**a**) no control case and (**b**) optimal control case; (**c**) population plots in x at the final time for both cases. The parameter values are from the Base Case list (7).

We varied parameters for the no control, constant control, and optimal control cases and show the changes in our objective functional J values. Starting with our Base Case parameter values in (7), we changed the values of D, T, r and K and investigated the resulting changes in J. Table 1 shows the J values for the Base Case along with cases with changes of D and T while Table 2 shows the variation of K and r. In all cases, the objective functional from the optimal control was smaller than the objective functional for the constant control and the no control cases, which is expected. In Figure 4, we plot the optimal controls for each of the parameters varied. Before changing any parameters for the Base Case, the optimal control is a 29% decrease in the objective functional from the constant control. This indicates that a time varying control can be significantly effective.

In Table 1, for the decrease in diffusion coefficient D, we see a 45% decrease in the objective functional from the constant control to the optimal control and a 74% decrease from the no control to optimal control. When we increase diffusion, there is almost no change in the J values for the constant control and optimal control. There is a 94% decrease from the objective functional value from the no control to the optimal control J value. Figure 4a shows the different optimal controls. When we increase the diffusion coefficient, the optimal control is at the upper bound for most of the time, which explains why the objective functional value was similar to the constant control one. Decreasing the diffusion creates a similar curve to the Base Case, but the optimal control values are slightly smaller. Returning to Table 1, decreasing the final time yields a 57% decrease in the objective functional from the constant control population to the optimal control population, and a 28% decrease from the no control to the optimal control. Meanwhile, increasing the final time gives a 20% decrease again from the constant control population to the optimal control population and a 97% decrease from no control to optimal control. In Figure 4b, increasing T causes the optimal control to be at the upper bound for about half the time, and to then steadily decrease to 0 at the final time. If we decrease T, the optimal control starts midway between the upper and lower bound and decreases down to 0 at the final time. The Base Case falls directly between these two cases.

Table 1. Objective functional outputs for the cases tested where we changed the parameter values for D and T (given below). The base values are from the list (7). Constant control was run with $Q = 10$.

	Base Case	$D = 0.05$	$D = 0.5$	$T = 5$	$T = 15$
No Control	239.52	111.96	5485.3	17.13	2373
Constant Control	56.63	53.72	316.89	28.39	84.87
Optimal Control	40.12	29.40	316.56	12.31	68.17

Table 2 shows a 32% decrease in the objective functional from the constant control to the optimal control when we decrease the carrying capacity, K. Between the no control population and the optimal control population, there is a 81% decrease in the optimal control objective functional value. Increasing K yields a 28% decrease of J from the constant control to the optimal control, while a 85% decrease in J is seen from the no control case to the optimal control case. Figure 4c illustrates that increasing and decreasing K has little change in the optimal control values.

For decreasing the growth rate, r, there is a 61% reduction in the objective functional from the constant control population to the optimal control population. A decrease of 63% occurs in the objective functional values from the no control case to the optimal control case. When we increased r, there is virtually no change in the objective functional values between the constant control and the optimal control, but there was a 95% reduction in the J value from the no control population to the optimal control population. Figure 4d shows the optimal control at the upper bound for most of the time when we increase r. Decreasing r gives the optimal control initially at 6 and steadily decreases to 0 at the final time.

Since our objective functional is meant to represent minimizing the population at upstream and the cost of removing the invasive species, in most cases, using an optimal control would give at least a 20% decrease in $J(Q^*)$ from the constant control. In addition, the optimal control would give a smaller objective functional values $J(Q^*)$ than applying no control at all.

Table 2. Objective functional outputs for the cases tested where we changed parameter values for K and r (given below). The base values are from the list (7). Constant control was run with $Q = 10$.

	Base Case	$K = 150$	$K = 250$	$r = 0.3$	$r = 1.6$
No Control	239.52	204.28	269.14	55.18	3837.3
Constant Control	56.63	56.16	56.97	52.44	176.67
Optimal Control	40.12	37.96	41.13	20.29	176.55

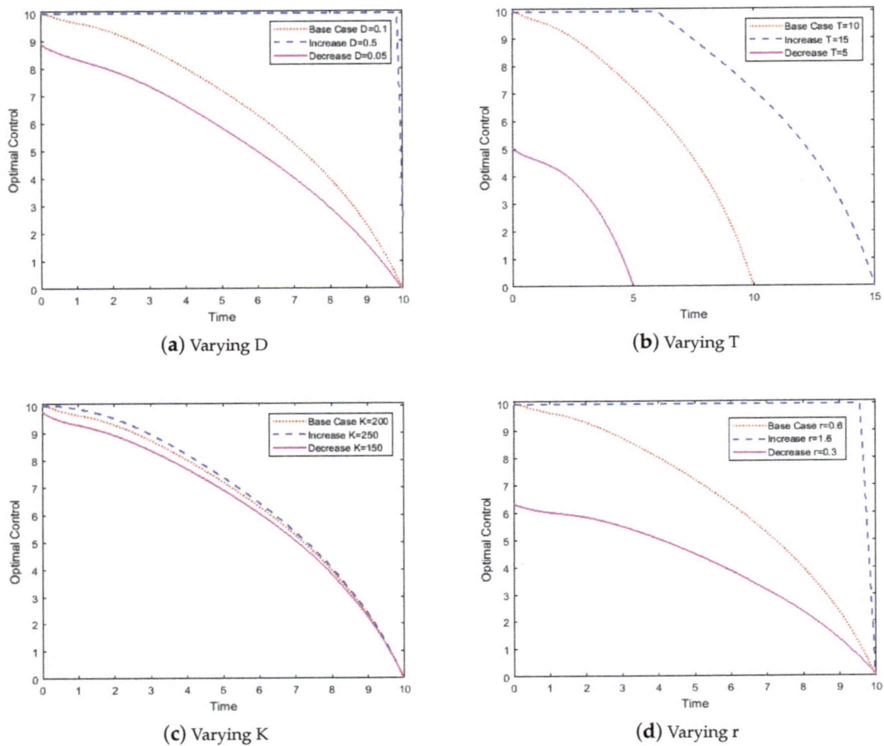

(**a**) Varying D (**b**) Varying T

(**c**) Varying K (**d**) Varying r

Figure 4. The optimal control plots for varying of the parameters D, T, K, and r. The base parameter values are from the list (7).

3.2. Cross-Sectional Area Is Not Constant

We now have the cross-sectional area dependent on space with a linear function,

$$A(x) = 0.5x + 25,$$

and ran all the same cases (varying D, T, K, and r) as for a constant cross-sectional area. The parameters are the same as before in the Base Case list (7). We do not show the specific graphs for this case, but we will compare the various location results later.

Finally, we let the cross-sectional area vary with time and space where

$$A(x,t) = (0.5x + 25) + (0.2t\,(10 - t)).$$

We chose a function so that the river was higher in the middle of the year. Again, the parameters are all the same as in the Base Case list (7). In Figure 5, the no control population starts at 8 and moves to 2.8 with an objective functional value of 238.66. The population with constant control moves from 8 to 5.5 with an objective functional value of 74.70. The population with the optimal control starts at 8 and ends at 4.9. The objective functional value for the optimal control is 66.55, which is again a 10.9% decrease from the constant control and a 72.1% decrease from the no control case. Figure 6 shows the population plots for the no control and optimal control cases, and, as expected, the population in the no control case is larger.

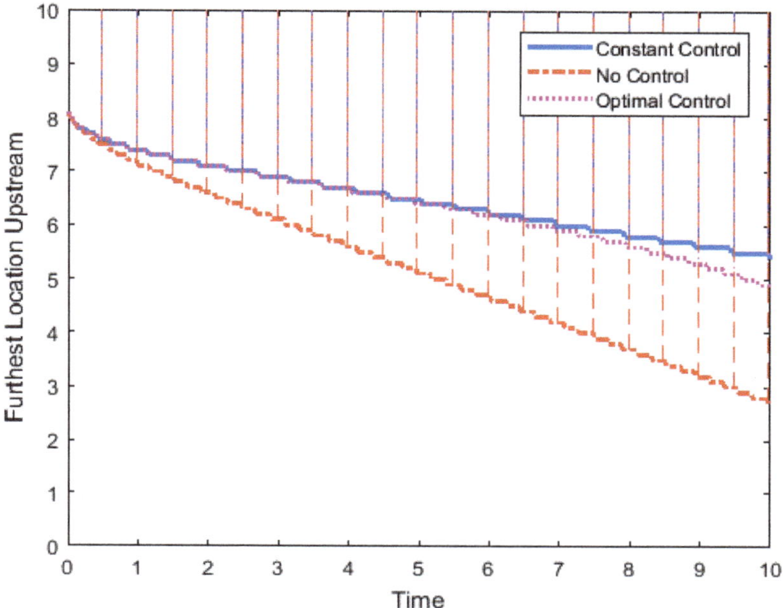

Figure 5. Location of the population in the river for the no control case (red), for the constant control case with $Q = 10$ (blue), and the optimal control case (magenta) over the time with the cross-sectional area is dependent on space and time, $A(x,t) = (0.5x + 25) + (0.2t\,(10 - t))$. The parameter values are $T = 10$, $L = 10$, $r = 0.6$, $K = 200$, $D = 0.1$, and $0 \leq Q(t) \leq 10$.

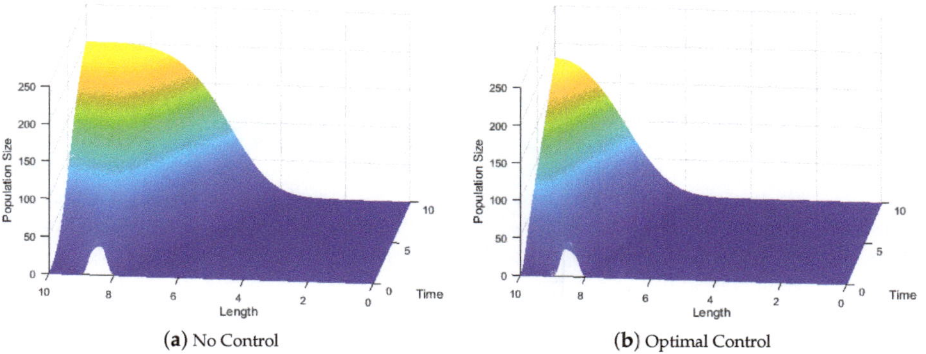

(**a**) No Control (**b**) Optimal Control

Figure 6. The population plots for the (**a**) no control case and (**b**) the optimal control case with the cross-sectional area is dependent on space and time, $A(x,t) = (0.5x + 25) + (0.2t\,(10 - t))$. The parameter values are from the Base Case list (7).

We compare the optimal control results for the three different cross-sectional area functions. We have that $25 \leq A(x) \leq 35$ and $25 \leq A(x,t) \leq 35$, while the constant $A = 20$. In Figure 7, in the constant cross-sectional area case, the population moved upstream the least, at 5.5 at the furthest point. In addition, notice that $A(x)$ and $A(x,t)$ caused similar movement upstream. When the cross-sectional area is not constant, the species is able to move further upstream, possibly due to larger cross-sectional area. The objective functional values are 40.12 for constant A, 60.55 for $A(x)$, and 66.55 for $A(x,t)$. The J values increase as our cross-sectional areas become larger and more complex. This makes sense as the effort to keep the population downstream would become more difficult with the varying larger areas.

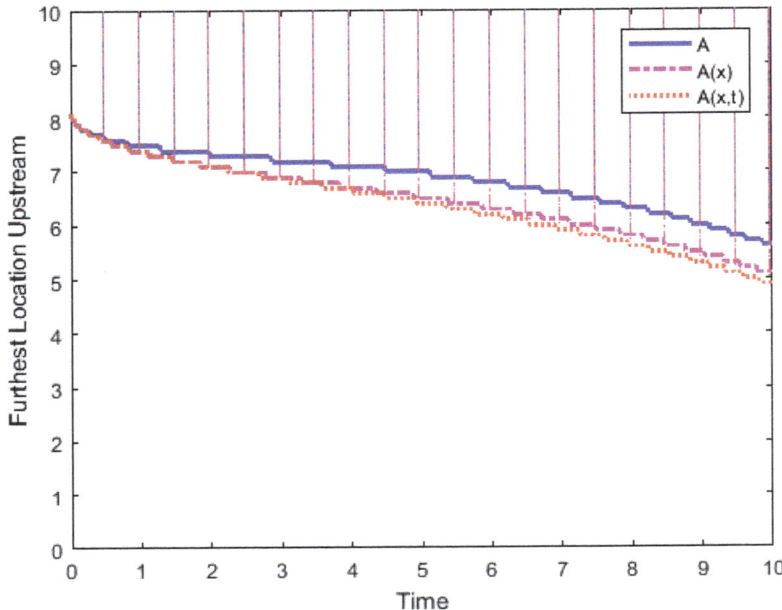

Figure 7. Location of the population in the river for the optimal control where the different cross-sectional area values are: constant $A = 20$ (blue), $A(x) = 0.5x + 25$ (magenta), and $A(x,t) = (0.5x + 25) + (0.2t\,(10 - t))$ (red). The parameter values are from the Base Case list (7).

3.3. Approximate Controls

Optimal controls may change at every time step. However, changing river flow at every time step may be difficult, so we may use an optimal control to design an approximate control that is easier to implement. For example, using the optimal control as a guide, we design an approximate control where we set the control at a high value at the beginning time and a low value towards the end of our time period. In Figure 8a, we set the control, Q, to 10 for $t \leq 5$ and then $Q = 0$ for $t > 5$, and plot this approximate control with the optimal control. For the approximate control, our objective functional value is 48.01, which is an increase from the 40.12 from the optimal control. In Figure 8b, another approximate control has $Q = 8$ for $t \leq 7$ and $Q = 0$ for $t > 7$. Here, our objective value is 42.49, which is again an increase compared to the Base Case value from the optimal control. This approximate control in Figure 8b is a good approximation since its corresponding J value is only 6% different from the optimal J.

We also changed the weight W in the objective function to be linear function and obtained quantitatively similar results. Changing the initial condition to be somewhat larger gives a larger objective functional values but with a similar optimal control.

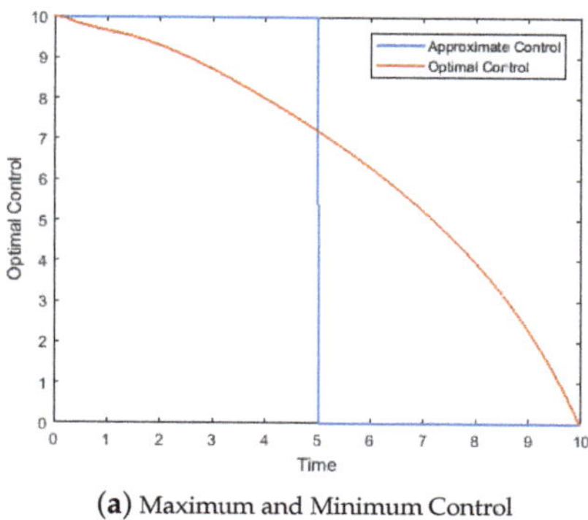

(a) Maximum and Minimum Control

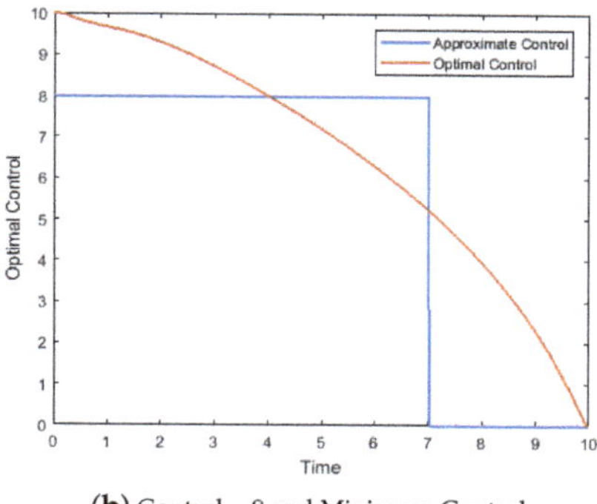

(b) Control = 8 and Minimum Control

Figure 8. Control plots for both the optimal control case and two constant case. (**a**) the two constant control case where $Q = 10$ for $t \leq 5$ and $Q = 0$ for $t > 5$, and (**b**) the two constant control case where $Q = 8$ for $t \leq 7$ and $Q = 0$ for $t > 7$. The parameter values are from the Base Case list (7).

4. Conclusions

To investigate how an invasive population moves upstream, we began with a PDE describing the movement of a species upstream, and then completed the needed analysis results for optimal control of this PDE using the flow rate as the control. We proved the differentiability of the control-to-state map as a directional derivative, which satisfies a linear sensitivity PDE. From the sensitivity PDE, we constructed our adjoint PDE that we used to characterize our optimal control. The control was chosen to keep the population downstream while minimizing the implementation cost.

To run numerical simulations to solve our optimality system, we used a finite difference method together with the forward-backward sweep method. In our simulations, we explored how changing the parameter values would affect the population's location in the river and the J values for cases with no control, constant control, and optimal control while leaving our initial conditions the same for all cases and parameter changes. We first did this for a constant cross-sectional area of the river. As expected, we saw the optimal control populations were always an improvement in J on the no control populations and the constant control populations. The time varying control allowed for an immediate response to the changing population size. We also observed that varying the final time greatly changed the population's location and size.

Changing the cross-sectional area to be dependent on space, and later on space and time, increased the J values from the constant cross-sectional area case with $A(x,t)$ being the largest. The non-constant A allowed the populations with constant and optimal controls to move further upstream than before (as seen in Figures 2, 5, and 7). The population probably moved farther upstream since there is more room for the species to move in the river for the $A(x)$ and $A(x,t)$ cases due to larger sizes of the cross-sections.

We investigated how our objective functional values changed if we implemented approximate controls that are easily implemented. We constructed an approximate control that gave a J value close to the optimal value.

Using optimal control theory can give strategies for controlling a flow rate to successfully manage a population in a river. Our future work will be to obtain some data for a particular invasive river species and apply these techniques. After approximating parameters and choosing appropriate cross-sectional area for a particular species in a river, we would explore how our optimal control strategy would affect the movements of the invasive population. We would also want to investigate more realistic cross-sectional area functions to better represent a river. Considering only allowing seasonal changes in controls would also be important for some situations.

Author Contributions: Investigation, analysis, computations and writing, S.L. and R.P.

Funding: This research received no external funding.

Acknowledgments: We acknowledge useful conservations with Mark Lewis and Yu Jin. This work was partially supported by the National Institute for Mathematical and Biological Synthesis, an Institute sponsored by the National Science Foundation through NSF Award #DBI-1300426, with additional support from the University of Tennessee, Knoxville.

Conflicts of Interest: The authors declare no conflict of interest.

References

1. Kolar, C.S.; Chapman, D.C.; Courtenay, W.R.; Housel, C.M.; Williams, J.D.; Jennings, D.P. *Bigheaded Carps: A Biological Synopsis and Environmental Risk Assessment*; American Fisheries Society: Bethesda, MD, USA, 2007.
2. Simberloff, D. *Invasive Species: What Everyone Needs to Know*; Oxford University Press: New York, NY, USA, 2013.
3. Lockwood, J.L.; Hoopes, M.F.; Marchetti, M.P. *Invasion Ecology*; Blackwell Publishing: Malden, MA, USA, 2007.
4. Jacobsen, J.; Jin, Y.; Lewis, M.A. Integrodifference models for persistence in temporally varying river environments. *J. Math. Biol.* **2015**, *70*, 549–590. [CrossRef] [PubMed]
5. Jin, Y.; Lewis, M.A. Seasonal influences on population spread and persistence in streams: critical domain size. *SIAM J. Appl. Math.* **2011**, *71*, 1241–1262. [CrossRef]
6. Jin, Y.; Lewis, M.A. Seasonal influences on population spread and persistence in streams: Spreading speeds. *J. Math. Biol.* **2012**, *65*, 403–439. [CrossRef]
7. Gagnon, K.; Peacock, S.J.; Jin, Y.; Lewis, M.A. Modelling the spread of the invasive alga *Codium fragile* driven by long-distance dispersal of buoyant propagules. *Ecol. Model.* **2015**, *316*, 111–121. [CrossRef]
8. Jin, Y.; Hilker, F.M.; Steffler, P.M.; Lewis, M.A. Seasonal Invasion Dynamics in a Spatially Heterogeneous River with Fluctuating Flows. *Bull. Math. Biol.* **2014**, *76*, 1522–1565. [CrossRef] [PubMed]
9. Clifford, N.J.; Harmar, O.P.; Harvey, G.; Petts, G.E. Physical habitat, eco-hydraulics and river design: A review and re-evaluation of some popular concepts and methods. *Aquat. Conserv. Mar. Freshw. Ecosyst.* **2006**, *16*, 389–408. [CrossRef]
10. Huang, Q.; Jin, Y.; Lewis, M.A. R_0 analysis of a benthic-drift model for a stream population. *SIAM J. Appl. Dyn. Syst.* **2016**, *15*, 287–321. [CrossRef]
11. Jin, Y.; Lutscher, F.; Pei, Y. Meandering rivers: How important is lateral variability for species persistence? *Bull. Math. Biol.* **2017**, *79*, 2954–2985. [CrossRef] [PubMed]
12. Poff, N.L.; Zimmerman, J.K. Ecological responses to altered flow regimes: A literature review to inform the science and management of environmental flows. *Freshw. Biol.* **2010**, *55*, 194–205. [CrossRef]
13. De Silva, K.R.; Phan, T.V.; Lenhart, S. Advection control in parabolic PDE systems for competitive populations. *Discret. Contin. Dyn. Syst. Ser. B* **2017**, *22*, 1049–1072. [CrossRef]
14. Kelly, M.; Xing, Y.; Lenhart, S. Optimal fish harvesting for a population modeled by a nonlinear parabolic partial differential equation. *Nat. Resour. Model. J.* **2016**, *29*, 36–70. [CrossRef]
15. Simon, J. Compact sets in the $L^p(0,t:b)$. *Ann. Mat. Pura Appl.* **1987**, *146*, 65–96. [CrossRef]
16. Lions, J.L. *Equations Différentielles Opérationnelles et Problèmes aux Limites*; Springer: Berlin, Germany, 1961.
17. Friedman, A. *Foundations of Modern Analysis*; Dover Publications: New York, NY, USA, 1970.
18. Evans, L. *Partial Differential Equations*; American Mathematical Society: Providence, RI, USA, 2010.
19. Lenhart, S. Optimal Control of a Convective Diffusive Fluid Problem. *Math. Models Methods Appl. Sci.* **1995**, *5*, 225–237. [CrossRef]
20. Ladyzenskaja, O.A.; Solonnikov, V.A.; Ural'ceva, N.N. *Linear and Quasi-Linear Equations of Parabolic Type*; American Mathematical Society: Providence, RI, USA, 1968.
21. Li, X.; Yong, J. *Optimal Control for Infinite Dimensional Systems*; Birkhauser: Boston, MA, USA, 1995.
22. Hackbush, W. A numerical method for solving parabolic equations with opposite orientations. *Computing* **1978**, *20*, 229–240. [CrossRef]

23. Lenhart, S.; Workman, J.T. *Optimal Control Applied to Biological Models*; Chapman and Hall/CRC: Boca Raton, FL, USA, 2007.
24. Iserles, A. *A First, Course in the Numerical Analysis of Differential Equations*; Cambridge University Press: New York, NY, USA, 2008.

Article

Directionally Correlated Movement Can Drive Qualitative Changes in Emergent Population Distribution Patterns

Jonathan R. Potts

School of Mathematics and Statistics, University of Sheffield, Hicks Building, Hounsfield Road, Sheffield S3 7RH, UK; j.potts@sheffield.ac.uk

Received: 26 June 2019; Accepted: 16 July 2019; Published: 18 July 2019

Abstract: A fundamental goal of ecology is to understand the spatial distribution of species. For moving animals, their location is crucially dependent on the movement mechanisms they employ to navigate the landscape. Animals across many taxa are known to exhibit directional correlation in their movement. This work explores the effect of such directional correlation on spatial pattern formation in a model of between-population taxis (i.e., movement of each population in response to the presence of the others). A telegrapher-taxis formalism is used, which generalises a previously studied diffusion-taxis system by incorporating a parameter T, measuring the characteristic time for directional persistence. The results give general criteria for determining when changes in T will drive qualitative changes in the predictions of linear pattern formation analysis for $N \geq 2$ populations. As a specific example, the $N = 2$ case is explored in detail, showing that directional correlation can cause one population to 'chase' the other across the landscape while maintaining a non-constant spatial distribution. Overall, this study demonstrates the importance of accounting for directional correlation in movement for understanding both quantitative and qualitative aspects of species distributions.

Keywords: animal movement; correlated random walk; movement ecology; population dynamics; taxis; telegrapher's equation

---

1. Introduction

Understanding spatial distributions of animal species is a key concern for ecology, being of fundamental importance for a wide range of applications including conservation efforts [1], quantifying biodiversity [2], and invasive species research [3,4]. For mobile animals, decisions about where to move drive their individual locations. Consequently there has been a huge amount of effort in recent years to understand the causes and consequences of animal movement [5–7]. However, these movement decisions have an effect not only on individuals' locations but also on the spatial distribution of the whole population [8]. To understand this individual-to-population upscaling in a non-speculative way requires mathematical models that are built from the underlying movement processes of individuals [9]. Such models exhibit emergent phenomena on the population level that can be then quantified and related concretely to the underlying mechanisms of individual movement (e.g., [10,11]).

One recent study in this general area examines how the movement responses between individuals from $N \geq 2$ different animal populations can drive spatio-temporal distribution patterns on temporal scales whereby births and deaths are negligible [12]. This takes spatial ecology in a slightly different direction to its tradition trajectory, whereby the combination of nonlinear birth-and-death terms (a.k.a. kinetics) combine with diffusive or cross-diffusive movement to drive spatio-temporal patterns [13–17]. Instead, the study of [12] shows that inter-population taxis (a form of cross-diffusion) can drive a wide range of complex patterns on its own, without the need for nonlinear kinetics.

However, the study by [12] assumes that individuals move diffusively in the absence of interactions. While this may be a reasonable approximation in many cases, in reality animals will always display some directional correlation in movement [18,19], if only due to the significant energetic costs of turning [20]. For some organisms, this correlation may only persist over quite small spatio-temporal scales, in which case diffusion is a reasonable model [21,22]. However, if the spatial scale of correlation is at a similar order of magnitude to the scale over which animals move in the time between successive interactions, diffusive assumptions may be less valid [23].

Here, I address this issue of directional correlation by extending the model of [12] from a diffusion-taxis system to a telegrapher-taxis system in 1D. The telegrapher's equation accurately models random movement with directional correlation in a 1D setting [24,25]. Moreover, it is a direct generalisation of the diffusion equation, simply requiring the introduction of a characteristic time-scale parameter, T, which is set to $T = 0$ in the diffusion limit.

The aim of this study is to examine the effect of introducing $T > 0$ on the linear pattern formation properties of the model in [12]. These patterning properties separate parameter space into three well-known categories [26]. First, patterns may not form at all from small non-constant perturbations of the steady state, i.e., the system is *stable* to such perturbations. Second, small non-constant perturbations of the steady state may grow over small times in a non-oscillatory fashion. This is classically known as a *Turing instability* after [27]. Third, these perturbations may both grow in magnitude and oscillate, which is sometimes known as a *Turing-Hopf instability*.

I give general criteria for when an increase T can cause a shift from one of these three patterning regions to another. When this shift occurs, it holds for all $T > T_*$ where T_* is a threshold persistence time, which is quantified. As an example, I examine in detail the case $N = 2$ in the absence of self-aggregation. Here, the diffusion-taxis model ($T = 0$) can either be stable to non-constant perturbations or exhibit a Turing instability, dependent on the parameter values of the model. Furthermore, the parameter regimes where the system falls into these two patterning regions is known precisely [12]. However, the $N = 2$ and $T = 0$ case is never susceptible to a Turing-Hopf instability unless a self-aggregation process is in play [28]. I show that when T is increased sufficiently high, Turing-Hopf instabilities are possible without self-aggregation. Furthermore, they occur when the populations are engaged in a "pursuit-and-avoid" situation, with one population exhibiting taxis towards the other, and the latter exhibiting taxis away from the first. In contrast, for $T = 0$, this pursuit-and-avoid situation is always linearly stable to non-constant perturbations.

2. The Modelling Framework and General Results

This work focusses on a system of N populations, each of fixed size (i.e., no births or deaths). Individuals from each population move on a 1D landscape as biased, correlated random walkers (i.e., those where there is both persistence in movement and bias in a particular direction). This bias depends on the density of the various populations in the system, and is represented by a taxis term up or down the density gradient of the various populations. The correlated aspect of movement is modelled using a telegrapher's equation formalism [24,25].

Denoting by $u_i(x,t)$ the spatial probability density function of population i at time t, the study system is given by the following telegrapher-taxis equation for each population i ($i \in \{1,\ldots,N\}$)

$$T\frac{\partial^2 u_i}{\partial t^2} + \frac{\partial u_i}{\partial t} = d_i \frac{\partial^2 u_i}{\partial x^2} - \frac{\partial}{\partial x}\left[u_i \sum_{j=1}^{N} \gamma_{ij} \frac{\partial}{\partial x}(\mathcal{K} * u_j)\right], \tag{1}$$

where $T \geq 0$, $d_i > 0$, $\gamma_{ij} \in \mathbb{R}$, and one can assume, without loss of generality, that the units are dimensionless and $d_1 = 1$. The case where $T = 0$ and $\gamma_{ii} = 0$ for all i was studied in [12], and the reader is referred there for details of the non-dimensionalisation process. The dynamics take place on a unit line segment, $[0,1]$, with periodic boundary conditions

$$u_i(0,t) = u_i(1,t), \tag{2}$$

so locations, x, live in the quotient space $[0,1]/\{0,1\}$. In Equation (1), $\mathcal{K}(x)$ is an integrable function, symmetric about $x = 0$ on the domain $[0,1]/\{0,1\}$, and $\mathcal{K} * u_j$ is the following spatial convolution

$$\mathcal{K} * u_j(x,t) = \int_0^1 \mathcal{K}(x-y)u_j(y,t)\mathrm{d}y. \tag{3}$$

The inter-population taxis term, in the right-hand summand of Equation (1), can come about through different biological mechanisms. The simplest is for the organisms in population i to sense directly the gradient of population density, $\frac{\partial}{\partial x}(\mathcal{K} * u_j)$, a mechanism that may be valid for very small individuals such as single-celled organisms or swarming insects. However, for larger organisms, this population gradient is more likely to be observed indirectly. For example, the deposition of marks in the environment from population j (e.g., through scenting; [29,30]) may indicate the population density, u_j, across space. Similarly, memory of past interactions with individuals from population j can act as a proxy for sensing the population density gradient [31,32]. In [12], the authors showed how all of these biological mechanism can be modelled via the same taxis term, given in Equation (1).

The linear pattern formation properties of Equation (1) are analysed by perturbing the system about the constant steady-state solution, $u_i(x,t) = 1$ for all i, x, t. Specifically, let $\mathbf{w}(x,t) = (u_1 - 1, \ldots, u_N - 1)' = (u_1^{(0)}, \ldots, u_N^{(0)})' \exp(\sigma t + i\kappa x)$, where $u_1^{(0)}, \ldots, u_N^{(0)} \in \mathbb{R}$, $\kappa \in \mathbb{R}_{\geq 0}$ and $\sigma \in \mathbb{C}$ are constants, and $'$ denotes matrix transpose. By neglecting non-linear terms, Equation (1) becomes

$$(T\sigma^2 + \sigma)\mathbf{w} = \kappa^2 M(\kappa)\mathbf{w}, \tag{4}$$

where $M(\kappa) = [M_{ij}(\kappa)]_{i,j}$ is an $N \times N$ matrix with

$$M_{ij}(\kappa) = \begin{cases} -d_i + \gamma_{ii}\hat{\mathcal{K}}(\kappa), & \text{if } i = j, \\ \gamma_{ij}\hat{\mathcal{K}}(\kappa), & \text{otherwise,} \end{cases} \tag{5}$$

and $\hat{\mathcal{K}}(\kappa)$ is the Fourier transform of $\mathcal{K}(x)$ on $[0,1]/\{0,1\}$.

Let $\lambda_1(\kappa), \ldots, \lambda_N(\kappa)$ be the eigenvalues of $M(\kappa)$ (which are not necessarily distinct). If $T = 0$ then $\sigma = \kappa^2\lambda_i(\kappa)$ gives a solution to Equation (4) for some non-trivial vector $\mathbf{w}$. From the perspective of pattern formation, there are three regimes of interest. These are well-documented [26] but it is valuable to summarise them briefly, for the purposes of introducing the key concepts and nomenclature used throughout this work:

1. **Stable.** All eigenvalues have negative real part: $\mathrm{Re}(\lambda_i(\kappa)) < 0$ for all $i \in \{1, \ldots, N\}$, $\kappa > 0$,
2. **Turing instability.** The dominant eigenvalue (i.e., the one with the largest real part) is positive and real, i.e., $\mathrm{argmax}_{\lambda_i(\kappa)}[\mathrm{Re}(\lambda_i(\kappa))] \in \mathbb{R}_{>0}$,
3. **Turing-Hopf instability.** The dominant eigenvalue is not real but has positive real part, i.e., $\mathrm{argmax}_{\lambda_i(\kappa)}[\mathrm{Re}(\lambda_i(\kappa))] \in \{z \in \mathbb{C} : \mathrm{Re}(z) > 0, z \notin \mathbb{R}\}$.

The first of these states that the linear perturbation, $\mathbf{w}(x,t)$, will decay back to the homogeneous steady state, the second that $\mathbf{w}(x,t)$ will grow at small times for certain wavenumbers, κ, in a non-oscillatory fashion, and the third that $\mathbf{w}(x,t)$ will grow and oscillate at small times. These regimes give an indication for the pattern formation properties of the system. In the Stable region, the expectation is that spatial patterns do not form. In the Turing instability region, stationary patterns are predicted to form, which are fixed over time. If there is a Turing-Hopf instability, patterns that are in perpetual flux are expected to emerge. However, it is important to note that these are merely predictions, arising from a linearisation of the system, and that it is possible for the non-linear terms to cause different pattern formation properties asymptotically.

The main aim of this work is to examine how the demarcation into the above three regimes changes as T is increased from 0. When $T > 0$, for each $i \in \{1, \ldots, N\}$, there are up to two values of σ such that $T\sigma^2 + \sigma = \kappa^2\lambda_i(\kappa)$. Thus I define

$$\sigma_i^{\pm}(\kappa, T) = \frac{-1 \pm \sqrt{1 + 4T\kappa^2\lambda_i(\kappa)}}{2T}, \tag{6}$$

for each $i \in \{1, \ldots, N\}$, and note that $\sigma_i^{\pm}(\kappa, T)$ solves Equation (4) for each i. The main results from this work are given in the following three theorems.

Theorem 1. *In the case where the system is linearly stable for $T = 0$ (i.e., $Re(\lambda_i(\kappa)) < 0$ for all $i \in \{1, \ldots, N\}$, $\kappa > 0$) one of two situations can occur:*

1. *If $\lambda_i(\kappa) \in \mathbb{R}$ for all i, κ, then $Re(\sigma_i^{\pm}(\kappa, T)) < 0$ for all i, κ, T, so the system stays in the Stable regime for all $T > 0$.*
2. *If there exist i and κ such that $\lambda_i(\kappa) \notin \mathbb{R}$ then there exists some $T_* > 0$ such that for all $T > T_*$, there is a Turing-Hopf instability. In other words, for this value of i and κ, $argmax_{\sigma_i(\kappa,T)}[Re(\sigma_i(\kappa, T)] \in \{z \in \mathbb{C} : Re(z) > 0, z \notin \mathbb{R}\}$ for all $T > T_*$. Furthermore, T_* is the minimum $T > 0$ such that there exist i, κ with $Re(\sqrt{1 + 4T\kappa^2\lambda_i(\kappa)}) > 1$.*

Proof. For part (1), if $\lambda_i(\kappa) \in \mathbb{R}$ for all i, κ then, since $\text{Re}(\lambda_i(\kappa)) < 0$, the inequality $1 + 4T\kappa^2\lambda_i(\kappa) < 1$ holds, so $\text{Re}(\sigma_i^{\pm}(\kappa, T)) < 0$ for all i, κ, T.

For part (2), let $i \in \{1, \ldots, N\}$ and $\kappa > 0$ such that $\lambda_i(\kappa) \notin \mathbb{R}$. Assume, without loss of generality, that $\text{Im}(\lambda_i(\kappa)) > 0$ (otherwise, pick the complex conjugate of $\lambda_i(\kappa)$). Then if $\text{Re}(\sqrt{1 + 4T\kappa^2\lambda_i(\kappa)}) > 1$, the inequality $\text{Re}(\sigma_i^{+}(\kappa, T)) > 0$ holds, so that there is a Turing-Hopf instability.

I now show that if T is arbitrarily large, the criterion $\text{Re}(\sqrt{1 + 4T\kappa^2\lambda_i(\kappa)}) > 1$ is always satisfied. Here,

$$\text{Re}(\sqrt{1 + 4T\kappa^2\lambda_i(\kappa)}) \approx \text{Re}(\sqrt{4T\kappa^2\lambda_i(\kappa)}). \tag{7}$$

Now, $\arg(\sqrt{4T\kappa^2\lambda_i(\kappa)}) = \arg(\lambda_i(\kappa))/2$. Since $\text{Im}(\lambda_i(\kappa)) > 0$ and $\text{Re}(\lambda_i(\kappa)) < 0$, the following holds

$$\pi/2 < \arg(\lambda_i(\kappa)) < \pi. \tag{8}$$

Therefore

$$0 < \arg(\sqrt{4T\kappa^2\lambda_i(\kappa)}) < \pi/2,$$

so that $\text{Re}(\sqrt{4T\kappa^2\lambda_i(\kappa)}) > 0$. Hence $\text{Re}(\sqrt{1 + 4T\kappa^2\lambda_i(\kappa)}) > 0$ whenever T is sufficiently large. Furthermore, $\text{Re}(\sqrt{1 + 4T\kappa^2\lambda_i(\kappa)}) \to \infty$ as $T \to \infty$, so there exists some $T_*^{i,\kappa}$ such that $\text{Re}(\sqrt{1 + 4T\kappa^2\lambda_i(\kappa)}) > 1$ for all $T > T_*^{i,\kappa}$. There may be more than one $j \in \{1, \ldots, N\}$ and $\kappa > 0$ such that $\lambda_j(\kappa) \notin \mathbb{R}$. Thus let T_* be the minimum of $T_*^{j,\kappa}$ over such j and all κ. Then T_* satisfies the requirements of the theorem. □

Theorem 2. *Consider the Turing instability case for $T = 0$ (i.e., $argmax_{\lambda_i(\kappa)}[Re(\lambda_i(\kappa)] \in \mathbb{R}_{>0}$). For a given κ, let $\lambda_i(\kappa)$ be the dominant eigenvalue of $M(\kappa)$. Then one of two situations can occur:*

1. *If $Re(\sqrt{1 + 4\kappa^2T\lambda_j(\kappa)}) \leq \sqrt{1 + 4\kappa^2T\lambda_i(\kappa)}$ for all j then there is a Turing instability at wavenumber κ and persistence time T.*
2. *If there is some j such that $Re(\sqrt{1 + 4\kappa^2T\lambda_j(\kappa)}) > \sqrt{1 + 4\kappa^2T\lambda_i(\kappa)}$ then there is a Turing-Hopf instability at wavenumber κ and persistence time T. Let T_* be the minimum $T > 0$ such that*

$Re(\sqrt{1+4\kappa^2 T\lambda_j(\kappa)}) > \sqrt{1+4\kappa^2 T\lambda_i(\kappa)}$ for some j. Then there is a Turing-Hopf instability for all $T > T_*$.

Proof. For part (1), let $\kappa, T > 0$. If $\mathrm{Re}(\sqrt{1+4\kappa^2 T\lambda_j(\kappa)}) \leq \sqrt{1+4\kappa^2 T\lambda_i(\kappa)}$ for all j then $\sigma_i^+(\kappa, T)$ is the dominant eigenvalue. Since $\lambda_i(\kappa) \in \mathbb{R}_{>0}$, $\sqrt{1+4\kappa^2 T\lambda_i(\kappa)} \in \mathbb{R}_{>1}$ so $\sigma_i^+(\kappa, T) \in \mathbb{R}_{>0}$ so there is a Turing instability for these values of T and κ.

For part (2), suppose there exist j, T, κ with $\mathrm{Re}(\sqrt{1+4\kappa^2 T\lambda_j(\kappa)}) > \sqrt{1+4\kappa^2 T\lambda_i(\kappa)}$ and suppose $\mathrm{Re}(\sqrt{1+4\kappa^2 T\lambda_j(\kappa)}) \geq \mathrm{Re}(\sqrt{1+4\kappa^2 T\lambda_k(\kappa)})$ for all k. Then $\sigma_j^+(\kappa, T)$ is the dominant eigenvalue. It is necessary to show that $\sigma_j^+(\kappa, T)$ is not real. Therefore, for a contradiction, suppose $\sqrt{1+4\kappa^2 T\lambda_j(\kappa)} \in \mathbb{R}$. Then $\lambda_j(\kappa) \in \mathbb{R}$. However, $\lambda_j(\kappa) \leq \lambda_i(\kappa)$, since $\lambda_i(\kappa)$ is the dominant eigenvalue of $M(\kappa)$, so $\sqrt{1+4\kappa^2 T\lambda_j(\kappa)} \leq \sqrt{1+4\kappa^2 T\lambda_i(\kappa)}$, which contradicts the assumption. Hence $\sigma_j^+(\kappa, T) \notin \mathbb{R}$ so there is a Turing-Hopf instability for these values of T and κ. □

Corollary. If there is some j such that $\mathrm{Re}(\sqrt{\lambda_j(\kappa)}) > \sqrt{\lambda_i(\kappa)}$ then there is a Turing-Hopf instability at wavenumber κ for sufficiently large T.

Theorem 3. *Consider the case where there is a Turing-Hopf instability for $T = 0$. Then there is a Turing-Hopf instability for all $T > 0$.*

Proof. Let i be such that $\lambda_i(\kappa)$ is the dominant eigenvalue of $M(\kappa)$ for some κ where there is a Turing-Hopf instability for $T = 0$. Assume, without loss of generality, that $\mathrm{Im}(\lambda_i(\kappa)) > 0$ (otherwise choose the complex conjugate of $\lambda_i(\kappa)$). Since $\mathrm{Re}(\lambda_i(\kappa)) > 0$, the inequality $\mathrm{Re}(1 + 4T\kappa^2\lambda_i(\kappa)) > 1$ holds, so $\sqrt{\mathrm{Re}(1+4T\kappa^2\lambda_i(\kappa))} > 1$. Since $1 + 4T\kappa^2\lambda_i(\kappa)$ has positive real and imaginary part, $\sqrt{\mathrm{Re}(1+4T\kappa^2\lambda_i(\kappa))} < \mathrm{Re}(\sqrt{1+4T\kappa^2\lambda_i(\kappa)})$. The latter follows from the following general calculation for $r > 0$ and $0 < \theta < \pi/2$

$$\sqrt{\mathrm{Re}(re^{\mathrm{i}\theta})} = \sqrt{r}\sqrt{\cos(\theta)} < \sqrt{r}\sqrt{\frac{\cos(\theta)+1}{2}} = \sqrt{r}\cos\left(\frac{\theta}{2}\right) = \mathrm{Re}(\sqrt{re^{\mathrm{i}\theta}}). \tag{9}$$

The inequality in (9) requires $0 < \cos(\theta) < 1$, which follows from $0 < \theta < \pi/2$.

It follows that $\mathrm{Re}(\sqrt{1+4T\kappa\lambda_i(\kappa)}) > 1$, and thus from Equation (6) that $\mathrm{Re}(\sigma_i^+(\kappa, T)) > 0$ for all $T > 0$. To show that this is within the Turing-Hopf instability region, it is necessary to check that there does not exist j such that both $\sigma_j^+(\kappa, T) \in \mathbb{R}_{>0}$ and $\sigma_j^+(\kappa, T) > \mathrm{Re}(\sigma_i^+(\kappa, T))$, since otherwise this is the Turing instability region. For a contradiction, suppose such a $\sigma_j^+(\kappa, T)$ exists. Then

$$\sqrt{1+4T\kappa^2\lambda_j(\kappa)} > \mathrm{Re}(\sqrt{1+4T\kappa^2\lambda_i(\kappa)}) > \sqrt{\mathrm{Re}(1+4T\kappa^2\lambda_i(\kappa))} \tag{10}$$

so $1 + 4T\kappa^2\lambda_j(\kappa) > \mathrm{Re}(1 + 4T\kappa^2\lambda_i(\kappa))$ so $\lambda_j(\kappa) > \mathrm{Re}(\lambda_i(\kappa))$, which contradicts the fact that $\lambda_i(\kappa)$ is the dominant eigenvalue for the $T = 0$ case. □

3. The Case of Two Interacting Populations ($N = 2$)

In this section, I look in detail at the specific example where $N = 2$ and $\gamma_{ii} = 0$ for all $i \in \{1, \ldots, N\}$, to show how persistent movement can drive qualitative changes in pattern formation properties even in this simple situation. In the case $T = 0$ (studied by [12]), the following holds

$$\lambda_{\pm}(\kappa) = \frac{-(1+d_2) \pm \sqrt{(1-d_2)^2 + 4\gamma_{12}\gamma_{21}\hat{\mathcal{K}}^2(\kappa)}}{2}, \tag{11}$$

and it follows that the system for $T = 0$ is linearly stable to wave perturbations whenever $\gamma_{12}\gamma_{21} < d_2$. When $\gamma_{12}\gamma_{21} > d_2$, the system can have a Turing instability, and always does in the limit $\hat{\mathcal{K}}(\kappa) \to 1$, where the spatial averaging given by $\mathcal{K}$ is arbitrarily small. The system never exhibits a Turing-Hopf instability for $T = 0$ [12].

In the Turing instability case for $T = 0$, both the eigenvalues are real, so the conditions of Theorem 2, part 2, are met. Thus there is a Turing instability for all $T > 0$. The case where $\gamma_{12}\gamma_{21} < d_2$, however, is more interesting. Here, Theorem 1 states that a Turing-Hopf instability will occur for sufficiently large T as long as $\lambda_{\pm}(\kappa) \notin \mathbb{R}$, that is

$$(1 - d_2)^2 + 4\gamma_{12}\gamma_{21}\hat{\mathcal{K}}^2(\kappa) < 0. \tag{12}$$

The first thing to notice about the condition given by (12) is that γ_{12} and γ_{21} need to be of opposite signs. This means that one population tends to advect up the density gradient of the other, while the second population retreats down the gradient of the first. This is termed a 'pursuit-and-avoid' situation in [12]. In this situation, for $T = 0$, no patterns form: the populations each settle to a steady-state distribution whereby they are uniformly distributed on the terrain. However, if T_* is defined to be the minimum positive real number such that there exists $\kappa > 0$ with $\mathrm{Re}(\sqrt{1 + 4T\kappa^2\lambda_+(\kappa)}) > 1$, then Theorem 1 implies there is a Turing-Hopf instability whenever $T > T_*$.

To understand how T_* depends on the parameters, I fix $d_2 = 1$ and set $\gamma = +\sqrt{-\gamma_{12}\gamma_{21}}$. (Notice that $\gamma \in \mathbb{R}_{>0}$ since γ_{12} and γ_{21} are of opposite signs.) It is also necessary to pick a particular functional form for $\mathcal{K}(x)$. This is a slightly delicate matter as it is not necessarily the case that the pattern formation problem is well-posed for an arbitrary $\mathcal{K}(x)$. For example, if $\mathcal{K}(x) = \delta(x)$ is used, where $\delta(x)$ is the Dirac delta function, then, by Equation (6), the dominant eigenvalue is

$$\sigma_i^+(\kappa, T) = \frac{-1 + \sqrt{1 - 4T\kappa^2 + 4T\mathrm{i}\kappa^2\gamma}}{2T}. \tag{13}$$

If T is fixed and κ is arbitrarily large, then $\mathrm{Re}(\sigma_i^+(\kappa, T)) \approx \kappa\mathrm{Re}(\sqrt{-1 + \mathrm{i}\gamma}/\sqrt{T}) \to \infty$ as $\kappa \to \infty$. Hence patterns can form for arbitrarily high wavenumbers, meaning the pattern formation problem is ill-posed.

However, suppose instead that

$$\mathcal{K}(x) = \begin{cases} \frac{\exp\left(-\frac{x^2}{\sigma^2}\right)}{\sigma\sqrt{\pi}\mathrm{erf}\left(\frac{1}{2\sigma}\right)}, & \text{if } -1/2 < x < 1/2 \\ 0, & \text{otherwise,} \end{cases} \tag{14}$$

for some $\sigma > 0$, so that

$$\hat{\mathcal{K}}(\kappa) = \frac{\mathrm{erf}\left(\frac{1}{2\sigma} - \frac{\mathrm{i}\kappa\sigma}{2}\right) + \mathrm{erf}\left(\frac{1}{2\sigma} + \frac{\mathrm{i}\kappa\sigma}{2}\right)}{2\mathrm{erf}\left(\frac{1}{2\sigma}\right)} \exp\left(\frac{-\kappa^2\sigma^2}{4}\right). \tag{15}$$

The case of interest is where σ is small, in which case the following approximation can be made

$$\hat{\mathcal{K}}(\kappa) \approx \exp\left(\frac{-\kappa^2\sigma^2}{4}\right). \tag{16}$$

Then Equation (6) gives

$$\sigma_i^+(\kappa, T) = \frac{-1 + \sqrt{1 - 4T\kappa^2 + 4T\mathrm{i}\gamma\kappa^2\exp(-\kappa^2\sigma^2/4)}}{2T}. \tag{17}$$

Now, notice that if T is fixed and κ is arbitrarily large, then

$$\text{Re}(\sigma_i^+(\kappa, T)) \approx \kappa\text{Re}\left(\frac{\sqrt{-1+\text{i}\gamma\exp(-\kappa^2\sigma^2/4)}}{\sqrt{T}}\right) \to 0, \tag{18}$$

as $\kappa \to \infty$. Therefore, with the Gaussian averaging kernel from Equation (14), the pattern formation problem is well-posed, in the sense that patterns cannot form at arbitrarily high wavenumbers.

By Theorem 1, there is some T_* such that there is a Turing-Hopf instability for all $T > T_*$. Furthermore, this Theorem states that T_* is the minimum T such that the following holds for some κ

$$f(\kappa) := Re[\sqrt{1 - 4T\kappa^2 + 4T\text{i}\gamma\kappa^2\exp(-\kappa^2\sigma^2/4)}] > 1. \tag{19}$$

In Figure 1a, $f(\kappa)$ is plotted against κ for $\gamma = 0.2$, $\sigma = 0.05$, and varying values of T. This reveals that T not only affects whether patterns form, but the range of wavenumbers for which patterns may form. In this example, there is a Turing-Hopf bifurcation for value of $T = T_*$ somewhere between $T = 0.05$ and $T = 0.1$.

The precise values of T_* for a range of γ- and σ-values are plotted in Figure 1b,c. The γ parameter encodes the strength of attraction/avoidance. Figure 1b,c shows that T_* decays for increasing γ and grows with the width of spatial averaging, σ.

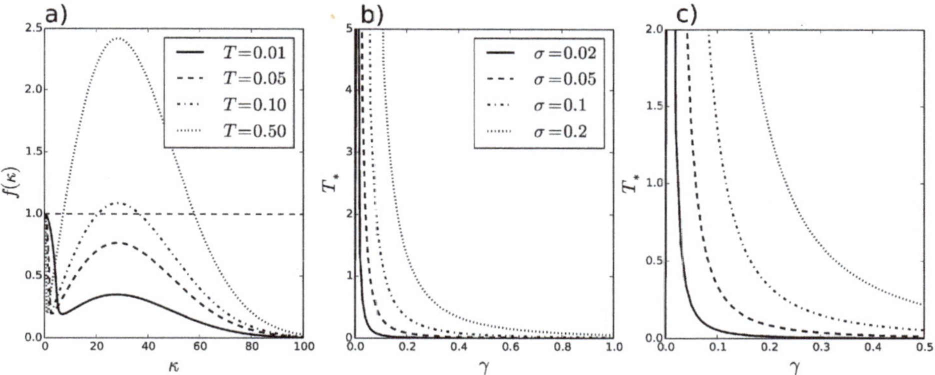

Figure 1. Critical value of T for pattern formation. In Panel (**a**), Equation (19) is plotted for $\gamma = 0.2$, $\sigma = 0.05$, and various values of T. Where $f(\kappa) > 1$, there is a Turing-Hopf instability. In Panels (**b**,**c**), the value, T_*, above which there is a Turing-Hopf instability for some κ and below which there is not, is plotted for various values of σ and γ.

To understand the qualitative properties of the patterns that emerge from increasing T_* past the Turing-Hopf bifurcation point, the system in Equations (1)–(3) is solved numerically, with $N = 2$, $d_2 = 1$, $\gamma = +\sqrt{-\gamma_{12}\gamma_{21}}$, and the Gaussian spatial averaging from Equation (14). In Figure 2, the results of these numerics are shown for certain values of γ, σ and T. Here, rather than decaying to the homogeneous steady state ($T = 0$), for higher T the populations move across the landscape while maintaining a non-constant population distribution.

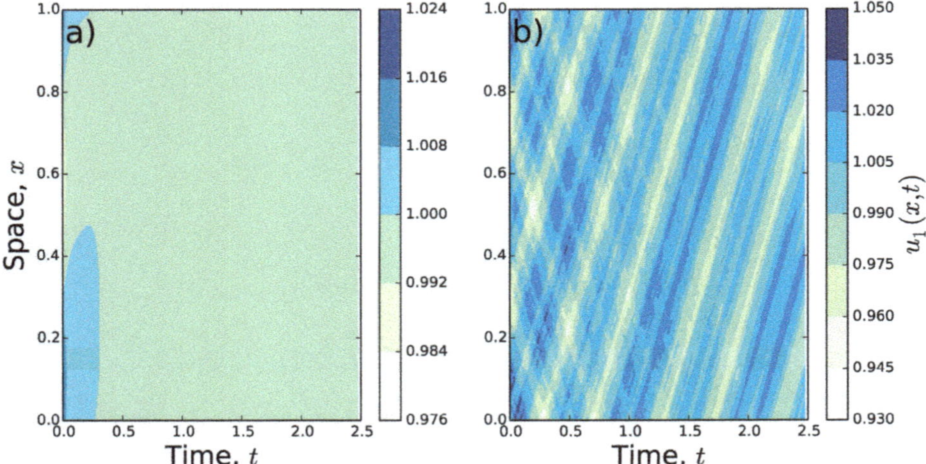

Figure 2. Example numerics for $N = 2$. Numerical solutions of Equation (1) for $N = 2$, $d_2 = 1$, $\gamma_{12} = 1$, $\gamma_{21} = -1$, and $\mathcal{K}(x)$ as in Equation (14) with $\sigma = 0.1$. Both examples have random continuous initial conditions, but forced to be symmetric about $x = 0.5$ to satisfy periodic boundary conditions. In Panel (**a**), $T = 0$, and the initial perturbation of the constant steady state decays to the solution $u_1(x, \infty) = 1$. In Panel (**b**), $T = 1$. Here, the population u_1 moves across the landscape, not settling to a constant distribution.

4. Discussion

This study examines a telegrapher-taxis system of N populations to demonstrate how directional correlation can drive pattern formation in systems of between-population taxis. General criteria are given for switches in pattern formation regime driven by directional correlation and the $N = 2$ case is examined in detail. These results demonstrate the importance of considering directional correlation when seeking to understanding the spatio-temporal population distributions of moving organisms. As our ability to capture data on animal movement is becoming increasingly sophisticated, better understanding of how the details of animal movement can affect population dynamics is becoming an ever-more pertinent question [6], with implications for a diverse range of ecological areas such as connectivity dynamics [33] and conservation [34].

The model of [12], on which the present model is based, is closely related to aggregation and chemotaxis models inspired by cell biology [35–38]. Although many such models only examine a single population, there are various examples of two-population models [28,39], and some that incorporate an arbitrary number of populations [40,41]. In the latter examples, the diffusion-taxis equations are coupled via interactions with a diffusive chemical, different to the model studied here. While cells lack the momentum of much larger organisms, directional correlation is known to be a factor in the movement of cells in certain circumstances [42–44]. Therefore the results presented here suggest that it is worth exploring how directional correlation may affect the pattern formation properties of such aggregation and chemotaxis models.

Here, ecosystems are modelled on a constrained timescale whereby births and deaths are negligible. However, it would be valuable to extend the model presented here to incorporate such effects, via competition and/or predation terms (i.e., kinetics), and so explore the effect of directional persistence over longer timescales. Adding directional persistence to models that incorporate kinetics is non-trivial, though, and does not simply involve adding a second temporal derivative to a reaction-diffusion-taxis model [45].

Nonetheless, building such a system of reaction-telegrapher-taxis equations would enable connection with cross-diffusion models (e.g., [16,17,46,47]). These are generalisations of reaction-diffusion systems, whereby terms are added that allow for motion driven by the presence of foreign populations. These terms include, but are not limited to, the taxis terms observed in the model from Equation (1). However, incorporating directional correlation via a telegrapher's term into a cross-diffusive setting is much rarer (but see [48]) and the pattern formation properties of such systems are not well-explored. Given the results presented here, it may be interesting to extend cross-diffusive models in this way, to show how directional correlation may affect pattern formation in such models.

In summary, the simple but general results shown here demonstrate that directional correlation of individuals' movement can have a great effect on the spatio-temporal distribution of species. While I have demonstrated this in a model of ecosystems relevant over relatively short timescales, where births and deaths are minimal, it highlights a general principle that is little-studied and may have much wider implications.

Acknowledgments: JRP acknowledges support from the Natural Environment Research Council (NERC) (grant number NE/R001669/1).

Conflicts of Interest: The author declares no conflict of interest.

References

1. Johnson, C.J.; Seip, D.R.; Boyce, M.S. A quantitative approach to conservation planning: using resource selection functions to map the distribution of mountain caribou at multiple spatial scales. *J. Appl. Ecol.* **2004**, *41*, 238–251. [CrossRef]
2. Ferrier, S.; Guisan, A. Spatial modelling of biodiversity at the community level. *J. Appl. Ecol.* **2006**, *43*, 393–404. [CrossRef]
3. Peterson, A.T. Predicting the geography of species invasions via ecological niche modeling. *Q. Rev. Biol.* **2003**, *78*, 419–433. [CrossRef] [PubMed]
4. Lewis, M.A.; Petrovskii, S.V.; Potts, J.R. *The Mathematics Behind Biological Invasions*; Springer: Berlin, Germany, 2016; Volume 44.
5. Nathan, R.; Getz, W.M.; Revilla, E.; Holyoak, M.; Kadmon, R.; Saltz, D.; Smouse, P.E. A movement ecology paradigm for unifying organismal movement research. *Proc. Natl. Acad. Sci. USA* **2008**, *105*, 19052–19059. [CrossRef]
6. Börger, L. Stuck in motion? Reconnecting questions and tools in movement ecology. *J. Anim. Ecol.* **2016**, *85*, 5–10. [CrossRef] [PubMed]
7. Hays, G.C.; Ferreira, L.C.; Sequeira, A.M.; Meekan, M.G.; Duarte, C.M.; Bailey, H.; Bailleul, F.; Bowen, W.D.; Caley, M.J.; Costa, D.P.; et al. Key questions in marine megafauna movement ecology. *Trends Ecol. Evol.* **2016**, *31*, 463–475. [CrossRef] [PubMed]
8. Mueller, T.; Fagan, W.F. Search and navigation in dynamic environments–from individual behaviors to population distributions. *Oikos* **2008**, *117*, 654–664. [CrossRef]
9. Lewis, M.A.; Maini, P.K.; Petrovskii, S.V. Dispersal, individual movement and spatial ecology. In *Lecture Notes in Mathematics (Mathematics Bioscience Series)*; Springer: Berlin/Heidelberg, Germany, 2013; Volume 2071. [CrossRef]
10. Moorcroft, P.R.; Lewis, M.A.; Crabtree, R.L. Mechanistic home range models capture spatial patterns and dynamics of coyote territories in Yellowstone. *Proc. R. Soc. B* **2006**, *273*, 1651–1659. [CrossRef]
11. Potts, J.R.; Mokross, K.; Lewis, M.A. A unifying framework for quantifying the nature of animal interactions. *J. R. Soc. Interface* **2014**, *11*, 20140333. [CrossRef]
12. Potts, J.R.; Lewis, M.A. Spatial Memory and Taxis-Driven Pattern Formation in Model Ecosystems. *Bull. Math. Biol.* **2019**. [CrossRef]
13. Shigesada, N.; Kawasaki, K.; Teramoto, E. Spatial segregation of interacting species. *J. Theor. Biol.* **1979**, *79*, 83–99. [CrossRef]
14. Durrett, R.; Levin, S. The importance of being discrete (and spatial). *Theor. Pop. Biol.* **1994**, *46*, 363–394. [CrossRef]

15. Sherratt, J.A.; Lewis, M.A.; Fowler, A.C. Ecological chaos in the wake of invasion. *Proc. Natl. Acad. Sci. USA* **1995**, *92*, 2524–2528. [CrossRef] [PubMed]
16. Lee, J.; Hillen, T.; Lewis, M. Pattern formation in prey-taxis systems. *J. Biol. Dyn.* **2009**, *3*, 551–573. [CrossRef] [PubMed]
17. Gambino, G.; Lombardo, M.C.; Sammartino, M. A velocity–diffusion method for a Lotka–Volterra system with nonlinear cross and self-diffusion. *Appl. Numer. Math.* **2009**, *59*, 1059–1074. [CrossRef]
18. Kareiva, P.; Shigesada, N. Analyzing insect movement as a correlated random walk. *Oecologia* **1983**, *56*, 234–238. [CrossRef] [PubMed]
19. Codling, E.A.; Plank, M.J.; Benhamou, S. Random walk models in biology. *J. R. Soc. Interface* **2008**, *5*, 813–834. [CrossRef]
20. Wilson, R.P.; Griffiths, I.W.; Legg, P.A.; Friswell, M.I.; Bidder, O.R.; Halsey, L.G.; Lambertucci, S.A.; Shepard, E.L.C. Turn costs change the value of animal search paths. *Ecol. Lett.* **2013**, *16*, 1145–1150. [CrossRef]
21. Patlak, C.S. Random walk with persistence and external bias. *Bull. Math. Biophys.* **1953**, *15*, 311–338. [CrossRef]
22. Turchin, P. Translating foraging movements in heterogeneous environments into the spatial distribution of foragers. *Ecology* **1991**, *72*, 1253–1266. [CrossRef]
23. Giuggioli, L.; Potts, J.; Harris, S. Predicting oscillatory dynamics in the movement of territorial animals. *J. R. Soc. Interface* **2012**, *9*, 1529–1543. [CrossRef] [PubMed]
24. Kac, M. A stochastic model related to the telegrapher's equation. *Rocky Mt. J. Math.* **1956**, *4*, 497–509. [CrossRef]
25. Weiss, G.H. Some applications of persistent random walks and the telegrapher's equation. *Physica A* **2002**, *311*, 381–410. [CrossRef]
26. Murray, J.D. *Mathematical Biology II: Spatial Models and Biomedical Applications*; Springer: New York, NY, USA, 2003.
27. Turing, A.M. The chemical basis of morphogenesis. *Philos. Trans. R. Soc. Lond. B* **1952**, *237*, 37–72.
28. Painter, K.; Bloomfield, J.; Sherratt, J.; Gerisch, A. A nonlocal model for contact attraction and repulsion in heterogeneous cell populations. *Bull. Math. Biol.* **2015**, *77*, 1132–1165. [CrossRef] [PubMed]
29. Lewis, M.; Moorcroft, P. *Mechanistic Home Range Analysis*; Princeton University Press: Princeton, NJ, USA, 2006.
30. Giuggioli, L.; Potts, J.R.; Rubenstein, D.I.; Levin, S.A. Stigmergy, collective actions, and animal social spacing. *Proc. Natl. Acad. Sci. USA* **2013**, *110*, 16904–16909. [CrossRef] [PubMed]
31. Fagan, W.F.; Lewis, M.A.; Auger-Méthé, M.; Avgar, T.; Benhamou, S.; Breed, G.; LaDage, L.; Schlägel, U.E.; Tang, W.w.; Papastamatiou, Y.P.; et al. Spatial memory and animal movement. *Ecol. Lett.* **2013**, *16*, 1316–1329. [CrossRef] [PubMed]
32. Potts, J.R.; Lewis, M.A. How memory of direct animal interactions can lead to territorial pattern formation. *J. R. Soc. Interface* **2016**, *13*, 20160059. [CrossRef]
33. Zeigler, S.L.; Fagan, W.F. Transient windows for connectivity in a changing world. *Mov. Ecol.* **2014**, *2*, 1. [CrossRef]
34. Thurfjell, H.; Ciuti, S.; Boyce, M.S. Applications of step-selection functions in ecology and conservation. *Mov. Ecol.* **2014**, *2*, 4. [CrossRef]
35. Alt, W. Degenerate diffusion equations with drift functionals modelling aggregation. *Nonlinear Anal. Theory Methods Appl.* **1985**, *9*, 811–836. [CrossRef]
36. Stevens, A.; Othmer, H.G. Aggregation, blowup, and collapse: The ABC's of taxis in reinforced random walks. *SIAM J. Math. Appl.* **1997**, *57*, 1044–1081. [CrossRef]
37. Mogilner, A.; Edelstein-Keshet, L. A non-local model for a swarm. *J. Math. Biol.* **1999**, *38*, 534–570. [CrossRef]
38. Topaz, C.M.; Bertozzi, A.L.; Lewis, M.A. A nonlocal continuum model for biological aggregation. *Bull. Math. Biol.* **2006**, *68*, 1601. [CrossRef] [PubMed]
39. Burger, M.; Francesco, M.D.; Fagioli, S.; Stevens, A. Sorting phenomena in a mathematical model for two mutually attracting/repelling species. *SIAM J. Math. Appl.* **2018**, *50*, 3210–3250. [CrossRef]
40. Wolansky, G. Multi-components chemotactic system in the absence of conflicts. *Eur. J. Appl. Math.* **2002**, *13*, 641–661. [CrossRef]

41. Horstmann, D. Generalizing the Keller–Segel model: Lyapunov functionals, steady state analysis, and blow-up results for multi-species chemotaxis models in the presence of attraction and repulsion between competitive interacting species. *J. Nonlinear Sci.* **2011**, *21*, 231–270. [CrossRef]
42. Hall, R. Amoeboid movement as a correlated walk. *J. Math. Biol.* **1977**, *4*, 327–335. [CrossRef]
43. Potdar, A.A.; Jeon, J.; Weaver, A.M.; Quaranta, V.; Cummings, P.T. Human mammary epithelial cells exhibit a bimodal correlated random walk pattern. *PLoS ONE* **2010**, *5*, e9636. [CrossRef]
44. Wadkin, L.; Orozco-Fuentes, S.; Neganova, I.; Swan, G.; Laude, A.; Lako, M.; Shukurov, A.; Parker, N. Correlated random walks of human embryonic stem cells in vitro. *Phys. Biol.* **2018**, *15*, 056006. [CrossRef]
45. Hillen, T. A Turing model with correlated random walk. *J. Math. Biol.* **1996**, *35*, 49–72. [CrossRef]
46. Chen, L.; Jüngel, A. Analysis of a multidimensional parabolic population model with strong cross-diffusion. *SIAM J. Math. Anal.* **2004**, *36*, 301–322. [CrossRef]
47. Zamponi, N.; Jüngel, A. Analysis of degenerate cross-diffusion population models with volume filling. In *Annales de l'Institut Henri Poincare (C) Non Linear Analysis*; Elsevier: Amsterdam, The Netherlands, 2017; Volume 34, pp. 1–29.
48. Hernandez-Martinez, E.; Puebla, H.; Perez-Munoz, T.; Gonzalez-Brambila, M.; Velasco-Hernandez, J.X. Spatiotemporal dynamics of telegraph reaction-diffusion predator-prey models. In *BIOMAT 2012*; World Scientific: Singapore, 2013; pp. 268–281.

Article

Continuum Modeling of Discrete Plant Communities: Why Does It Work and Why Is It Advantageous?

Ehud Meron [1,2,*], Jamie J. R. Bennett [1], Cristian Fernandez-Oto [3], Omer Tzuk [1], Yuval R. Zelnik [4] and Gideon Grafi [5]

1 Department of Solar Energy and Environmental Physics, Blaustein Institutes for Desert Research, Ben-Gurion University of the Negev, Sede Boqer Campus 84990, Israel; jjrbenn@gmail.com (J.J.R.B.); omertz@post.bgu.ac.il (O.T.)
2 Department of Physics, Ben-Gurion University, Beer Sheva 84105, Israel
3 Complex Systems Group, Facultad de Ingenieria y Ciencias Aplicadas, Universidad de los Andes, Av. Mon. Alvaro del Portillo 12.455, Santiago 7620086, Chile; cristianfernandezoto@gmail.com
4 Centre for Biodiversity Theory and Modelling, Theoretical and Experimental Ecology Station, CNRS and Paul Sabatier University, 09200 Moulis, France; yuval.zelnik@sete.cnrs.fr
5 French Associates Institute for Agriculture and Biotechnology of Drylands, Jacob Blaustein Institutes for Desert Research, Ben-Gurion University of the Negev, Sede Boqer Campus 84990, Israel; ggrafi@bgu.ac.il
* Correspondence: ehud@bgu.ac.il

Received: 16 September 2019; Accepted: 15 October 2019; Published: 17 October 2019

Abstract: Understanding ecosystem response to drier climates calls for modeling the dynamics of dryland plant populations, which are crucial determinants of ecosystem function, as they constitute the basal level of whole food webs. Two modeling approaches are widely used in population dynamics, individual (agent)-based models and continuum partial-differential-equation (PDE) models. The latter are advantageous in lending themselves to powerful methodologies of mathematical analysis, but the question of whether they are suitable to describe small discrete plant populations, as is often found in dryland ecosystems, has remained largely unaddressed. In this paper, we first draw attention to two aspects of plants that distinguish them from most other organisms—high phenotypic plasticity and dispersal of stress-tolerant seeds—and argue in favor of PDE modeling, where the state variables that describe population sizes are not discrete number densities, but rather continuous biomass densities. We then discuss a few examples that demonstrate the utility of PDE models in providing deep insights into landscape-scale behaviors, such as the onset of pattern forming instabilities, multiplicity of stable ecosystem states, regular and irregular, and the possible roles of front instabilities in reversing desertification. We briefly mention a few additional examples, and conclude by outlining the nature of the information we should and should not expect to gain from PDE model studies.

Keywords: continuum models; partial differential equations; individual based models; plant populations; phenotypic plasticity; vegetation pattern formation; desertification; homoclinic snaking; front instabilities

1. Introduction

Global warming and the concomitant increased frequency of intense droughts threaten the viability of plant populations and communities throughout the world [1–3]. As plants are primary producers that constitute the basal levels of whole food webs, that threat extends to ecosystem function as well. Understanding ecosystem response to drier climates, therefore, calls for unraveling mechanisms by which plant populations tolerate water stress. Such mechanisms operate at the organism level through various forms of phenotypic changes, but also at higher organization levels through self-organization in spatial patterns. These response forms are particularly relevant in dryland ecosystems, where both phenotypic changes and spatial self-organization are highly significant.

Empirical studies of plant-community dynamics are hampered by the long time-scales of plant growth and the yet longer time scales of self-organization in space. A powerful methodology that complements empirical studies and compensates for the time-scale limitation is the construction and study of dynamic mathematical models. These models can be divided into two major groups. The first group consists of individual-based models (IBM) (also called agent-based models), which are stochastic computational algorithms for interacting individual organisms. Each individual is described by a set of time-dependent attributes, often including its spatial location on a grid and various physiological and behavioral traits [4–6]. The second group of models consists of continuum partial differential equations (PDE) that do not address discrete individual organisms, but rather deterministic processes at small spatial scales. The population is then described by a variable that represents the population size, such as the population number density, and is considered to be continuous in time and space [7–9].

The advantage of PDE models over IBM is that they lend themselves to the powerful methods of dynamical-system and pattern-formation theories, and therefore can provide deeper insights into the mechanisms that drive ecosystem dynamics and ecosystem response to environmental variability [8,10,11]. However, the application of the PDE modeling approach to plant populations in drylands should first be justified, as these populations are typically small, especially in arid regions, where the vegetation is sparse. This inherent character of drylands questions the use of continuous variables to describe population sizes; variables, such as number densities, become highly discrete, and demographic noise and extinction may become important aspects of the dynamics [12,13].

In this paper, we first argue that the discrete nature of small plant populations may still be discarded in modeling their dynamics owing to the high phenotypic plasticity of plants. That calls for a different description of the population size; rather than describing it by a discrete number density, we describe it by a biomass variable, which remains continuous even at the level of a single plant. We further demonstrate with a few examples the advantage of continuum PDE modeling over discrete IBM and discuss the ecological significance of this advantage. The examples include the use of linear stability analysis to calculate thresholds for the emergence of periodic patterns, the use of numerical continuation to calculate bifurcation diagrams, and a study of a transverse instability of desertification fronts. We conclude the paper with a brief discussion of additional examples that demonstrate the utility of continuum PDE models in gaining mechanistic information and insights about large, landscape-scale behaviors.

2. Distinctive Aspects of Plant Populations

Population size is often described by the number density of the individuals that constitute the population. We describe here two distinctive properties of plants, not shared by most other organisms: high phenotypic plasticity and seed dispersal, that will be used in Section 3 to motivate a better way of describing plant population size, especially in drylands where the vegetation is often sparse.

2.1. Phenotypic Plasticity and Plant Adaptation to Variable Environments

Phenotypic plasticity is defined as the ability of an organism (or a given genotype) to give rise to distinct observable traits (phenotypes) when exposed to variable environmental conditions [14]. Two basic, distinct developmental approaches have been evolved in plants and animals reflecting different degrees of phenotypic plasticity. Animals have a relatively short period of embryonic phase prior to birth, whereby all major organs are formed and postnatal development is essentially restricted to expansion and growth of existing organs. By contrast, plants spend most of their life, from weeks and months to hundreds and often thousands of years, at the embryonic phase, in the sense that they can produce new organs (stems, leaves, flowers) throughout their life, depending on environmental conditions. The higher phenotypic plasticity of plants can be attributed to their sessile nature and their inability to migrate to less stressful conditions as mobile animals do.

A plant embryo has only a small fraction of the organs an adult body has, consisting of two apical meristems located at the tip of the shoot and the root, which are responsible for growth and expansion of the shoot and the root systems. Both meristems consist of pluripotent cells (cells capable of differentiation into multiple cell types that make up the plant body) that are functionally equivalent to animal stem cells [15]. The shoot apical meristem (SAM) gives rise to the production of the aerial part of the plant including leaves and axillary buds (from which new stems emerge) as well as flowers. The root apical meristem (RAM) allows the growth and development of the root system and its expansion below ground. Thus, while the SAM activity leads to the expansion of the photosynthetic activity, which is carbon fixation by light energy, the RAM activity enhances uptake and mobilization of water and minerals to the canopy.

The proportion between aboveground (shoot system) biomass and belowground (root system) biomass, the so-called root-to-shoot ratio, is dynamic and needs to be balanced to achieve optimal performance (and therefore survival) under variable environmental conditions. Commonly there is a positive correlation between shoot growth and root growth, and both are interconnected [16]. Accordingly, shoot growth provides ample energy for the expansion of the root system, which in turn increases water and nutrient uptake from a larger belowground domain to feed and enable the expansion of the shoot system, a mechanism that plays an important role in the development of vegetation patterns (see root-augmentation feedback in Section 3.1.2).

The capacity of plants to dynamically change the allocation of biomass to different organs is central to plant response to variable environments and crucial for plant survival. A well-known example of periodic change in above-ground (shoot) biomass is the shedding of leaves in temperate deciduous forests during the winter (when leaves are susceptible to cold and are not photosynthetically active) as a mechanism to tolerate low winter temperature, e.g., saving energy via remobilization and storage of leaf nutritional constituents in stems. Shedding of leaves may also occur during the dry season in tropical and subtropical deciduous forests as a mechanism to tolerate seasonal drought by reducing the loss of water through transpiration.

Plants thriving in variable desert environments show many additional mechanisms to cope with seasonal climate variations that involve changes in above-ground biomass. A good example of such mechanisms is found in *Zygophyllum dumosum* Boiss (bushy bean caper), which is well adapted to a variable, desert environment [17,18]. Its root system is composed mostly of lateral, rapidly growing roots in the upper soil layer that constitute the major active elements in absorbing water [19], but also a few roots that can extend several meters in depth [20]. The shrub develops new stems with compound leaves during the wet season, where each leaf consists of two leaflets carried on a thick, fleshy petiole (Figure 1A). On entry into the summer, the above-ground biomass is altered due to the shedding of leaflets (to reduce whole plant transpiration) while the fleshy, wax-covered petioles remain alive (Figure 1B,C) [18].

Another strategy employed by *Z. dumosum*, particularly in successive drought years, is splitting the main axis (Figure 1D,E), a common phenomenon in desert shrubs [21] that enables the survival of certain units at the expense of others [22]. Also, certain desert shrubs such as *Artemisia sieberi* Besser (synonym name: *A. herba-alba*) and *Encelia farinosa* can change their mode of development and produce new leaf types on the transition from the wet to dry season, which is better adapted to dry conditions [23,24].

Figure 1. Phenology of *Zygophyllum dumosum* Boiss growing on a southeast-facing slope at Sede Boqer research area (30o 51′ N 34o 46′ E; elevation 498 m). (**A**) A typical branch of *Zygophyllum* plant during the wet season (March 2008). Note the coumpond leaf composed of two leaflets (LL) carried on a fleshy, cylindrical petiole (P). 1YOP, 1-year-old petiole. (**B**) A typical *Z. dumosum* branch during the dry season carrying petioles (P). (**C**) A closer look at emerging new bud from the axil of a 1YOP at the beginning of the winter. (**D**) Axis splitting resulting in a dual appearance of a *Z. dumosum* plant showing healthy branches carrying new leaves and flowers (right from the broken line) and unhealthy branches (left from the broken line). (**E**) A *Z. dumosum* plant with the main axis divided into several distinct units.

2.2. Seed Persistence in the Soil

Seeds persist in the soil until they germinate or die as a result of aging, predation or decay by fungi or bacteria [25]. Based on their longevity in soil, soil seed banks are divided into transient and persistent types, whereby the latter refers to seeds that remain viable for more than one year (Thompson and Grim, 1979) and often persist for decades [26] and even centuries [27]. There are several attributes that assist in seed persistence and longevity in the soil including dormancy, the capacity to repair accumulated damages and to possess active and/or passive defense mechanisms against potential predators and pathogens (reviewed in Ref. [28]).

Seed persistence is an important factor controlling the survival of a species long after the death of the mother plant. It avoids germination under unfavorable conditions (bet-hedging strategy, [29]) and allows for genetic preservation and distribution in time and space [30,31]. The capacity of seed persistence in the soil has implications for weed management, flora restoration and for understanding plant community dynamics, particularly in light of global climate change [31,32]. Persistence in the soil varies significantly between plant species and is dependent on the physical, physiological, chemical and biochemical properties of the dispersal unit and its capacity to withstand variable biotic and abiotic conditions [28].

These persistence characteristics may be altered upon exposure of mother plants to various biotic and abiotic stress conditions during seed development and prior to dispersal. Some dispersal unit characteristics underlying seed persistence in the soil are of maternal origin, embedded within the dead organs enclosing the embryos (DOEEs) including the seed coat (dry, dehiscent fruits), the pericarp (dry, indehiscent fruits) and dead floral bracts (glumes, lemmas, paleas) in grasses [33]. DOEEs are thought to function in seed dispersal and in protecting the embryo from mechanical and physical damages. However, detailed study of DOEEs revealed their capacity to store substances such as proteins (e.g., hydrolytic enzymes), growth factors (e.g., phytohormones) and various metabolites that might affect various aspects of seed biology including longevity, germination, and seedling vigor [33].

3. Modeling Dryland Vegetation

The high phenotypic plasticity of plants and the consequent wide range of above-ground biomass values a single plant can assume, suggest a description of plant-population sizes in terms of a biomass density variable, continuously varying in time and space, rather than in terms of a discrete number-density variable, as is often done in studies of animal populations. Since overland water flow and soil–water diffusion are continuous processes too, a natural way of describing dryland-vegetation dynamics is in terms of systems of partial differential equations (PDE) for biomass and water variables. This continuum modeling approach avoids the introduction of factitious demographic noise, which is a concern in small, low-plasticity animal populations. It is also consistent with the little relevance of extinction events in small plant populations, unlike animal populations. Such events are unlikely because of dispersed seeds that can remain viable long after plant mortality takes place, and are capable of reviving the plant population once favorable environmental conditions resume. The residual exponentially-small biomass that follows vegetation decay and convergence to bare soil in model solutions can be viewed as representing long-lived seeds.

PDE models of dryland vegetation describe the size of a plant population by a biomass-density variable, $B(X,Y,T)$, which stands for the above-ground biomass of plants per unit area. Here, X,Y are the spatial coordinates in the plane (in units of meters), T is time (in units of years), and B has units of $\mathrm{kg/m^2}$. The biomass $BdXdY$ in a small area element $dXdY$ may represent the contribution of a single, several or many plants, depending on the particular plant species and on the spatial scale over which B varies. Several PDE models for dryland vegetation have been proposed. The simplest of which is a single-variable model for the vegetation biomass [34], while more detailed models also include a water variable [35–37], or two water variables representing below-ground water per unit area, $W(X,Y,T)$, and above-ground water per unit area, $H(X,Y,T)$ (both in units of $\mathrm{kg/m^2}$) [38,39].

In order to account for the emergence of vegetation patterns from uniform vegetation, the models should capture positive feedback loops that are capable of inducing nonuniform instabilities of a uniform-vegetation state. The more detailed models are advantageous in that they capture several pattern-forming feedbacks of this kind and thus allow us to study the interplay between different feedbacks [40–42]. These models also introduce better defined and measurable parameters. When applied to particular ecological contexts these models often simplify considerably and allow further mathematical analysis [43]. We refer the reader to Ref. [43] for a detailed description of a three-variable vegetation model that captures three distinct pattern-forming feedbacks. Here, we briefly describe this feedbacks and presents two simplified versions of the model that will be used later on to demonstrate the advantages of PDE models in studies of dryland ecosystems.

3.1. Pattern-Forming Feedbacks

Pattern-forming feedbacks in flat terrains follow the general scheme illustrated in Figure 2, that is, a positive feedback loop between local vegetation growth and water transport towards the growing vegetation [11,44]. They differ from one another in the mechanism of water transport: overland water flows, conduction of water by laterally extended root systems, and soil–water transport, as explained below. The transport of water towards denser vegetation patches accelerates vegetation growth there, and, at the same time, inhibits the growth in their surroundings. It, therefore, favors the growth of nonuniform perturbations. This is a short-range activation—long-range inhibition mechanism of pattern formation [45], also termed "scale-dependent feedback" [46], where biomass is the activator and lack of water is the inhibitor. Associated with the three water-transport mechanisms are three different positive feedback loops as described below.

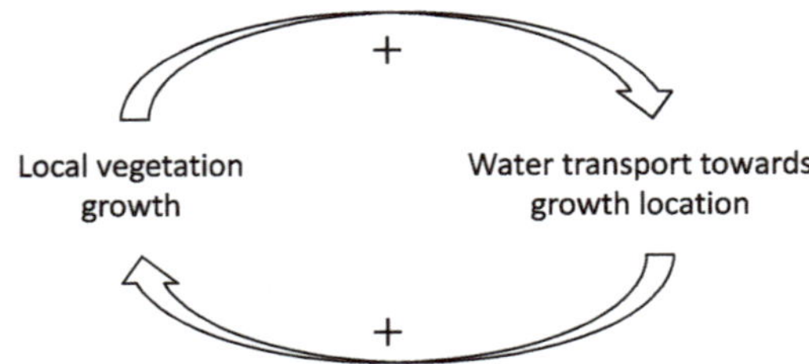

Figure 2. A general positive feedback loop that drives the formation of vegetation patterns in drylands. The feedback concomitantly accelerates vegetation growth in patches of denser vegetation and inhibits the growth in adjacent sparser patches. The combined processes favor the growth of nonuniform perturbations and the formation of vegetation patterns. From [11].

3.1.1. Infiltration Feedback

Infiltration rates of surface-water into the soil in sparsely vegetated areas are typically lower than those in densely vegetated areas. Two main factors contribute to that effect: soil crusts in bare-soil that reduce the infiltration rate [47,48] and denser roots in denser vegetation that make the soil more porous and increase the infiltration rate. The infiltration contrast that builds up as the vegetation becomes denser in a given location induces overland water flow towards that location, which accounts for the upper arrow in Figure 2, i.e., enhancement of water transport by local vegetation growth. The increased soil moisture in the growth location further increases the rate of vegetation growth (lower arrow in Figure 2), which completes the positive feedback loop and sets the ground for a yet stronger positive-feedback loop. This pattern-forming feedback is referred to as the infiltration feedback.

To capture the infiltration feedback in the model equations we assume a monotonically increasing dependence of the infiltration rate I on the biomass variable B [38,39,49],

$$I = A\frac{B+Qf}{B+Q}. \tag{1}$$

The parameter Q controls how fast the asymptote $I = A$ is approached. The parameter $f \in [0,1]$ controls the infiltration contrast, and, thus, the strength of the infiltration feedback. The value $f = 1$ corresponds to a constant infiltration rate, $I = A$, or no infiltration contrast, while values $f \ll 1$ correspond to high infiltration contrasts, for which the infiltration rate in bare soil is significantly higher than in vegetated soil. This feedback may apply to ecosystems in which bare soil is covered by soil crusts, as these crusts typically reduce the infiltration rate relative to vegetation patches [48]. It is not expected to apply to uncrusted sandy soil, where the infiltration rate is high everywhere.

3.1.2. Root-Augmentation Feedback

As noted in Section 2.1, there is a positive correlation between root growth and shoot growth that is quantified by the so-called root-to-shoot ratio. This property constitutes another mechanism by which water transport is enhanced by vegetation growth, as the lateral extension of the roots as the shoot grows enables water uptake and conduction from a larger volume. These processes and the consequent accelerated vegetation growth define the root-augmentation feedback.

The root-augmentation feedback is modeled by introducing a biomass-dependent kernel function,

$$G(\mathbf{X},\mathbf{X}') = \tilde{G}\left(\frac{|\mathbf{X}-\mathbf{X}'|}{S[B(\mathbf{X})]}\right). \tag{2}$$

This function describes the spatial distribution of the roots in the horizontal directions X and Y, where $\mathbf{X} = (X, Y)$ represents the plant (shoot) location and $\mathbf{X}'$ a distant point. The function $S(B)$ determines the width of the kernel function and provides a measure for the horizontal extension of the roots. The root-augmentation feedback is captured by letting S increase monotonically with the biomass variable $B(\mathbf{X})$, which represents the shoot mass. Figure 3 shows an example of a root kernel and its lateral extension as the plant grows and the above-ground biomass increases. The mathematical form of the kernel function is given by Equations (A1) and (A2) in Appendix A.

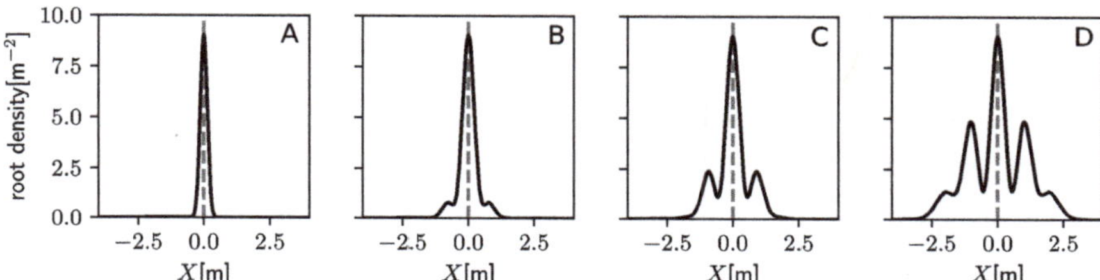

Figure 3. Root system growth as shoot biomass increases. A plot of the axisymmetric root kernel, G (see (A1) in Appendix A), as a function of the space coordinate X when **(A)** $B(X,T) = 0$, **(B)** $B(X,T) = 0.2$, **(C)** $B(X,T) = 0.3$, **(D)** $B(X,T) = 0.4 =: K$. The kernel is a Gaussian function multiplied by a polynomial factor, that mimics adventitious root branching observed in some plant species [50]. Parameters in (A1) and (A2): $S_0 = 0.15\text{m}$, $c_0 = 1$, $c_2 = -0.135$, $c_4 = 7.09 \times 10^{-3}$, $c_6 = -9.11 \times 10^{-5}$, $c_8 = 3.84 \times 10^{-7}$.

3.1.3. Soil–Water Diffusion Feedback

Water depletion due to water uptake by the plants' roots, followed by soil–water diffusion from the patch surroundings, is a third mechanism of water transported that is enhanced by vegetation growth [51]. The associated feedback loop is referred to as the lsoil—water diffusion feedback. Like the root-augmentation feedback, this feedback relies on the root-to-shoot property of plants, except that here the role of the roots is to create soil–water gradients by local water uptake. This pattern-forming feedback can possibly apply to plants with vertical roots and strong water uptake, and to soil types for which lateral water diffusion is fast relative to the rate of vegetation spread.

3.2. *Mathematical Models*

For the sake of simplicity, we confine ourselves in this paper to ecosystems with sandy soil for which the infiltration rate is high also in bare-soil areas and, therefore, overland water flow can be neglected. Mathematically, this amount to assuming $f = 1$ (no infiltration contrast) and to the elimination of the equation for surface water. The full model then reduces to the following pair of equations for B and W [43]:

$$\begin{aligned}\partial_T B &= G_B B\,(1 - B/K) - MB + D_B \Delta B\,,\\ \partial_T W &= P - L_W W - G_W W + D_W \Delta W\,.\end{aligned} \tag{3}$$

Here, Δ is the Laplacian operator in the X, Y plane, L_W is a biomass dependent evaporation rate,

$$L_W = \frac{N}{(1 + RB/K)}, \tag{4}$$

where N is the evaporation rate in bare soil and R quantifies the reduction in evaporation rate in vegetation patches, and G_B and G_W are the rates of biomass growth and water uptake by plants' roots, respectively. These rates are given by the integrals

$$G_B = \Lambda \int_\Omega G(\mathbf{X}, \mathbf{X}', T) W(\mathbf{X}', t) \mathrm{d}\mathbf{X}', \tag{5a}$$

$$G_W = \Gamma \int_\Omega G(\mathbf{X}', \mathbf{X}, T) B(\mathbf{X}', t) \mathrm{d}\mathbf{X}', \tag{5b}$$

over the system domain Ω, where the kernel function $G(\mathbf{X}, \mathbf{X}', t)$ satisfies the general form (2), and Λ and Γ are rate constants in units of $(\mathrm{kg/m^2})^{-1}\mathrm{y}^{-1}$. These nonlocal forms represent the effects of laterally extended roots; the growth of a plant at point $\mathbf{X}$ depends on water availability at all points $\mathbf{X}'$ within the reach of the plant's roots, and likewise, the water uptake at point $\mathbf{X}$ is due to plants at all points $\mathbf{X}'$ whose roots extend to $\mathbf{X}$. Two forms for the root kernel, $G(\mathbf{X}, \mathbf{X}', T)$, will be used in this study. The first form represents vertical roots with negligible lateral extension. In that case the expressions in Equation (5a) for G_B and G_W simplify to algebraic expressions as shown in Appendix B. The second form represents laterally extended roots, as described in Appendix A.

Equation (3) capture several processes that affect biomass dynamics: water-dependent plant growth ($G_B B$), mortality ($-MB$), and short-distance seed dispersal ($D_B \Delta B$). In addition, the late biomass growth phase is slowed down by species-specific constraints that dictate a maximal standing biomass per unit ground area K. These may represent self-shading, limited stem strength of a woody plant, etc. The processes that affect soil–water dynamics, according to Equation (3) are precipitation (P), water loss due to biomass dependent evaporation ($-LW$), water uptake by plants' roots ($-G_W W$), and soil–water diffusion ($D_W \Delta W$). The simplified model equations do not capture the infiltration feedback, as no overland water flow takes place, but still capture the root-augmentation feedback and the soil–water diffusion feedback.

4. Advantages of Continuum PDE Models

Continuum PDE models are advantageous over discrete models, such as individual based models (IBM), in that they lend themselves to mathematical analysis, which results in deeper insights into the phenomena in question and the ecological implications they have. We chose to demonstrate this aspect using three examples, as described in the following subsections.

4.1. Instability Thresholds

The simplest solutions of Equation (3) are stationary uniform solutions obtained by solving the algebraic equations that result by setting the time and space derivatives of the state variables to zero. Two such solutions are found, $(B, W) = (0, W_0)$ representing bare soil and existing for all positive precipitation values, P, and $(B, W) = (B_1, W_1)$ representing uniform vegetation and existing only beyond some precipitation threshold (see Figure 4). The stability of these solutions to infinitesimal perturbations of all forms can be calculated using linear stability analysis [11,52]. Such an analysis reveals that the bare-soil state is stable in the precipitation range $0 < P < P_0$, where P_0 is an instability threshold at which a uniform mode (characterized by zero wavenumber) begins to grow, as the dispersion curves in Figure A1A show. This is a uniform stationary instability [43] where the first mode to grow is spatially uniform and the growth is monotonic in time. Spatial patterning, induced by the root-augmentation feedback or the soil–water diffusion feedback, cannot arise from a nonuniform instability of the bare-soil state as this would involve the growth of a periodic mode from a zero-biomass state and therefore imply negative biomass values, which are unphysical. Failing to obtain a uniform instability of the bare-soil state would indicate a modeling flaw. Such a flaw would be harder to detect in discrete models, such as IBM.

Stationary periodic vegetation patterns may arise from a nonuniform stationary instability of the uniform vegetation state. A linear stability analysis of that state indeed reveals such instability as the precipitation rate P decreases below a threshold value P_T. At that threshold a periodic mode with a characteristic wavenumber k_T begins to grow monotonically in time while all other modes still decay, as the dispersion curves in Figure A1B shows. This analysis can be performed even for quite complicated laterally extended root kernels G as shown in Figure 3, see Appendix A. The analysis further confirms the expectation that each of the pattern-forming feedbacks alone, root-augmentation and soil–water diffusion, can induce a nonuniform stationary instability that culminates in stationary periodic patterns.

Linear stability analysis not only unravels possible instabilities and the nature of the modes that grow beyond instability points, but it also provides information about the parameters that control the instability threshold. For example, the instability of the bare-soil state occurs at $P = P_0 = MN/\Lambda$, indicating that species with low ratios of growth rates, Λ, to mortality rates, M, will grow from bare soil at higher precipitation thresholds, or that high evaporation rates, N, act to stabilize the bare soil state. Such relations are less apparent in IBM.

4.2. Bifurcation Diagrams

Linear stability analysis does not provide information about the new state that the system converges to following an instability. This kind of information can be obtained by nonlinear analysis of the model equations. A common approach, valid close enough to the instability point, involves the derivation of amplitude equations [11,53–55]. These are nonlinear differential equations for the amplitudes of the modes that grow beyond the instability point. Amplitude equations are, in general, much easier to analyze than the original model equations. Moreover, they are dictated by the instability type, rather than by the specific model equations, and therefore are universal; different systems that go through the same instability type (e.g., nonuniform stationary instability) will behave similarly near the instability point. Since all pattern-forming feedbacks lead to the same instability type they will also induce similar dynamics and patterns, although fine details, such as the relative spatial distribution of biomass and water, can reflect differences in the these feedbacks [56].

A convenient way to summarize the outcomes of nonlinear analysis is to draw a bifurcation diagram. Such a diagram shows the existence and stability ranges of various model solutions. In the context of dryland vegetation it often shows graphs of spatial biomass average (or L2 norm) vs. precipitation rate for selected model solutions, using the convention that solid (dashed) lines represent stable (unstable) solutions. If amplitude equations are known, a bifurcation diagram can often be calculated analytically [11,57,58]. When the derivation of amplitude equations is not easy to perform, or when there is an interest in dynamical behaviors far from instability points, bifurcation diagrams can be calculated by numerical continuation methods using software packages such as AUTO [59], pde2path [60], and others. Since these packages solve the equations using iterative solution methods, they converge to unstable solutions as well and, thereby, can provide complete solution branches in the bifurcation diagram.

The utility of bifurcation diagrams can be demonstrated with the example shown in Figure 4, a bifurcation diagram for a vegetation model (see Appendix B) in one space dimension. We focus here on one property that bifurcation diagrams often reveal, namely, parameter ranges where multiple stable states coexist. The (partial) bifurcation diagram of Figure 4 shows a bistability range of bare soil and periodic patterns, a bistability range of bare soil and uniform vegetation, and a tristability subrange of bare soil, periodic patterns and uniform vegetation. The stability of the bare soil state at relative high precipitations values, where uniform vegetation is stable too, can be achieved assuming high evaporation rates (N) in bare soil. Within the tristability range there exists many more stable states, which describe spatial hybrids of periodic pattern and uniform vegetation. These hybrid states appear as confined domains of an increasing size of the patterned state in an otherwise uniform vegetation state, as the insets in Figure 4 show. Hybrid states in bistability ranges of uniform and patterned states have been studied extensively, both mathematically and in particular physical systems. The confined pattern domains are homoclinic solutions that snake back and forth in the bifurcation diagram, a behavior termed "homoclinic snaking" [61–66]. They are tightly related to front pinning, that is, to the existence of stationary front solutions between uniform and patterned states over a range of the control parameter. This is in contrast to fronts between distinct uniform states that propagate in general, except, possibly, for a particular parameter value, the so-called Maxwell point at which the fronts are stationary [67,68], and unlike fronts that are pinned by external heterogeneities [69].

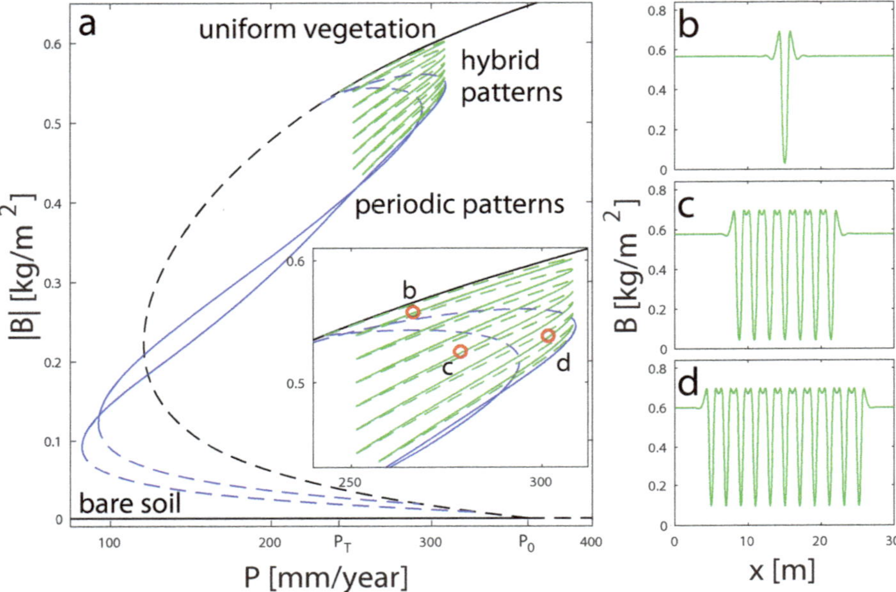

Figure 4. Bifurcation diagram for the model presented in Appendix B (Equation (A17)) in 1D (panel **a**). The vertical axis is the spatial biomass average, while the horizontal axis is the precipitation rate. Solid (dashed) lines represent stable (unstable) solutions. The diagram shows bistability precipitation ranges of bare soil and periodic patterns, and of bare soil and uniform vegetation, as well as a tristability range of bare soil, uniform vegetation and periodic patterns. Within the latter range there exists a subrange of hybrid patterns, consisting of confined pattern domains of increasing size in an otherwise uniform vegetation, examples of which are shown in the three panels (**b–d**). Parameters: $\Lambda = 0.1$ [(m^2/kg)/y], $\Gamma = 4$ [(m^2/kg)/y], $E = 7$ [m^2/kg], $K = 1$ [kg/m^2], $M = 4.5$ [1/y], $N = 8$ [1/y], $R = 0.1$, $D_B = 0.05$ [m^2/y], $D_W = 30$ [m^2/y].

Homoclinic snaking has been used to explain fairy-circle dynamics in Namibian grasslands [56,70–72]. Fairy circles are circular bare-soil gaps in grasslands that often form nearly periodic hexagonal patterns, where each gap is surrounded by nearly equidistant six other gaps on average [73]. Empirical studies indicate that fairy circles are occasionally "born" or "die" locally [74]. Taking into account the high rainfall inter-annual variability in fairy-circles sites, such dynamical events have been interpreted as transitions between different hybrid states (homoclinic solutions) induced by rainfall fluctuations that are strong enough to drive the ecosystem temporarily outside the existence range of these states (snaking range) [51]. The existence of a multitude of stable hybrid states makes dryland landscapes more plastic in the sense that their response to varying environmental conditions and disturbances can involve temporal convergence to different hybrid states, rather than direct convergence to a single alternative stable state [75].

In order for homoclinic snaking to occur in models of dryland landscapes bistability of uniform vegetation and periodic vegetation patterns are generally sufficient [75,76]. The snaking range in the bifurcation diagram of Figure 4, however, lies within a tristability range of uniform vegetation, periodic pattern and (uniform) bare soil. The existence of that tristability range may have interesting consequences. In particular, the organization of hybrid states within a homoclinic snaking structure can break down as it meets a Maxwell point, where fronts connecting the bare soil and uniform vegetation states are stationary [77]. Another implication of the tristability of two uniform states and a periodic pattern state is the emergence of a family of complex front structures that involve all three states [77].

An additional important aspect that bifurcation diagrams uncover is related to the existence of unstable solution branches. Tracking unstable solutions in bifurcation diagrams can be highly significant when the response of ecosystems to disturbances or human intervention is studied, as the appearance or disappearance of unstable solutions can dramatically affect the flow in phase space and, thus, the response [78]. These subtle aspects, which may have significant ecological implications, become apparent once a bifurcation diagram is calculated. Bifurcation diagrams showing stable and unstable uniform, periodic and localized solutions can be calculated using PDE models [60] but, practically, not with IBM. Bifurcation diagrams have been calculated using IBM but the information they contain is very limited [79,80].

4.3. Front Dynamics

Bistable ecosystems can go through state transitions, or regime shifts [81], in various ways, including a passage through a bifurcation point (B-tipping), as a result of environmental fluctuations (N-tipping), or as a result of fastly varying environmental conditions (R-tipping) [82,83]. Such transitions are generally discussed as whole-system responses, occurring simultaneously at all points in space. Spatially confined disturbances, however, can induce local state transitions. The dynamics that follow such disturbances are dictated by the motion of the fronts that bound the transition area. Of particular interest are degradation fronts, where a dysfunctional state gradually displaces a functional state by front propagation. An example of such a front in the context of dryland vegetation is a desertification front, where a bare-soil domain displaces a vegetated domain [68,84].

PDE models are highly valuable in analyzing desertification fronts and addressing questions such as how to reverse the process of desertification, as we briefly discuss below. We consider again the bifurcation diagram shown in Figure 4, and focus on a bistability precipitation range of bare soil and uniform vegetation, which may or may not include the tristability range where periodic patterns are stable too. In that range we consider precipitation values below the Maxwell point, where desertification fronts exist (bare soil displacing uniform vegetation). In the vicinity of the bare-soil instability to uniform vegetation ($P = P_C$ in Figure 4) the two-variable model presented in Appendix B can be reduced to an amplitude equation for the uniform mode that begins to grow at this instability [85]. Analysis of this equation reveals a transverse front instability [86,87], whereby small bulges along the front line are first enhanced, then develop into growing fingers that avoid one another, and ultimately fill up the system domain with a stationary labyrinthine pattern.

According to the analysis of the amplitude equation, the instability occurs as the soil–water diffusion coefficient, D_W, exceeds a threshold value, and that threshold is inversely related to the root-to-shoot ratio, E, which controls water uptake by plants' roots [85]. This result uncovers the mechanism of the instability—fast soil–water diffusion (relative to biomass expansion) towards incidental bulges along the front line that locally deplete the soil–water content. The fast diffusion acts to enhance the growth of these bulges and, at the same time, to inhibit vegetation growth on both sides of any bulge. Fast diffusion can be obtained by increasing the diffusion coefficient, D_W, which is consistent with the finding of a threshold value above which the instability develops. Fast diffusion is also obtained by steepening the soil–water gradients, which can be achieved by strong water uptake. This is consistent with the finding that the threshold value decreases as the parameter E that controls water uptake, is increased.

Figure 5 shows a desertification front that develops a transverse instability. While the bare soil area initially expands into the vegetated area, the instability results in vegetation fingers that grow backward into the bare-soil area and thereby reverse the desertification process. The resulting state is a productive vegetation pattern that prevents further irreversible degradation processes (not captured by the model), such as soil erosion, and maintains the ecosystem in a reversible state capable of forming uniform vegetation when favorable rainfall conditions resume. But how can a transverse instability be induced in stable desertification fronts? One possible answer to this question is the introduction of a species with sufficiently high root-to-shoot ratio so as to reduce the threshold value of D_W above which the instability sets in, and thereby induce a transverse instability. There is, however, another possibility associated with the tristability range of uniform vegetation, bare soil and periodic patterns. In this range, desertification fronts that are stable to small perturbations (linearly stable) may still be unstable to larger perturbations (nonlinearly unstable), which drive the system to the periodic-pattern state through finger growth [88] as Figure 6 demonstrates.

These results are very appealing from the point of view of ecosystem management. First, they imply local manipulations in the front zone only, rather than extensive intervention across the whole ecosystem. The manipulations may involve planting a different species capable of inducing a linear front instability, or modulating the front line strongly enough in order to induce a nonlinear front instability. The latter intervention form may involve periodic grazing management, clear-cutting or irrigation, and is relevant in cases when indications for the existence of periodic patterns exist (e.g., scattered patches of periodic vegetation). Second, such manipulations are limited in time, as they are needed only to trigger the instability. Once the instability sets in, a process of self-recovery begins. These conclusions, which are based on the mathematical analysis of front solutions, could not have been obtained in studies of IBM.

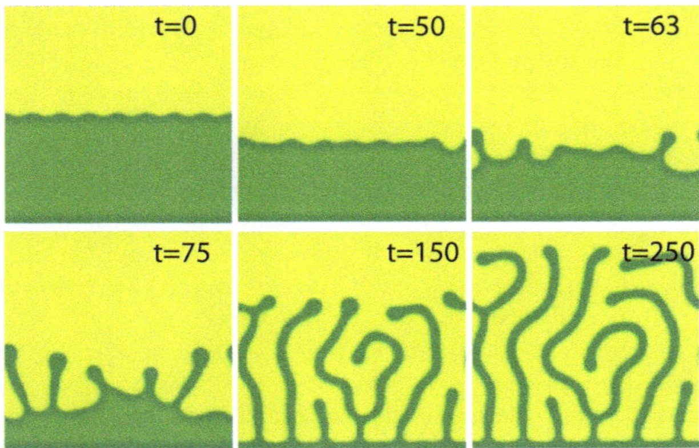

Figure 5. Snapshots of model solutions (see Appendix B) that demonstrate a linear front instability of a desertification front. Following a short phase in which the bare-soil domain (yellow) expands into the uniform-vegetation domain (green), transverse perturbations begin to grow and form vegetation fingers that grow back into the bare-soil domain. The time indicated in every snapshot is in units of years. From [85] .

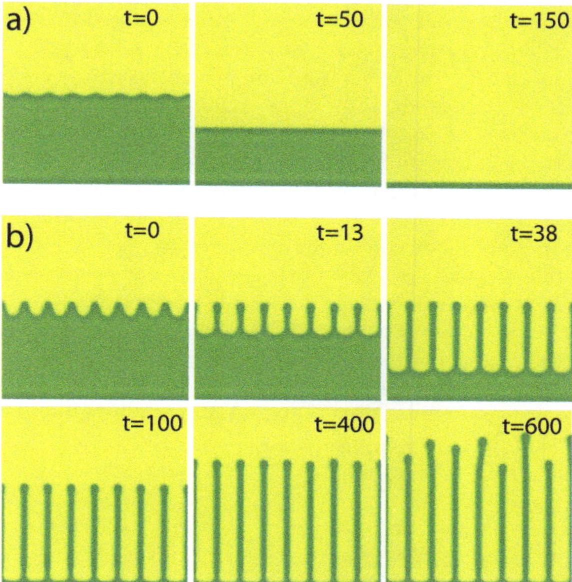

Figure 6. Nonlinear front instability in a tristability range. Snapshots of numerical model solutions showing (**a**) the stability of a planar desertification front to small transverse modulation, (**b**) instability to transverse modulations that are sufficiently large, and the development of vegetation fingers that grow back into bare soil. The time indicated in every snapshot is in units of years. From [85].

5. Discussion

Despite their typically small size, discrete plant populations in dryland ecosystems can still be described by continuum PDE models because of the high phenotypic plasticity of plants and the dispersal of stress-tolerant seeds, as discussed in Sections 2 and 3. Since the state variables that describe plant population sizes should reflect that plasticity, biomass densities are more appropriate choices for plant populations than the often used number densities in population dynamics studies. Unlike IBM, where the smallest entity is a single individual, in PDE models with biomass densities as state variables, the smallest entity is a small area element and the processes that take place there. That area element can be significantly smaller than the scale of a single plant and its roots.

We presented here several examples that demonstrate the utility of continuum PDE models in gaining mechanistic information and insights about large, landscape-scale behaviors, such as the onset of a nonuniform stationary instability that culminates in periodic vegetation patterns, uncovering precipitation ranges of bistability, tristability and multistability of uniform states, periodic patterns and localized patterns (hybrid states), and dynamics of desertification fronts, where we focused on front instabilities that may reverse gradual desertification. Many more examples exist, a few of them are briefly discussed below.

On a slope, a linear stability analysis of uniform vegetation has revealed a nonuniform oscillatory instability that results in traveling-wave solutions. These solutions describe periodic vegetation stripes migrating uphill [89,90], as observed in empirical studies [91]. The migration mechanism is easy to understand; while plants at the top part of a vegetation stripe receive runoff and grow, plants at the bottom part loose runoff and die. PDE models have been instrumental in clarifying the relations between migration speed, pattern wavelength, and slope [92,93].

Studies of PDE models in one space dimension have identified many periodic solutions along the rainfall gradient, the wavenumbers of which decrease as precipitation decreases, down to zero in a solution that represents a single vegetation patch [94–96]. These solutions reflect two manners by which patchy vegetation responds to water stress. The first is an increase of bare soil areas at the expense of vegetation-patch areas, keeping the wavenumber unchanged. This response occurs along each solution branch [97]. The second response is a transition to a periodic solution with a lower wavenumber, quite often half the original wavenumber [95–97]. In this response, the area of each vegetation patch does not necessarily decrease, but the number of vegetation patches decreases. In both response forms, the increase in bare-soil area compensates for the reduced rainfall through increased water transport to adjacent vegetation patches by one of the transport mechanisms discussed in Section 3.1. Empirical indications for the existence of periodic patterns with different wavenumbers have been found in studies of banded vegetation in Somalia [98].

In two space dimensions, patchy vegetation can respond to water stress by yet another mechanism of increasing bare-soil areas, namely, morphological changes, first from hexagonal gap patterns to stripe patterns, and, at lower precipitation, from stripe patterns to hexagonal spot patterns [36,43,99]. In cases where the instability of bare soil to uniform vegetation is supercritical, weak nonlinear analyses of PDE models in two space dimensions have produced bifurcation diagrams that unfold the basic periodic vegetation patterns along the rainfall gradient: hexagonal gap patterns, stripe patterns and hexagonal spot patterns [57,58]. Empirical indications for the existence of these vegetation patterns in nature are abundant, although not yet with a single plant species along a rainfall gradient [56,73,100–102].

PDE models have also been used to study the interactions between two distinct plant species. One interesting context of such interactions is the bistability of a uniform state of one species and a periodic-pattern state of the other species. In this range, homoclinic snaking can result in the multistability of hybrid states that involve the two species, and therefore constitutes a mechanism for species coexistence [103]. Another interesting context is woody-herbaceous systems, where the woody species is pattern forming. Studies of a PDE model that captures both the infiltration feedback and the root-augmentation feedback reveal changes in the relative importance of the two feedbacks along the rainfall gradient. At high precipitation dominance of the root augmentation feedback results in a strong depletion of the soil–water content and the exclusion of herbaceous vegetation. At low precipitation, the dominance of the infiltration feedback results in soil–water concentration at woody patches and the consequent facilitation of herbaceous-vegetation growth [41]. These results are consistent with the stress-gradient hypothesis in ecology [104].

While these results are hard to obtain using IBM, we should not regard PDE models as substitutes to IBM, but rather as complementary means to gain deeper mechanistic understanding. IBM typically go into much more details, and attempts to include these details in PDE models would render them mathematically intractable too. PDE models should rather be motivated by specific questions of interest and by judicious assessments of the processes that are most relevant to these questions. The vegetation models described in Section 3 have been motivated by empirical observations of regular vegetation patterns in drylands and by the understanding that the most relevant processes to these phenomena are positive feedback loops involving vegetation growth and water availability that can induce pattern-forming instabilities. Various other processes, considered to be of secondary significance, have been left aside, such as the effect of transpiration on the atmosphere, soil erosion and deposition, and various plant physiology processes. Such processes are likely to have quantitative rather than qualitative effects on vegetation pattern formation; they may affect, for example, instability thresholds, but are not likely to affect the occurrence of instabilities or change their nature. While providing deep mechanistic insights into ecological processes, and predicting possible qualitative responses to environmental changes, PDE models should not be regarded as tools for making quantitative forecasts.

Author Contributions: Conceptualization, E.M.; methodology, all authors; software, Y.R.Z. and O.T.; formal analysis, J.J.R.B. and C.F.-O.; investigation, all authors; writing—original draft preparation, E.M. and G.G.; writing—review and editing, all authors; visualization, all authors; funding acquisition, E.M.

Funding: This study was funded by the Israel Science Foundation under grant number 1053/17, and by the European Union's Horizon 2020 research and innovation programme under grant agreement No 641762.

Conflicts of Interest: The authors declare no conflict of interest.

Appendix A. Model for Laterally Extended Roots

We consider here the model Equation (3) in two space dimensions for a laterally extended root kernel (2) of the form

$$G(\mathbf{X},\mathbf{X}',T) = \frac{F(\mathbf{X},\mathbf{X}')}{2\pi S_0^2} \mathrm{e}^{-\frac{|\mathbf{X}-\mathbf{X}'|^2}{2S_0^2(1+EB(\mathbf{X},T))^2}}, \tag{A1}$$

where S_0 is the lateral root length of a seedling and F is a polynomial given by

$$F(\mathbf{X},\mathbf{X}') = (c_0 + c_2 S_0^{-2}|\mathbf{X}-\mathbf{X}'|^2 + c_4 S_0^{-4}|\mathbf{X}-\mathbf{X}'|^4 + c_6 S_0^{-6}|\mathbf{X}-\mathbf{X}'|^6 + c_8 S_0^{-8}|\mathbf{X}-\mathbf{X}'|^8)/\psi. \tag{A2}$$

Here $\psi = c_0 + 2c_2 + 8c_4 + 48c_6 + 384c_8$ is a normalization factor that ensures the integral of $G(\mathbf{X},\mathbf{X}',T)$ over the entire domain is unity when $B(\mathbf{X},T) = 0$. Since G has units of of m^{-2} (see Equation (5a)), F is a dimensionless quantity. The "shape parameters" c_0, c_2, c_4, c_6, $c_8 \in \mathbb{R}$ and the normalization factor ψ are then dimensionless too. For proper choices of these parameters the root kernel can describe root branching as Figure 3 shows.

Non-dimensionalisation: Using the scalings

$$b = \frac{B}{K},\ w = \frac{W\Lambda}{K\Gamma},\ t = MT,\ \mathbf{x} = \frac{\mathbf{X}}{S_0}, \tag{A3}$$

one can non-dimensionalise model (3) to obtain

$$\frac{\partial b}{\partial t} = G_b b(1-b) - b + \delta_b \Delta b, \tag{A4a}$$

$$\frac{\partial w}{\partial t} = p - \nu(1-\rho b)w - G_w w + \delta_w \Delta w. \tag{A4b}$$

The non-dimensional biomass growth rate and water uptake rate are given by

$$G_b = \lambda \int_\Omega g(\mathbf{x},\mathbf{x}',t) w(\mathbf{x}',t) \mathrm{d}\mathbf{x}', \tag{A5a}$$

$$G_w = \lambda \int_\Omega g(\mathbf{x}',\mathbf{x},t) b(\mathbf{x}',t) \mathrm{d}\mathbf{x}', \tag{A5b}$$

with

$$g(\mathbf{x},\mathbf{x}',t) = \frac{f(\mathbf{x},\mathbf{x}')}{2\pi} \mathrm{e}^{-\frac{|\mathbf{x}-\mathbf{x}'|^2}{2(1+\eta b(\mathbf{x},t))^2}}, \tag{A6}$$

where

$$f(\mathbf{x},\mathbf{x}') = (c_0 + c_2|\mathbf{x}-\mathbf{x}'|^2 + c_4|\mathbf{x}-\mathbf{x}'|^4 + c_6|\mathbf{x}-\mathbf{x}'|^6 + c_8|\mathbf{x}-\mathbf{x}'|^8)/\psi. \tag{A7}$$

The relations between non-dimensional and dimensional quantities are given by

$$\lambda = \frac{K\Gamma}{M},\ \eta = EK,\ p = \frac{\Lambda P}{K\Gamma M},\ \nu = \frac{N}{M},\ \rho = R,\ \delta_b = \frac{D_B}{MS_0^2},\ \delta_w = \frac{D_W}{MS_0^2}.$$

Linear stability: We use here linear stability analysis to study the stability properties of the solution of (3) that represents steady uniform vegetation, which we denote by $\mathbf{U}_0 = (b_0, w_0)^{\mathrm{T}}$. In this method one considers (infinitesimally) small nonuniform perturbations about the state in question and study whether all perturbations decay to zero, or some perturbations grow, indicating an instability. We denote a small perturbation of this kind by $\tilde{\mathbf{U}}$ of $\mathbf{U}_0$, so that the perturbed state is given by

$$\mathbf{U}(\mathbf{x},t) = \mathbf{U}_0 + \tilde{\mathbf{U}}(\mathbf{x},t), \tag{A8}$$

where $\mathbf{U} = (b, w)^{\mathrm{T}}$ and $\tilde{\mathbf{U}} = (\tilde{b}, \tilde{w})^{\mathrm{T}}$. In the standard way [11] we let $\tilde{\mathbf{U}}(\mathbf{x}, t) = \mathbf{a}(t)\mathrm{e}^{\mathrm{i}\mathbf{k}\cdot\mathbf{x}} + c.c$ where $\mathbf{a}(t) = (a_1, a_2)^{\mathrm{T}}$ and $\mathbf{k} = (k_x, k_y)$. Substitution of (A8) into (3) gives

$$\frac{\partial \tilde{b}}{\partial t} = G_b\big|_{\mathbf{U}_0+\tilde{\mathbf{U}}}(b_0 + \tilde{b})(1 - (b_0 + \tilde{b})) - (b_0 + \tilde{b}) + \delta_b \Delta \tilde{b}, \tag{A9a}$$

$$\frac{\partial \tilde{w}}{\partial t} = p - \nu(1 - \rho(b_0 + \tilde{b}))(w_0 + \tilde{w}) - G_w\big|_{\mathbf{U}_0+\tilde{\mathbf{U}}}(w_0 + \tilde{w}) + \delta_w \Delta \tilde{w}. \tag{A9b}$$

To evaluate $G_b\big|_{\mathbf{U}_0+\tilde{\mathbf{U}}}$ and $G_w\big|_{\mathbf{U}_0+\tilde{\mathbf{U}}}$ we expand the kernels (A6) as

$$g(\mathbf{x}, \mathbf{x}', t) = g_0(\mathbf{x}, \mathbf{x}') + g_1(\mathbf{x}, \mathbf{x}')\tilde{b}(\mathbf{x}, t) + \mathcal{O}(\tilde{b}^2) \tag{A10a}$$

$$g(\mathbf{x}', \mathbf{x}, t) = g_0(\mathbf{x}', \mathbf{x}) + g_1(\mathbf{x}', \mathbf{x})\tilde{b}(\mathbf{x}', t) + \mathcal{O}(\tilde{b}^2) \tag{A10b}$$

where

$$g_0(\mathbf{x} - \mathbf{x}') = g_0(\mathbf{x}' - \mathbf{x}) = g(\mathbf{x} - \mathbf{x}', t)\big|_{\mathbf{U}=\mathbf{U}_0} = \frac{f(\mathbf{x}, \mathbf{x}')}{2\pi}\mathrm{e}^{-\frac{|\mathbf{x}-\mathbf{x}'|^2}{2\sigma_0^2}} \tag{A11a}$$

$$g_1(\mathbf{x} - \mathbf{x}') = g_1(\mathbf{x}' - \mathbf{x}) = \frac{\partial g}{\partial b}(\mathbf{x} - \mathbf{x}', t)\bigg|_{\mathbf{U}=\mathbf{U}_0} = \frac{\eta|\mathbf{x} - \mathbf{x}'|^2 f(\mathbf{x}, \mathbf{x}')}{2\pi\sigma_0^3}\mathrm{e}^{-\frac{|\mathbf{x}-\mathbf{x}'|^2}{2\sigma_0^2}}, \tag{A11b}$$

with $\sigma_0 = 1 + \eta b_0$. Considering linear terms only we can then calculate

$$\begin{aligned} G_b\big|_{\mathbf{U}_0+\tilde{\mathbf{U}}} &\approx \lambda w_0 \int_\Omega g_0(\mathbf{x}, \mathbf{x}')\mathrm{d}\mathbf{x}' + \lambda \int_\Omega g_0(\mathbf{x}, \mathbf{x}')\tilde{w}(\mathbf{x}', t)\mathrm{d}\mathbf{x}' + \lambda w_0 \tilde{b}(\mathbf{x}, t) \int_\Omega g_1(\mathbf{x}, \mathbf{x}')\mathrm{d}\mathbf{x}' \\ &= \frac{\lambda w_0 \sigma_0^2 \psi_0}{\psi} + \lambda \mathcal{F}_0(k)\tilde{w}(\mathbf{x}, t) + \frac{\lambda \eta w_0 \psi_1}{\sigma_0 \psi}\tilde{b}(\mathbf{x}, t), \end{aligned} \tag{A12a}$$

$$\begin{aligned} G_w\big|_{\mathbf{U}_0+\tilde{\mathbf{U}}} &\approx \lambda b_0 \int_\Omega g_0(\mathbf{x}', \mathbf{x})\mathrm{d}\mathbf{x}' + \lambda \int_\Omega g_0(\mathbf{x}', \mathbf{x})\tilde{b}(\mathbf{x}', t)\mathrm{d}\mathbf{x}' + \lambda b_0 \int_\Omega g_1(\mathbf{x}', \mathbf{x})\tilde{b}(\mathbf{x}', t)\mathrm{d}\mathbf{x}' \\ &= \frac{\lambda b_0 \sigma_0^2 \psi_0}{\psi} + \lambda \mathcal{F}_0(k)\tilde{b}(\mathbf{x}, t) + \lambda b_0 \mathcal{F}_1(k)\tilde{b}(\mathbf{x}, t), \end{aligned} \tag{A12b}$$

where $k = |\mathbf{k}|$ is the wavenumber of the perturbation, $\psi_0 = c_0 + 2c_2\sigma_0^2 + 8c_4\sigma_0^4 + 48c_6\sigma_0^6 + 384c_8\sigma_0^8$, and $\psi_1 = 2c_0\sigma_0^2 + 8c_2\sigma_0^4 + 48c_4\sigma_0^6 + 384c_6\sigma_0^8 + 3840c_8\sigma_0^{10}$. $\mathcal{F}_0(k)$ and $\mathcal{F}_1(k)$ are the Fourier transforms of $g_0(\mathbf{x}, \mathbf{x}')$ and $g_1(\mathbf{x}, \mathbf{x}')$, respectively, and are expressed in terms of the quantities

$$\begin{aligned} F_0 &= \sigma_0^2 \exp(-\sigma_0^2 k^2/2), \\ F_2 &= F_0\sigma_0^2(2 - \sigma_0^2 k^2), \\ F_4 &= F_0\sigma_0^4(8 - 8\sigma_0^2 k^2 + \sigma_0^4 k^4), \\ F_6 &= F_0\sigma_0^6(48 - 72\sigma_0^2 k^2 + 18\sigma_0^4 k^4 - \sigma_0^6 k^6), \\ F_8 &= F_0\sigma_0^8(384 - 768\sigma_0^2 k^2 + 288\sigma_0^4 k^4 - 32\sigma_0^6 k^6 + \sigma_0^8 k^8), \\ F_{10} &= F_0\sigma_0^{10}(3840 - 9600\sigma_0^2 k^2 + 4800\sigma_0^4 k^4 - 800\sigma_0^6 k^6 + 50\sigma_0^8 k^8 - \sigma_0^{10} k^{10}), \end{aligned}$$

as

$$\begin{aligned} \mathcal{F}_0(k) &= \frac{1}{\psi}(c_0F_0 + c_2F_2 + c_4F_4 + c_6F_6 + c_8F_8), \\ \mathcal{F}_1(k) &= \frac{\eta}{\sigma_0^3\psi}(c_0F_2 + c_2F_4 + c_4F_6 + c_6F_8 + c_8F_{10}). \end{aligned}$$

The uniform steady states of (3) satisfy

$$\frac{\lambda\psi_0}{\psi}w_0b_0(1-b_0)\sigma_0^2 - b_0 = 0, \tag{A13a}$$

$$p - \nu w_0(1-\rho b_0) - \frac{\lambda\psi_0}{\psi}w_0b_0\sigma_0^2 = 0, \tag{A13b}$$

which yield two homogeneous steady states: the bare soil state $(0, p/\nu)$ and the uniform vegetation state, the expression of which we omit here for brevity. Substitution of (A12) into (A9), retaining only first order terms, and using (A13) yields the system of equations

$$\frac{\mathrm{d}\mathbf{a}}{\mathrm{d}t} = \mathbf{J}\mathbf{a} \tag{A14}$$

where the components of $\mathbf{J} \in \mathbb{R}^{2\times 2}$ are given by

$$\mathbf{J}_{11} = \frac{\lambda w_0}{\psi}\left(\frac{\eta\psi_1 b_0(1-b_0)}{\sigma_0} + \sigma_0^2\psi_0(1-2b_0)\right) - 1 - \delta_b k^2, \tag{A15a}$$

$$\mathbf{J}_{12} = \lambda\mathcal{F}_0(k)b_0(1-b_0), \tag{A15b}$$

$$\mathbf{J}_{21} = \rho\nu w_0 - \lambda w_0\mathcal{F}_0(k) - \lambda b_0 w_0\mathcal{F}_1(k), \tag{A15c}$$

$$\mathbf{J}_{22} = -\nu(1-\rho b_0) - \frac{\lambda b_0\sigma_0^2\psi_0}{\psi} - \delta_w k^2. \tag{A15d}$$

Assuming a_1, $a_2 \propto \exp(\mu t)$ we can calculate the dispersion relation μ=$\mu(k)$. Substitution of the steady states into μ reveal their stability at a given precipitation rate (Figure A1). For $P > P_0$ the bare soil state undergoes a uniform instability (Figure A1A) that drives the system towards a uniform vegetation state, which may not necessarily be stable. For $P < P_T$ the uniform vegetation state undergoes a non-uniform instability (Figure A1B) which generates a periodic vegetation state. Both instabilities are stationary (perturbations grow monotonically in time) as $Im(\mu) = 0$ for both of them. Furthermore, we verified that the root-augmentation feedback alone can induce a non-uniform instability by setting $\delta_w = 0$ (i.e., no soil–water diffusion feedback).

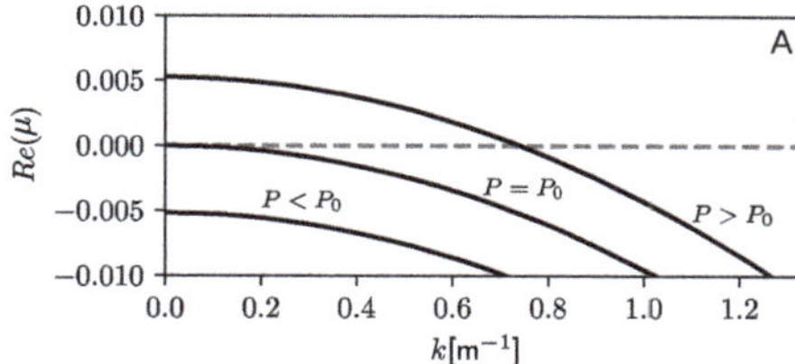

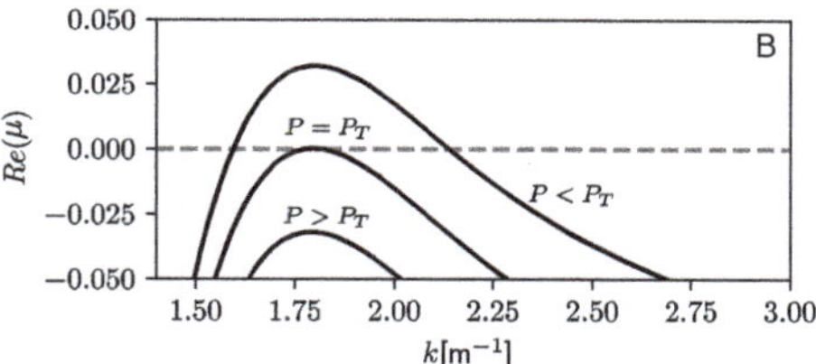

Figure A1. Dispersion relations for stable, marginally stable and unstable steady states. (**A**) Instability of the bare soil state to the growth of a uniform mode at $P_0 = 175.3$ mm/y (or $(\mathrm{kg/m^2})\mathrm{y}^{-1}$). (**B**) Instability of the uniform vegetation state to the growth of a non-uniform mode (of finite wavenumber) at $P_T = 296.2$ mm/y. Precipitation values: (**A**) $P = 299 > P_T$, $P = 293.4 < P_T$, (**B**) $P = 175.8 > P_0$, $P = 174.2 < P_0$. Other parameters are fixed: $E = 7[\mathrm{m^2/kg}]$, $K = 0.4[\mathrm{kg/m^2}]$, $M = 10.5[1/\mathrm{y}]$, $N = 15[1/\mathrm{y}]$, $\Lambda = 0.9[(\mathrm{m^2/kg})/\mathrm{y}]$, $\Gamma = 12[(\mathrm{m^2/kg})/\mathrm{y}]$, $R = 0.7$, $D_B = 0.1[\mathrm{m^2/y}]$, $D_W = 50[\mathrm{m^2/y}]$, $S_0 = 0.15$m, $c_0 = 1$, $c_2 = 0.0125$, $c_4 = 5.06 \times 10^{-4}$, $c_6 = -5.70 \times 10^{-6}$, $c_8 = 3.33 \times 10^{-7}$.

Appendix B. Model for Laterally Confined Roots

Laterally confined roots are modeled by considering the limit $S_0 \rightarrow 0$ in Equation (2). This amounts to replacing $G(\mathbf{X}, \mathbf{X}', T)$ in Equation (5a) by a delta function, which leads to the following model:

$$\begin{aligned} \frac{\partial B}{\partial T} &= \Lambda BW(1+EB)^2(1-\frac{B}{K}) - MB + D_B\Delta B\,, \\ \frac{\partial W}{\partial T} &= P - \frac{NW}{1+RB/K} - \Gamma BW(1+EB)^2 + D_W\Delta W\,. \end{aligned} \quad (A16)$$

This model is used in the studies described in Sections 4.2 and 4.3, with the nondimensional form

$$\begin{aligned} \partial_t b &= bw(1+\eta b)^2(1-b) - b + \Delta b\,, \\ \partial_t w &= p - \frac{nw}{1+\rho b} - \gamma bw(1+\eta b)^2 + \delta\Delta w\,, \end{aligned} \quad (A17)$$

obtained by introducing the non-dimensional variables:

$$b = \frac{B}{K},\ w = \frac{W\Lambda}{M},\ x = X\sqrt{M/D_B}.\ y = Y\sqrt{M/D_B},\ t = MT\,, \quad (A18)$$

and the non-dimensional parameters

$$p = \frac{P\Lambda}{M^2},\ n = \frac{N}{M},\ \gamma = \frac{\Gamma K}{M},\ \rho = R,\ \eta = EK,\ \delta = \frac{D_W}{D_B}\,. \quad (A19)$$

References

1. Hansen, J.; Sato, M.; Ruedy, R. Perception of climate change. *Proc. Natl. Acad. Sci. USA* **2012**, *109*, E2415–E2423. [CrossRef] [PubMed]
2. Hoerling, M.; Eischeid, J.; Perlwitz, J.; Quan, X.; Zhang, T.; Pegion, P. On the Increased Frequency of Mediterranean Drought. *J. Clim.* **2012**, *25*, 2146–2161. [CrossRef]
3. Cook, B.I.; Ault, T.R.; Smerdon, J.E. Unprecedented 21st century drought risk in the American Southwest and Central Plains. *Sci. Adv.* **2015**, *1*, e1400082. [CrossRef] [PubMed]
4. DeAngelis, D.; Grimm, V. Individual-based models in ecology after four decades. *F1000Prime* **2014**, *6*, 39. [CrossRef] [PubMed]
5. Grimm, V.; Railsback, S.F. *Individual-Based Modeling and Ecology*; Princeton University Press: Princeton, NJ, USA, 2005.
6. DeAngelis, D.L.; Yurek, S. Spatially Explicit Modeling in Ecology: A Review. *Ecosystems* **2016**. [CrossRef]
7. Holmes, E.E.; Lewis, M.A.; Banks, J.E.; Veit, R.R. Partial Differential Equations in Ecology: Spatial Interactions and Population Dynamics. *Ecology* **1994**, *75*, 17–29. [CrossRef]
8. Petrovskii, S.V.; Li, B.L. *Exactly Solvable Models of Biological Invasion*; Chapman and Hall/CRC Mathematical and Computational Biology, CRC Press: Boca Raton, FL, USA, 2005.
9. Petrovskii, S.V.; Morozov, A.Y.; Venturino, E. Allee effect makes possible patchy invasion in a predator–prey system. *Ecol. Lett.* **2002**, *5*, 345–352. [CrossRef]
10. Cantrell, R.S.; Cosner, C. *Spatial Ecology via Reaction-Diffusion Equations (Wiley Series in Mathematical & Computational Biology)*; Wiley: San Francisco, CA, USA, 2003.
11. Meron, E. *Nonlinear Physics of Ecosystems*; CRC Press, Taylor & Francis Group: Boca Raton, FL, USA, 2015.
12. Ovaskainen, O.; Meerson, B. Stochastic models of population extinction. *Trends Ecol. Evol.* **2010**, *25*, 643–652. [CrossRef]
13. Bonachela, J.A.; Muñoz, M.A.; Levin, S.A. Patchiness and demographic noise in three ecological examples. *J. Stat. Phys.* **2012**, *148*, 723–739. [CrossRef]
14. Pigliucci, M. Evolution of phenotypic plasticity: Where are we going now? *Trends Ecol. Evol.* **2005**, *20*, 481–486. [CrossRef]
15. Aichinger, E.; Kornet, N.; Friedrich, T.; Laux, T. Plant stem cell niches. *Annu. Rev. Plant Biol.* **2012**, *63*, 615–636. [CrossRef] [PubMed]
16. Mokany, K.; Raison, J.; Prokushkin, A. Critical analysis of root-shoot rations in terrestrial biomes. *Glob. Chang. Biol.* **2006**, *12*, 84–96. [CrossRef]

17. Khadka, J.; Yadav, N.; Granot, G.; Grafi, G. Summer dormancy is associated with loss of the permissive epigenetic marker dimethyl H3K4 and extensive reduction in proteins involved in basic cell functions. *Plants* **2018**, *7*, 59. [CrossRef] [PubMed]
18. Grafi, G. A "mille-feuilles" of stress tolerance in the desert plant Zygophyllum dumosum Boiss. *Isr. J. Plant Sci.* **2019**, *66*, 52–59. [CrossRef]
19. Evenari, M.; Shanan, L.; Tadmor, N. *The Negev: The Challenge of a Desert*; Harvard University Press: Cambridge, MA, USA, 1982.
20. Terwilliger, V.; Zeroni, M. Gas exchange of a desert shrub (Zygophyllum dumosum Boiss.) under different soil moisture regimes during summer drought. *J. Plant Ecol.* **1994**, *115*, 133–144.
21. Schenk, J.H.; Espino, S.; Goedhart, C.M.; Nordenstahl, M.; Cabrera, H.I.; Jones, C.S. Hydraulic integration and shrub growth form linked across continental aridity gradients. *Proc. Natl. Acad. Sci. USA* **2008**, *105*, 11248–11253. [CrossRef] [PubMed]
22. Ginzburg, C. Some anatomic features of splitting of desert shrubs. *Annu. Rev. Plant Biol.* **1963**, *13*, 92–97.
23. Orshan, G.; Zand, G. Seasonal body reduction of certain desert halfshrubs. *Bull. Res. Counc. Isr.* **1962**, *11D*, 35–42.
24. Cunningham, G.; Strain, B. An ecological significance of seasonal leaf variability in a desert shrub. *Ecology* **1969**, *50*, 400–408. [CrossRef]
25. Dalling, J.; Davis, A.; Schutte, B.; Elizabeth, A.A. Seed survival in soil: Interacting effects of predation, dormancy and the soil microbial community. *J. Ecol.* **2011**, *99*, 89–95. [CrossRef]
26. Kivilaan, A.; Bandurski, R. The one hundred-year period for Dr. Beal's seed viability experiment. *Am. J. Bot.* **1981**, *68*, 1290–1292. [CrossRef]
27. Shen-Miller, J.; Mudgett, M.; Schopf, J.; Clarke, S.; Berger, R. Exceptional seed longevity and robust growth: Ancient sacred lotus from China. *Am. J. Bot.* **1995**, *82*, 1367–1380. [CrossRef]
28. Long, R.; Gorecki, M.; Renton, M.; Scott, J.; Colville, L.; Goggin, D.; Commander, L.; Westcott, D.; Cherry, H.; Finch-Savage, W. The ecophysiology of seed persistence: A mechanistic view of the journey to germination or demise. *Biol. Rev. Camb. Philos. Soc.* **2015**, *90*, 31–59. [CrossRef] [PubMed]
29. Cohen, D. A general model of optimal reproduction in a randomly varying environment. *Am. J. Bot.* **1966**, *56*, 219–228. [CrossRef]
30. Venable, D.; Brown, J. The selective interactions of dispersal, dormancy, and seed size as adaptations for reducing risk in variable environments. *Am. Nat.* **1988**, *131*, 360–384. [CrossRef]
31. Ooi, M. Seed bank persistence and climate change. *Seed Sci. Res.* **2012**, *22*, S53–S60. [CrossRef]
32. Walck, J.; Hidayati, S.; Dixon, K.; Thompson, K.; Poschlod, P. Climate change and plant regeneration from seed. *Glob. Chang. Biol.* **2011**, *17*, 2145–2161. [CrossRef]
33. Raviv, B.; Godwin, J.; Granot, G.; Grafi, G. The dead can nurture: Novel insights into the function of dead organs enclosing embryos. *Int. J. Mol. Sci.* **2018**, *19*, 2455. [CrossRef]
34. Lefever, R.; Lejeune, O. On the origin of tiger bush. *Bull. Math. Biol.* **1997**, *59*, 263–294. [CrossRef]
35. Klausmeier, C. Regular and irregular patterns in semiarid vegetation. *Science* **1999**, *284*, 1826–1828. [CrossRef]
36. Von Hardenberg, J.; Meron, E.; Shachak, M.; Zarmi, Y. Diversity of Vegitation Patterns and Desertification. *Phys. Rev. Lett.* **2001**, *87*, 198101. [CrossRef] [PubMed]
37. Shnerb, N.; Sarah, P.; Lavee, H.; Solomon, S. Reactive glass and vegetation patterns. *Phys. Rev. Lett.* **2003**, *90*, 0381011. [CrossRef] [PubMed]
38. HilleRisLambers, R.; Rietkerk, M.; Van den Bosch, F.; Prins, H.; de Kroon, H. Vegetation pattern formation in semi-arid grazing systems. *Ecology* **2001**, *82*, 50–61. [CrossRef]
39. Gilad, E.; Von Hardenberg, J.; Provenzale, A.; Shachak, M.; Meron, E. Ecosystem Engineers: From Pattern Formation to Habitat Creation. *Phys. Rev. Lett.* **2004**, *93*, 098105. [CrossRef] [PubMed]
40. Gilad, E.; Von Hardenberg, J.; Provenzale, A.; Shachak, M.; Meron, E. A Mathematical Model for Plants as Ecosystem Engineers. *J. Theor. Biol.* **2007**, *244*, 680. [CrossRef]
41. Gilad, E.; Shachak, M.; Meron, E. Dynamics and Spatial Organization of Plant Communities in Water Limited Systems. *Theor. Popul. Biol.* **2007**, *72*, 214–230. [CrossRef] [PubMed]
42. Kinast, S.; Zelnik, Y.R.; Bel, G.; Meron, E. Interplay between Turing Mechanisms can Increase Pattern Diversity. *Phys. Rev. Lett.* **2014**, *112*, 078701. [CrossRef]
43. Meron, E. From Patterns to Function in Living Systems: Dryland Ecosystems as a Case Study. *Annu. Rev. Condens. Matter Phys.* **2018**, *9*, 79–103. [CrossRef]

44. Meron, E. Pattern formation—A missing link in the study of ecosystem response to environmental changes. *Math. Biosci.* **2016**, *271*, 1–18. [CrossRef]
45. Meinhardt, H. Pattern formation in biology: A comparison of models and experiments. *Rep. Prog. Phys.* **1992**, *55*, 797. [CrossRef]
46. Rietkerk, M.; van de Koppel, J. Regular pattern formation in real ecosystems. *Trends Ecol. Evol.* **2008**, *23*, 169–175. [CrossRef] [PubMed]
47. Verrecchia, E.; Yair, A.; Kidron, G.J.; Verrecchia, K. Physical properties of the psammophile cryptogamic crust and their consequences to the water regime of sandy softs, north-western Negev Desert, Israel. *J. Arid Environ.* **1995**, *29*, 427–437. [CrossRef]
48. Eldridge, D.J.; Zaady, E.; Shachak, M. Infiltration through three contrasting biological soil crusts in patterned landscapes in the Negev, Israel. *J. Stat. Phys.* **2012**, *148*, 723–739. [CrossRef]
49. Walker, B.; Ludwig, D.; Holling, C.; Peterman, R. Stability of semi-arid savana grazing systems. *J. Ecol.* **1981**, *69*, 473–498. [CrossRef]
50. Atkinson, J.A.; Rasmussen, A.; Traini, R.; Voß, U.; Sturrock, C.; Mooney, S.; Wells, D.M.; Bennett, M.J. Branching Out in Roots: Uncovering Form, Function, and Regulation. *Plant Physiol.* **2014**, *166*, 538–550. [CrossRef] [PubMed]
51. Zelnik, Y.R.; Meron, E.; Bel, G. Gradual Regime Shifts in Fairy Circles. *Proc. Natl. Acad. Sci. USA* **2015**, *112*, 12327–12331. [CrossRef]
52. Strogatz, S.H. *Nonlinear Dynamics and Chaos: With Applications to Physics, Biology, Chemistry, and Engineering*; Westview Press: Pittsburgh, PA, USA, 2001.
53. Newell, A.C.; Passot, T.; Lega, J. Order parameter equations for patterns. *Ann. Rev. Fluid Mech.* **1993**, *25*, 399–453. [CrossRef]
54. Elphick, C.; Tirapegui, E.; Brachet, M.E.; Coullet, P.; Iooss, G. A simple global characterization for normal forms of singular vector fields. *Phys. D* **1987**, *29*, 95–127. [CrossRef]
55. Cross, M.; Greenside, H. *Pattern Formation and Dynamics in Nonequilibrium Systems*; Cambridge University Press: Cambridge, UK, 2009.
56. Getzin, S.; Yizhaq, H.; Bell, B.; Erickson, T.E.; Postle, A.C.; Katra, I.; Tzuk, O.; Zelnik, Y.R.; Wiegand, K.; Wiegand, T.; et al. Discovery of fairy circles in Australia supports self-organization theory. *Proc. Natl. Acad. Sci. USA* **2016**, *113*, 3551–3556. [CrossRef]
57. Lejeune, O.; Tlidi, M.; Lefever, R. Vegetation spots and stripes: Dissipative structures in arid landscapes. *Int. J. Quantum Chem.* **2004**, *98*, 261–271. [CrossRef]
58. Gowda, K.; Riecke, H.; Silber, M. Transitions between patterned states in vegetation models for semiarid ecosystems. *Phys. Rev. E* **2014**, *89*, 022701. [CrossRef] [PubMed]
59. Doedel, E.; Paffenroth, R.C.; Champneys, A.R.; Fairgrieve, T.F.; Kuznetsov, Y.A.; Oldeman, B.E.; Sandstede, B.; Wang, X. *AUTO2000*; Technical report; Concordia University: Montréal, QC, Canada, 2002.
60. Uecker, H.; Wetzel, D.; Rademacher, J.D. pde2path-A Matlab package for continuation and bifurcation in 2D elliptic systems. *Numer. Math. Theory Methods Appl.* **2014**, *7*, 58–106. [CrossRef]
61. Kozyreff, G.; Chapman, S.J. Asymptotics of Large Bound States of Localized Structures. *Phys. Rev. Lett.* **2006**, *97*, 044502. [CrossRef] [PubMed]
62. Burke, J.; Knobloch, E. Snakes and ladders: Localized states in the Swift-Hohenberg equation. *Phys. Lett. A* **2007**, *360*, 681–688. [CrossRef]
63. Dawes, J. Localized Pattern Formation with a Large-Scale Mode: Slanted Snaking. *SIAM J. Appl. Dyn. Syst.* **2008**, *7*, 186–206. [CrossRef]
64. Lloyd, D.J.B.; Sandstede, B.; Avitabile, D.; Champneys, A.R. Localized Hexagon Patterns of the Planar Swift-Hohenberg Equation. *SIAM J. Appl. Dyn. Syst.* **2008**, *7*, 1049–1100. [CrossRef]
65. Avitabile, D.; Lloyd, D.; Burke, J.; Knobloch, E.; Sandstede, B. To Snake or Not to Snake in the Planar Swift–Hohenberg Equation. *SIAM J. Appl. Dyn. Syst.* **2010**, *9*, 704–733. [CrossRef]
66. Knobloch, E. Spatial Localization in Dissipative Systems. *Annu. Rev. Condens. Matter Phys.* **2015**, *6*, 325–359. [CrossRef]
67. Pomeau, Y. Front motion, metastability and subcritical bifurcations in hydrodynamics. *Physica D* **1986**, *23*, 3. [CrossRef]
68. Bel, G.; Hagberg, A.; Meron, E. Gradual regime shifts in spatially extended ecosystems. *Theor. Ecol.* **2012**, *5*, 591–604. [CrossRef]

69. Clerc, M.G.; Fernandez-Oto, C.; García-Ñustes, M.A.; Louvergneaux, E. Origin of the Pinning of Drifting Monostable Patterns. *Phys. Rev. Lett.* **2012**, *109*, 104101. [CrossRef] [PubMed]
70. Juergens, N. The biological underpinnings of Namib Desert fairy circles. *Science* **2013**, *339*, 1618–1621. [CrossRef] [PubMed]
71. Fernandez-Oto, C.; Tlidi, M.; Escaff, D.; Clerc, M.G. Strong interaction between plants induces circular barren patches: Fairy circles. *Philos. Trans. A Math. Phys. Eng. Sci.* **2014**, *372*, 20140009. [CrossRef] [PubMed]
72. Tarnita, C.; Bonachela, J.A.; Sheffer, E.; Guyton, J.A.; Coverdale, T.C.; Long, R.A.; Pringle, R.M. A theoretical foundation for multi-scale regular vegetation patterns. *Nature* **2017**, *541*, 398–401. [CrossRef] [PubMed]
73. Getzin, S.; Wiegand, K.; Wiegand, T.; Yizhaq, H.; von Hardenberg, J.; Meron, E. Adopting a spatially explicit perspective to study the mysterious fairy circles of Namibia. *Ecography* **2015**, *38*, 1–11. [CrossRef]
74. Tschinkel, W. The Life Cycle and Life Span of Namibian Fairy Circles. *PLoS ONE* **2012**, *7*, e38056. [CrossRef]
75. Zelnik, Y.R.; Meron, E.; Bel, G. Localized states qualitatively change the response of ecosystems to varying conditions and local disturbances. *Ecol. Complex.* **2016**, *25*, 26–34. [CrossRef]
76. Dawes, J.; Williams, J. Localised pattern formation in a model for dryland vegetation. *J. Math. Biol.* **2016**, *73*, 63–90. [CrossRef]
77. Zelnik, Y.R.; Gandhi, P.; Knobloch, E.; Meron, E. Implications of tristability in pattern-forming ecosystems. *Chaos Interdiscip. J. Nonlinear Sci.* **2018**, *28*, 033609. [CrossRef]
78. Mau, Y.; Haim, L.; Meron, E. Reversing desertification as a spatial resonance problem. *Phys. Rev. E* **2015**, *91*, 012903. [CrossRef]
79. Xu, C.; Van Nes, E.H.; Holmgren, M.; Kéfi, S.; Scheffer, M. Local Facilitation May Cause Tipping Points on a Landscape Level Preceded by Early-Warning Indicators. *Am. Nat.* **2015**, *186*, E81–E90. [CrossRef] [PubMed]
80. Van Strien, M.J.; Huber, S.H.; Anderies, J.M.; Grêt-Regamey, A. Resilience in social-ecological systems: Identifying stable and unstable equilibria with agent-based models. *Ecol. Soc.* **2019**, *24*, 8. [CrossRef]
81. Scheffer, M.; Carpenter, S.; Foley, J.; Folke, C.; Walkerk, B. Catastrophic shifts in ecosystems. *Nature* **2001**, *413*, 591–596. [CrossRef] [PubMed]
82. Ashwin, P.; Wieczorek, S.; Vitolo, R.; Cox, P. Tipping points in open systems: Bifurcation, noise-induced and rate-dependent examples in the climate system. *Philos. Trans. Math. Phys. Eng. Sci.* **2012**, *370*, 1166–1184. [CrossRef] [PubMed]
83. Feudel, U.; Pisarchik, A.N.; Showalter, K. Multistability and tipping: From mathematics and physics to climate and brain—Minireview and preface to the focus issue. *Chaos Interdiscip. J. Nonlinear Sci.* **2018**, *28*, 033501. [CrossRef] [PubMed]
84. Zelnik, Y.R.; Meron, E. Regime shifts by front dynamics. *Ecol. Indic.* **2018**, *94*, 544–552. [CrossRef]
85. Fernandez-Oto, C.; Tzuk, O.; Meron, E. Front Instabilities Can Reverse Desertification. *Phys. Rev. Lett.* **2019**, *122*, 048101. [CrossRef]
86. Hagberg, A.; Meron, E. Complex Patterns in reaction Diffusion Systems: A Tale of Two Front Instabilities. *Chaos* **1994**, *4*, 477–484. [CrossRef]
87. Goldstein, R.E.; Muraki, D.J.; Petrich, D.M. Interface proliferation and the growth of labyrinths in a reaction-diffusion system. *Phys. Rev. E* **1996**, *53*, 3933–3957. [CrossRef]
88. Hagberg, A.; Yochelis, A.; Yizhaq, H.; Elphick, C.; Pismen, L.; Meron, E. Linear and nonlinear front instabilities in bistable systems. *Phys. D Nonlinear Phenom.* **2006**, *217*, 186–192. [CrossRef]
89. Ursino, N. The influence of soil properties on the formation of unstable vegetation patterns on hillsides of semiarid catchments. *Adv. Water Resour.* **2005**, *28*, 956–963. [CrossRef]
90. Sherratt, J.A. An Analysis of Vegetation Stripe Formation in Semi-Arid Landscapes. *J. Math. Biol.* **2005**, *51*, 183–197. [CrossRef] [PubMed]
91. Deblauwe, V.; Couteron, P.; Bogaert, J.; Barbier, N. Determinants and dynamics of banded vegetation pattern migration in arid climates. *Ecol. Monogr.* **2012**, *82*, 3–21. [CrossRef]
92. Sherratt, J.A. Pattern Solutions of the Klausmeier Model for Banded Vegetation in Semiarid Environments V: The Transition from Patterns to Desert. *SIAM J. Appl. Math.* **2013**, *73*, 1347–1367. [CrossRef]
93. Sherratt, J.A. Using wavelength and slope to infer the historical origin of semiarid vegetation bands. *Proc. Natl. Acad. Sci. USA* **2015**, *112*, 4202–4207. [CrossRef]
94. Sherratt, J.A.; Lord, G.J. Nonlinear dynamics and pattern bifurcations in a model for vegetation stripes in semi-arid environments. *Theor. Popul. Biol.* **2007**, *71*, 1–11. [CrossRef]

95. Zelnik, Y.R.; Kinast, S.; Yizhaq, H.; Bel, G.; Meron, E. Regime shifts in models of dryland vegetation. *Philos. Trans. R. Soc. A* **2013**, *371*, 20120358. [CrossRef]
96. Siteur, K.; Siero, E.; Eppinga, M.B.; Rademacher, J.D.M.; Doelman, A.; Rietkerk, M. Beyond Turing: The response of patterned ecosystems to environmental change. *Ecol. Complex.* **2014**, *20*, 81–96. [CrossRef]
97. Yizhaq, H.; Gilad, E.; Meron, E. Banded Vegetation: Biological Productivity and Resilience. *Physica A* **2005**, *356*, 139. [CrossRef]
98. Bastiaansen, R.; Jaïbi, O.; Deblauwe, V.; Eppinga, M.B.; Siteur, K.; Siero, E.; Mermoz, S.; Bouvet, A.; Doelman, A.; Rietkerk, M. Multistability of model and real dryland ecosystems through spatial self-organization. *Proc. Natl. Acad. Sci. USA* **2018**, *115*, 11256–11261. [CrossRef]
99. Rietkerk, M.; Dekker, S.C.; de Ruiter, P.C.; van de Koppel, J. Self-Organized Patchiness and Catastrophic Shifts in Ecosystems. *Science* **2004**, *305*, 1926–1929. [CrossRef] [PubMed]
100. Valentine, C.; d'Herbes, J.; Poesen, J. Soil and water components of banded vegetation patterns. *Catena* **1999**, *37*, 1–24. [CrossRef]
101. Deblauwe, V.; Barbier, N.; Couteron, P.; Lejeune, O.; Bogaert, J. The global biogeography of semi-arid periodic vegetation patterns. *Glob. Ecol. Biogeogr.* **2008**, *17*, 715–723. [CrossRef]
102. Borgogno, F.; D'Odorico, P.; Laio, F.; Ridolfi, L. Mathematical models of vegetation pattern formation in ecohydrology. *Rev. Geophys.* **2009**, *47*, RG1005. [CrossRef]
103. Kyriazopoulos, P.; Jonathan, N.; Meron, E. Species coexistence by front pinning. *Ecol. Complex.* **2014**, *20*, 271–281. [CrossRef]
104. Malkinson, D.; Tielbörger, K. What does the stress-gradient hypothesis predict? Resolving the discrepancies. *Oikos* **2010**, *119*, 1546–1552. [CrossRef]

Article

Pattern Formation and Bistability in a Generalist Predator-Prey Model

Vagner Weide Rodrigues [1,†], Diomar Cristina Mistro [2,†] and Luiz Alberto Díaz Rodrigues [2,*,†]

1 Instituto Federal do Rio Grande do Sul–Campus Bento Gonçalves, Bento Gonçalves RS 95700-206, Brazil; vagner.rodrigues@bento.ifrs.edu.br

2 Departamento de Matemática, Universidade Federal de Santa Maria, Santa Maria RS 97105-900, Brazil; dcmistro@gmail.com

* Correspondence: ladiazrodrigues@gmail.com

† These authors contributed equally to this work.

Received: 14 November 2019; Accepted: 15 December 2019; Published: 20 December 2019

Abstract: Generalist predators have several food sources and do not depend on one prey species to survive. There has been considerable attention paid by modellers to generalist predator-prey interactions in recent years. Erbach and collaborators in 2013 found a complex dynamics with bistability, limit-cycles and bifurcations in a generalist predator-prey system. In this paper we explore the spatio-temporal dynamics of a reaction-diffusion PDE model for the generalist predator-prey dynamics analyzed by Erbach and colleagues. In particular, we study the Turing and Turing-Hopf pattern formation with special attention to the regime of bistability exhibited by the local model. We derive the conditions for Turing instability and find the region of parameters for which Turing and/or Turing-Hopf instability are possible. By means of numerical simulations, we present the main types of patterns observed for parameters in the Turing domain. In the Turing-Hopf range of the parameters, we observed either stable patterns or homogeneous periodic distributions. Our findings reveal that movement can break the effect of hysteresis observed in the local dynamics, what can have important implication in pest management and species conservation.

Keywords: generalist predator; pattern formation; Turing instability; Turing-Hopf bifurcation; bistability; regime shift

1. Introduction

The relevance of considering species movement and spatial distribution is nowadays well known and recognized in mathematical ecology [1]. In some cases, such as the study of biological invasions of an exotic species or the spread of epidemics, the phenomena is essentially spatial. In other situations, however, the explicit introduction of space can change the forecasts of the nonspatial counterpart model. A classical example where the spatial model change the conclusions of the local one, is dynamics of hosts and parasitoids whose coexistence is extensively observed in nature: the nonspatial Nicholson-Bailey host-parasitoid model [2] forecasts extinction of one or both species; however when space and local movement are considered along with this dynamics, the persistence of hosts and parasitoids is obtained and different spatial configurations are observed [3,4].

There are several ways of including space in the model, depending on how the variables are considered, discrete or continuous. If the spatial and temporal variables are assumed continuous and the population is described in terms of densities, Partial Differential Equations (PDE) are the common framework. The so called reaction-diffusion models formulated in terms of second-order PDE have now a long history in theoretical ecology since the analogy between the movement of molecules and the random dispersal of individuals of a population was proposed by Skellam (1951) [5–8]. Travelling waves of population invasions [9–12] for studying the spread of epidemics and exotic species; the critical patch size problems which study the minimum region necessary for a population to persist [5,13] and pattern formation in ecology [8,14–16] are among the relevant biological phenomena that have been being addressed through reaction-diffusion models.

Biological populations are rarely homogeneously distributed in space what makes the study of pattern formation a very important issue in ecology. Briefly speaking, Turing showed in his seminal paper [17] that, under appropriate conditions, a stationary stable homogeneous distribution of two chemical substances in the absence of diffusion can be destabilized by heterogeneous perturbations when diffusion is present. In their pioneer work Segel and Jakson (1972) [16] first presented the Turing's ideas of instability induced by diffusion in an ecological context while and Gierer and Meinhardt (1972) [18] identified the "activator-inhibitor mechanism". Since then the literature abounds with research on the Turing mechanism of pattern formation in Biology [19] and, in particular, in predator-prey systems. When the stable equilibrium of the local dynamics undergoes a Hopf bifurcation as some parameter varies, a limit cycle, very common in predator-prey systems, appears. For parameters close to the bifurcation point, it is known that heterogeneous perturbations can lead the spatial system to complex spatio-temporal dynamics including spatio-temporal chaos. This mechanism of pattern formation is known as Turing-Hopf bifurcation [20,21].

Models focusing on pattern formation in predator-prey systems, in most cases, consider that the predator is a specialist [20–22]. It means that predators feed itself on a single species of prey, so that in the absence of it the predator goes extinct. Generalist predators, in their turn, have more than one food source and do not depend exclusively on one prey to survive. Although generalist predators have been receiving considerable attention in the context of invasions and biological control in the last decades, there are few studies analysing the influence of a generalist predator in a spatially distributed system [23,24]. A recent study analyse how a generalist predator may influence the stable heterogeneous spatial pattern formation [25]. In this research, the author considered a reaction-diffusion generalist predator-prey system with constant alternative food source and Holling type II functional response [26]. Chakraborty (2015) [25] analysed the case in which the model presents just one coexistence equilibrium. As a result of Turing instability, patterns such as cold spots, hot spots and stripes were observed. For parameters near the Turing-Hopf region, the model exhibits a chaotic behavior in space and time.

The model by Erbach and collaborators (2013) [27] consists of a prey species, which grows logistically, and a predator that has a constant alternative food source, so that it does not become extinct in the absence of the focal prey. Predation is described by the functional response Holling type III. The local model explored by Erbach and collaborators shows a rich and complex dynamics which can exhibit one stable interior equilibrium, bistability of two stable equilibria, limit cycles (stable and unstable) and bistability between a stable equilibrium and a stable limit-cycle with plenty of bifurcations. The concomitant existence of two stable equilibria creates basins of attraction which means that the initial condition determines the fate of the system solution. It also makes possible sudden and abrupt state transitions that may be dramatic from the ecological, social or economic point of view [28].

Erbach and collaborators (2013) say, in the discussion of the above mentioned paper, that they were taken by a great curiosity about the effects of diffusion on the spatiotemporal dynamics of the species conducted by this dynamics. Hence, here we explore the effects of spatial distribution and movement on the generalist predator-prey dynamics studied by Erbach et al. (2013). Our goal is to investigate the role of space on the complex dynamics of their model. In particular, we intend to analyze the Turing and Turing-Hopf pattern formation with special attention to the effects of dispersal on the bistability regime. We start describing the model and giving a brief review of the main results by Erbach and collaborators in Section 2. In what follows, we derive the Turing instability conditions for diffusive instability in the model in Section 3 and then, by means of numerical simulations, presented in Section 4, we illustrate the patterns obtained. Finally, we discuss the implications of our results for the dynamics of the species.

2. The Model

The proposed model is a reaction-diffusion system with a prey and a generalist predator in the presence of a constant alternative food for the predator. Predation is described by the functional response Holling type III, which means that the rate in which predators attack the prey is small when the prey density is low. In this scenario, the predator prefers to capture more abundant alternative prey species and abandon the focal prey. However, as the focal prey density increases, the rate at which predators capture it, also increases, in a phenomenon known as prey switching. Since for many biological species the individuals disperse randomly in space [8], we adopt the classical diffusion as paradigma for modelling the spatial dispersal. Hence, the spatio-temporal dynamics is described by the following system:

$$\begin{aligned}\frac{\partial N}{\partial T} &= RN\left(1-\frac{N}{K}\right) - \frac{AN^2P}{1+HN^2} + D_1\left(\frac{\partial^2 N}{\partial X^2}+\frac{\partial^2 N}{\partial Y^2}\right),\\ \frac{\partial P}{\partial T} &= \frac{BP}{1+EP} - MP + C\frac{AN^2P}{1+HN^2} + D_2\left(\frac{\partial^2 P}{\partial X^2}+\frac{\partial^2 P}{\partial Y^2}\right),\end{aligned} \tag{1}$$

where $N(X,Y,T)$ and $P(X,Y,T)$ denote the densities of the prey and the predator, respectively, at $(X,Y) \in \mathbb{R}^2$ and time $T \geq 0$.

In the absence of predator, the prey population reproduces according to the logistic function. Parameters R and K denote the intrinsic growth rate and the carrying capacity of the prey. In its turn, in the absence of the focal prey, the predator population reproduces following a Beverton-Holt function. The parameter B denotes the per capita reproduction rate of the predator while E represents the strength of density-dependence. The death rate of the predator is denoted by M. In the term corresponding to the type-III functional response, A represents the searching efficiency and H, the handling time [26]. The efficiency of converting prey into predator biomass is denoted by C. D_1 and D_2 are the diffusion coefficients of prey and predator, respectively. All the parameters are positive and since we are considering that predator can survive in the absence of the focal prey, we assume $B > M$.

We introduce the following dimensionless variables: $n = N/K$, $p = (AKP)/R$, $x = X\sqrt{R/D_2}$, $y = Y\sqrt{R/D_2}$ and $t = RT$, so that predator and prey population dynamics are governed by the nondimensional system

$$\begin{aligned}\frac{\partial n}{\partial t} &= n(1-n) - \frac{n^2p}{1+an^2} + D\left(\frac{\partial^2 n}{\partial x^2}+\frac{\partial^2 n}{\partial y^2}\right),\\ \frac{\partial p}{\partial t} &= \frac{cp}{1+dp} - ep + \frac{bn^2p}{1+an^2} + \left(\frac{\partial^2 p}{\partial x^2}+\frac{\partial^2 p}{\partial y^2}\right),\end{aligned} \tag{2}$$

where $a = HK^2$, $b = (CAK^2)/R$, $c = B/R$, $d = (ER)/(AK)$, $e = M/R$ and $D = D_1/D_2$. Model (2) now contains only six parameters (against ten in its dimensional form). Since $B > M$, we have now $c > e$.

Let Ω be a two-dimensional bounded square domain and η the outward unit normal vector of the boundary $\partial\Omega$. We investigate the model (2) under the following initial conditions:

$$\begin{aligned}n_0 &= n(x,y,0) > 0,\\ p_0 &= p(x,y,0) > 0,\end{aligned} \tag{3}$$

where $(x,y) \in \Omega$, with zero-flux boundary conditions

$$\left.\frac{\partial n}{\partial \eta}\right|_{(x,y)} = \left.\frac{\partial p}{\partial \eta}\right|_{(x,y)} = 0, \tag{4}$$

where $(x,y) \in \partial\Omega$.

Local Stability

For completeness, we briefly summarize the main results of the local stability. A detailed analysis can be found in [27].

The corresponding non-diffusive model is

$$\begin{aligned} \frac{dn}{dt} &= n(1-n) - \frac{n^2 p}{1+an^2}, \\ \frac{dp}{dt} &= \frac{cp}{1+dp} - ep + \frac{bn^2 p}{1+an^2}. \end{aligned} \tag{5}$$

Moreover $n = 0$ and $p = 0$, the n-nullclines and p-nullclines are given, respectively, by

$$f(n) = \frac{(1+an^2)(1-n)}{n} \quad \text{and} \quad g(n) = \frac{1}{d}\left(\frac{c(1+an^2)}{e(1+an^2) - bn^2} - 1\right). \tag{6}$$

The shape of function f depends on parameter a. If $a \leq 27$ then $f(n)$ is monotonically decreasing, whereas if $a > 27$, $f(n)$ has two local extrema ([27]; see also [22]). Moreover, the n-nullcline always has a singularity at $n = 0$ and a root at $n = 1$. Function $g(n)$, on the other hand, is always monotonically increasing. When $b < ae$, $g(n)$ is bounded. However, for $b > ae$, there is a vertical asymptote at $n = \sqrt{e/(b-ae)}$ [27]. Figure 1 shows a typical shape of both nullclines for $a > 27$ and $b < ae$.

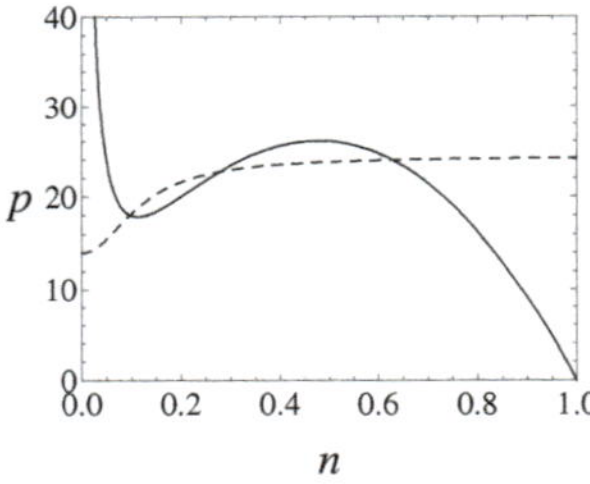

Figure 1. Possible shape of the local model nullclines. The continuous and dashed curves represent the graphic of f and g, respectively. f has two local extreme ($a > 27$) and g has an upper limit ($b < ae$).

The model always has the trivial equilibrium $(0,0)$ and the two semi-trivial equilibria $(1,0)$ and $(0,(c-e)/ed)$. Furthermore, depending on the parameter values, the system can have one or three coexistence equilibria.

Linearization of the system around the coexistence equilibrium (n^*, p^*) produces the Jacobian matrix

$$J(n^*, p^*) = \begin{pmatrix} a_{11} & a_{12} \\ a_{21} & a_{22} \end{pmatrix}, \tag{7}$$

where $a_{11} = 1 - 2n^* - \frac{2n^* p^*}{(1+an^*)^2}$, $a_{12} = -\frac{n^{*2}}{1+an^{*2}}$, $a_{21} = \frac{2bn^* p^*}{(1+an^{*2})^2}$ and $a_{22} = \frac{bn^{*2}}{1+an^{*2}} + \frac{c}{(1+dp^*)^2} - e$.

By applying the Routh-Hurwitz criterion, the equilibrium $(0,0)$ is always an unstable node and $(1,0)$ and $(0,(c-e)/ed)$ are saddles. To analyze the stability of the coexistence equilibria, Erbach and colleagues [27] found two relations between the slope of f and g with the elements of matrix J, as follows:

$$f'(n^*) = -a_{11}/a_{12}, \tag{8}$$

$$g'(n^*) = -a_{21}/a_{22}. \tag{9}$$

These conditions lead to the following conclusions:

(i) If $f'(n^*) < 0$, then the equilibrium is stable;
(ii) The equilibrium is a saddle point if and only if $f'(n^*) > g'(n^*)$.

In other words, conclusion (i) establishes that if an equilibrium point is located where the n-nullcline is decreasing, then it is stable. From this, Erbach and colleagues conclude that for $a < 27$ the unique coexistence equilibria is always stable. Conclusion (ii) states that if an equilibrium point stands where the slope of the n-nullcline is greater than the slope of the p-nullcline, then this equilibrium is a saddle point. This scenario only occurs for $a > 27$ and a proper choice for the other parameters. The authors found a complex dynamical behaviour including biestability, limit cycles and several bifurcations related to the positive slope of n-nullcline at the equilibrium. Pattern formation in the spatial model will be explored in these different scenarios.

3. Turing Instability

The Turing mechanism [17] shows that, under appropriate conditions, two chemical substances that react and diffuse can lead to stable heterogeneous spatial patterns. Turing considered Fickian classical diffusion for the substances movement and the law of mass action for the interaction. For diffusive instability to occur it is assumed that the reaction achieved a stable steady state. The classical diffusion is a macroscopic description (or on the population level) of the random movement of the molecules (or biological individuals) [29]. This process results in a net movement from regions of high concentrations towards regions of low concentrations; it means that it has a stabilizing effect that lead to an uniform spatial distribution. Counter-intuitively, Turing showed that the combination of two stabilizing processes (diffusion and stable interaction) can have a destabilizing effect that can promote heterogeneous distributions of the substances.

In this section, we present the conditions that lead to diffusive instability for model (2), following the usual analysis (see, e.g., [6,15,16,30]). Briefly speaking, Turing instability consists in two assumptions: (1) initially, in the absence of diffusion, the system must exhibit a stationary homogeneous state and (2) in the presence of diffusion, the stationary homogeneous state is unstable to small perturbations, what can lead to spatially stable heterogeneous patterns.

Therefore, let us consider the existence of a stationary uniform solution $E^* = (n^*, p^*)$ for the system (2). To examine the stability of E^* with respect to perturbations

$$\varepsilon(x,y,t) = \alpha_1 \cos(k_1 x + k_2 y) e^{\lambda t}, \tag{10}$$

$$\delta(x,y,t) = \alpha_2 \cos(k_1 x + k_2 y) e^{\lambda t}, \tag{11}$$

where α_i, k_i $(i = 1,2)$ and λ are constants, we linearize system (2) and analyze the eigenvalues λ of the matrix

$$J_D = \begin{pmatrix} a_{11} - Dk^2 & a_{12} \\ a_{21} & a_{22} - k^2 \end{pmatrix}, \tag{12}$$

where $k^2 = k_1^2 + k_2^2$ is the wave number (see [16] for a detailed analysis).

Perturbations (10) and (11) decay if and only if $Re(\lambda) < 0$. In other words, the homogeneous equilibrium (n^*, p^*) of system (2) is stable if $tr J_D(n^*, p^*) < 0$ and $\det J_D(n^*, p^*) > 0$, i.e.

$$a_{11} + a_{22} - (D+1)k^2 < 0, \tag{13}$$

$$(a_{11} - Dk^2)(a_{22} - k^2) - a_{12}a_{21} > 0. \tag{14}$$

Note that for $k = 0$ (which means perturbations of zero wavenumber), inequalities (13) and (14) become

$$a_{11} + a_{22} < 0, \tag{15}$$
$$a_{11}a_{22} - a_{12}a_{21} > 0. \tag{16}$$

These conditions have already been satisfied since we assume the existence of a stationary homogeneous state in the absence of diffusion.

Now, by violating (13) or (14), the stationary homogeneous state becomes unstable to small perturbations in the presence of diffusion, satisfying the second assumption of diffusive instability. But inequality (13) is always true, because $a_{11} + a_{22} < 0$. So inequality (14) is the only one that can be "broken".

Expanding the left hand side of (14) and reversing the inequality, we find $H(k^2)$ given by

$$H(k^2) = Dk^4 - (Da_{22} + a_{11})k^2 + a_{11}a_{22} - a_{12}a_{21} < 0. \tag{17}$$

Function $H(k^2)$ is a quadratic function and its graph is a parabola. Since we are looking for a positive k^2 such that $H(k^2) < 0$, it is necessary that

$$Da_{22} + a_{11} > 0. \tag{18}$$

Since $D = D_1/D_2$, from (18) follows that $D \neq 1$ (or $D_1 \neq D_2$). Furthermore, the minimum of H is given by

$$H(k^2_{min}) = a_{11}a_{22} - a_{12}a_{21} - \frac{(Da_{22} + a_{11})^2}{4D}, \tag{19}$$

where $k^2_{min} = \dfrac{Da_{22} + a_{11}}{2D}$.

In order to satisfy inequality (17) is sufficient that H be negative at its minimum. Therefore, imposing $H(k^2_{min}) < 0$ we obtain the last condition for diffusive instability:

$$(Da_{22} + a_{11})^2 - 4D(a_{11}a_{22} - a_{12}a_{21}) > 0. \tag{20}$$

In summary, from (15), (16) and (20), the conditions for Turing instability are:

$$a_{11} + a_{22} < 0, \tag{21}$$
$$a_{11}a_{22} - a_{12}a_{21} > 0, \tag{22}$$
$$(Da_{22} + a_{11})^2 - 4D(a_{11}a_{22} - a_{12}a_{21}) > 0. \tag{23}$$

As mentioned above, the Turing instability starts with a homogeneous locally stable steady state in the absence of diffusion what means that the system returns to the spatially homogeneous equilibrium distribution after being homogeneously perturbed. However, if heterogeneous small perturbations of a certain wave number are applied, the system can evolve to heterogeneous spatial distributions. If the homogeneous equilibrium stability is broken via a Hopf bifurcation, the local dynamics becomes oscillatory and homogeneous small perturbations do not vanish with time. The Turing-Hopf bifurcation is then characterized as the point in the parameter space for which the system steady state is destabilized by both homogeneous (Hopf) as well as heterogeneous (Turing) perturbations [20,21,31,32]. Many authors have reported that in the vicinity or inside the Turing-Hopf domain heterogeneous stationary patterns [21] as well as spatially irregular non-stationary patterns and spatiotemporal chaos [20,31] are expected.

4. Numerical Simulations

Our goal in this section is to investigate the effects of species movement on the dynamics, in particular, we intend to investigate the existence of spatial patterns in the system (2). We perform extensive numerical simulations for the model (2) in a bounded spatial habitat $\Omega = [0, 12.5] \times [0, 12.5] \subset \mathbb{R} \times \mathbb{R}$ with reflective zero-flux boundary conditions. To solve the system, we discretized the model in time and space using FTCS (forward-time central-space) scheme in a personal code developed in Mathematica 10.0, in which we take $\Delta x = \Delta y = 0.25$ and the time-step is $\Delta t = 0.01$. From an initial random perturbations of the uniform distribution of the species at the local equilibrium, we run the simulations until the system reach a steady state. In all the cases, predator and prey species assumed the same type of spatial distribution so that we will only refer to the prey one.

The simplest case occurs when we take $a \leq 27$ so that f (n-nullcline) is always decreasing and thus the model has only one coexistence equilibrium. In this case, although conditions (21) and (22) are satisfied, inequality (23) never happens. Indeed, since $a \leq 27$, the slope of f is negative and from conclusion **(i)**, the equilibrium state is stable. Now, noting that a_{12} is always negative, from equation (8) we conclude that a_{11} must be negative too. Moreover, as g is always increasing ($g' > 0$) and $a_{21} > 0$, equation (9) implies that a_{22} is negative. Therefore, as we have $a_{11} < 0$, $a_{22} < 0$ and $D > 0$, condition (23) is not satisfied, which means that there is no diffusive instability for $a \leq 27$.

From now on, we focus the analysis on the case $a > 27$ and study the following three sets of parameters a, c, d and e for which the dynamics has been explored by Erbach and collaborators (2013):

- Set I: $a = 100$, $c = 1$, $d = 0.9$ and $e = 0.5$;
- Set II: $a = 100$, $c = 1$, $d = 0.1$ and $e = 0.5$;
- Set III: $a = 80$, $c = 0.7$, $d = 0.6$ and $e = 0.1$.

For each set, we present the corresponding bifurcation diagram with b as controlling parameter. After applying the Turing conditions for diffusive instability, we show in the space of parameters b, D, the regions for which Turing and/or Turing-Hopf instability can occurs, referred to as Turing or Turing-Hopf region, accordingly.

We also present the density plots of the species distribution to illustrate the patterns obtained. We will refer to the heterogeneous distributions obtained as Turing or Turing-Hopf patterns, accordingly. We adopted the classification used by Baurmann and collaborators [20] for the patterns obtained: *Homogeneous Distribution and Stationary Patterns*. The first one obviously refers to a homogeneous distribution of the species over the whole habitat. If the system evolves to a stationary heterogeneous spatial distribution of prey and predators, we will use the second classification. The resulting pattern can have different spatial configurations. Isolated areas with high population densities surrounded by areas with very low population densities are called *hot spots*. However, we make a distinction when high picks of density are observed and call the pattern *spikes*. On the other hand, when the distribution of the species presents isolated areas with low densities, the resulting pattern is called *cold spots*. When high and low population densities are alternated, the pattern is called *stripes*. Patterns that do not fit any of these characteristics will be simply called *heterogeneous patterns*.

5. Results

5.1. Set I

For parameters' set I the system presents only one coexistence equilibrium which can lose stability and gives place to a stable limit cycle depending on the values of b. Figure 2a illustrates a bifurcation diagram of the prey equilibrium with respect to b while Figure 2b shows one nullcline for the prey and several predator-nullclines. In both figures the values of b for which we obtain Turing as well as Turing-Hopf spatial patterns can be observed.

From our previous comments, Turing instability can only be possible when the equilibrium occurs in an increasing region of the n-nullcline. For parameters' set I, this happens for b in the gray interval of

Figure 2a. For b in the dark gray region, the equilibrium is stable and Turing instability condition (23) can be satisfied for some values of D. For parameters in light gray interval, the coexistence equilibrium is unstable and limit cycles are observed what suggests possible Turing-Hopf instabilities for suitable values of D.

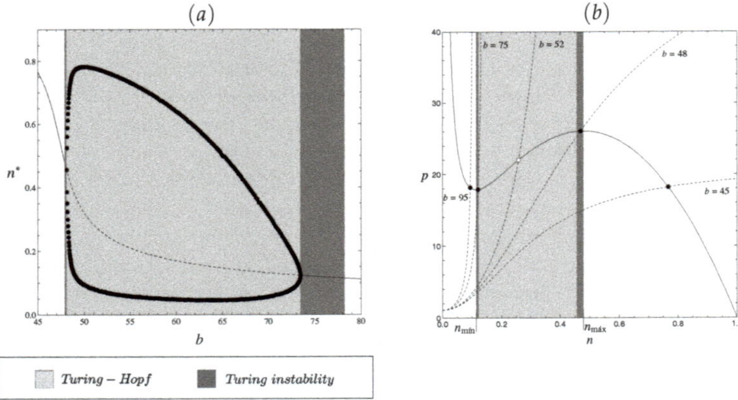

Figure 2. (**a**) Bifurcation diagram for prey species with b as the controlling parameter. The continuous (dashed) line indicates the stable (unstable) steady state. Black dots are the maximum and minimum prey density in the stable limite cycle. (**b**) Phase plane for different values of b. The continuous curve corresponds to one nullcline for prey while the dashed curves represent p-nullclines for different values of b. Black (white) dots represent the stable (unstable) coexistence equilibrium. In both images, other parameters are given by Set I.

Through this analysis we restricted our attention to the range of b for which diffusion driven instabilities are possible. We then, numerically verify the values of D for which condition (23) is satisfied. The only restriction for D is $0 < D < 1$ since $D_2 < D_1$. Figure 3 illustrates the region in the b, D parameters space where Turing and/or Turing-Hopf instabilities take place while Figure 4 indicates the types of pattern obtained for different combinations of the parameters b and D inside these regions.

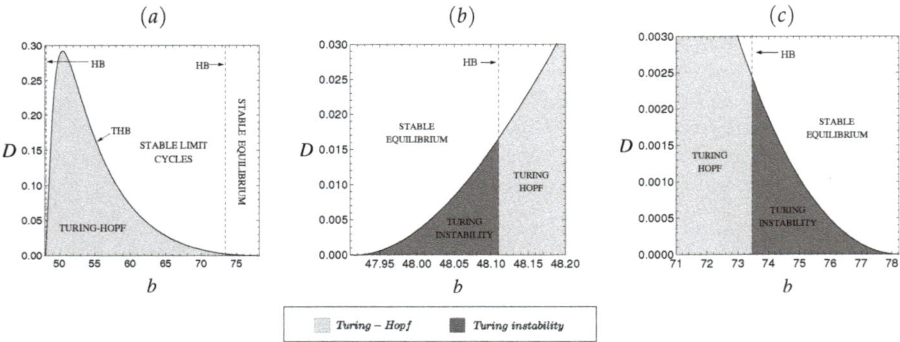

Figure 3. (**a**) Stability diagram in the $b \times D$ parameters space. (**b**,**c**) represent a zoom for a better visualization of the Turing instability region. HB stands for Hopf Bifurcation and THB indicates Turing Hopf Bifurcation.

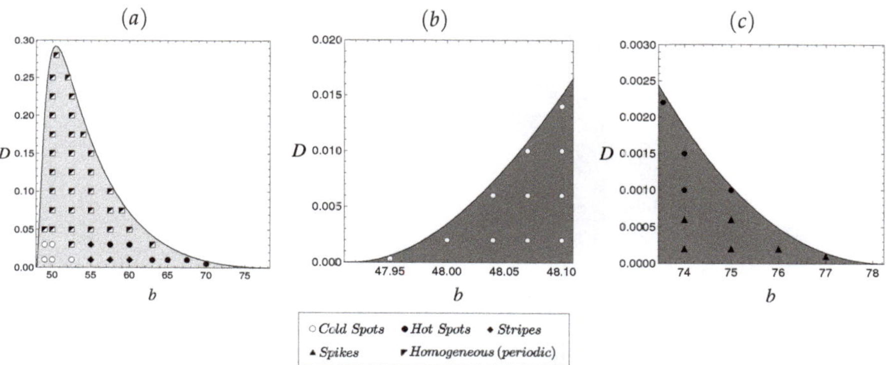

Figure 4. Pattern types obtained for different combinations of b and D inside the Turing-Hopf region (**a**) and inside Turing regions, illustrated in a zoom in (**b**,**c**).

Examples of patterns obtained for parameters inside the Turing-Hopf region are illustrated in Figure 5. We can observe that increasing b and keeping $D = 0.01$, the resulting pattern evolves from *cold spots* to *hot spots* after passing by a transition of stripes patterns. For grater values of D even though inside the Turing-Hopf region, the system evolves to a homogeneous population distribution. Since the equilibrium of the local dynamics is not stable for the parameters Set I with b inside this region; the system undergoes oscillations according to the corresponding limit cycle. Hence, the spatial dynamics exhibit time-oscillating homogeneous distributions.

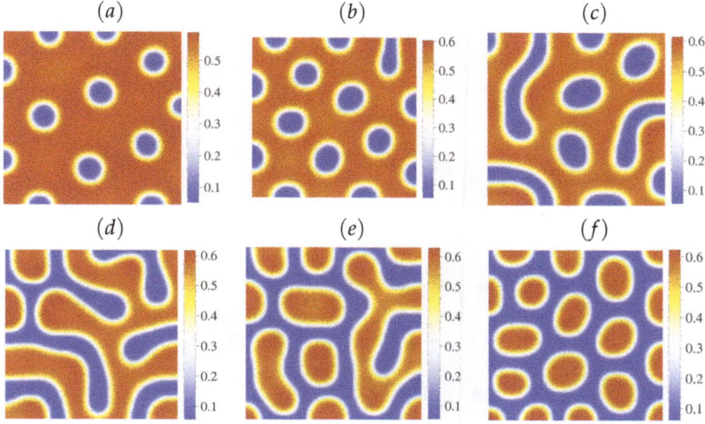

Figure 5. Prey spatial distribution in Turing-Hopf region for $D = 0.01$ and different values of b: (**a**) $b = 50$, (**b**) $b = 52.5$, (**c**) $b = 55$, (**d**) $b = 57.5$, (**e**) $b = 60$ and (**f**) $b = 65$.

For parameters taken inside the Turing instability region of the parameters illustrated in Figure 4b we obtained *cold spots* as resulting patterns (see Figure 6a,b for examples). On the other hand, with parameters inside the Turing region illustrated in Figure 4c, *spikes* and *hot spots* were observed as, for example, those showed in Figure 6c,d, respectively.

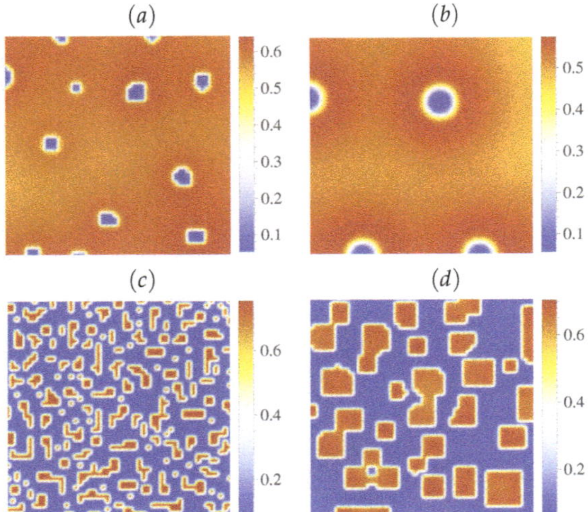

Figure 6. Example of patterns obtained in the two different Turign regions: (**a**) $b = 48.07$ and $D = 0.002$; (**b**) $b = 48.07$ and $D = 0.01$; (**c**) $b = 74$ and $D = 0.0002$ and (**d**) $b = 74$ and $D = 0.0015$.

5.2. Set II

For parameters set II, the system exhibits bistability and unstable limit cycles as illustrated in Figure 7. It suggests that it would be possible, for the same parameter range, two different pattern regimes depending on whether the initial perturbation is made around one or another equilibrium. Furthermore, one can ask about the influence of one equilibrium on the other. We will refer to the equilibrium with small prey density as E_1 and to the equilibrium with high prey density as E_2.

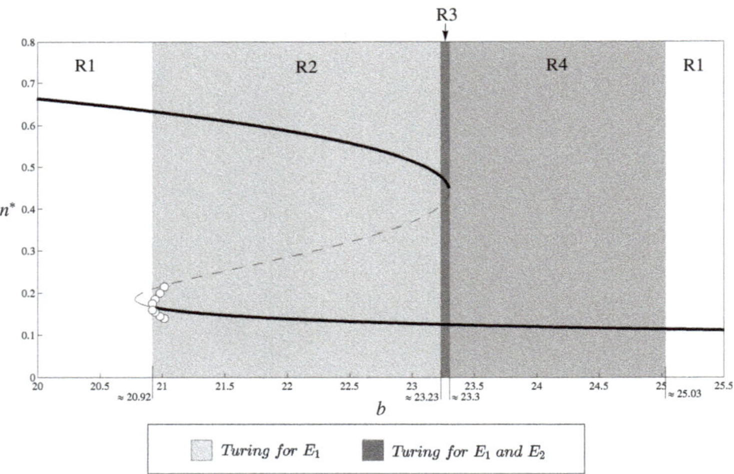

Figure 7. Bifurcation diagram for the prey equilibrium with regard to b with the remainder parameters fixed as in Set II. The black bold parts of the curve indicate stable equilibria while thin and dashed curves stand for unstable equilibria. The white dots represent unstable limit-cycles. In the white regions labeled R_1 no Turing pattern is possible. Turing instabilities are feasible in the light and dark gray regions as explained in the text.

For $b < 20.23$, E_2 is located where the n-nullcline is decreasing. Hence, condition (23) is not satisfied and Turing pattern is not possible for any perturbation of this equilibrium. For $20.92 < b < 25.03$ approximately, which includes regions R_2, R_3 and R_4 of Figure 7, the diffusive instability conditions (21)–(23) are satisfied for equilibrium E_1. For b in the small interval R_3 Turing instability is possible for both equilibria.

We then applied condition (23) for both equilibria in order to find the combinations of b and D for which Turing instability can happen. Figure 8a,b illustrate these regions for the equilibrium E_2 and E_1, respectively. For b in R_3 and D taken in the dark gray region of Figure 8a, small perturbation of equilibrium E_2, lead the species distributions to *cold spots*. For the equilibrium E_1, on the other hand, we obtained different results. For very small values of D ($D < 0.005$ approximately) regardless the value of b in the interval $20.92 < b < 25.03$ which includes regions $R2$ to $R4$, small heterogeneous initial perturbations lead to non-homogeneous patterns as those illustrated in Figure 9. On the other hand, as D increases, cold spots and heterogeneous stripes like patterns as illustrated in Figure 10 are obtained in regions $R2$ and $R3$.

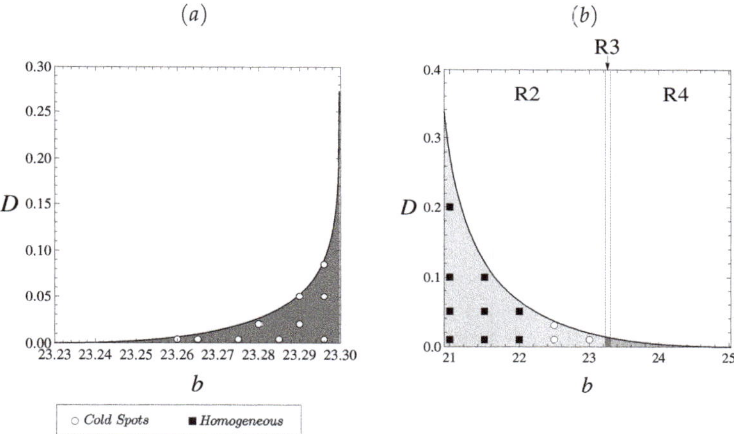

Figure 8. Regions in the $b - D$ parameters space where all the diffusive instability conditions are satisfied for (**a**) E_2 and (**b**) E_1.

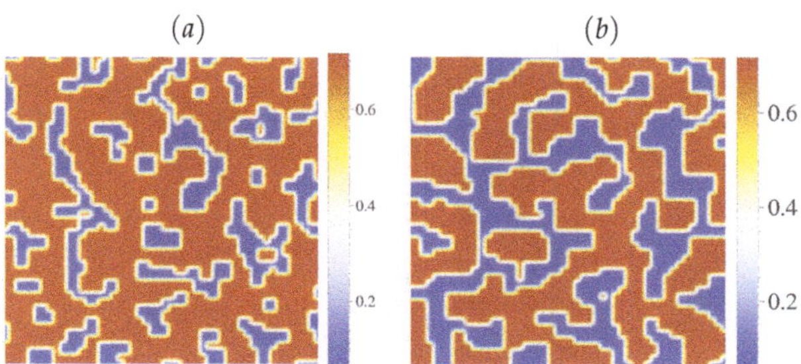

Figure 9. Examples of Turing patterns obtained for equilibrium E_1 in different regions of the parameters b with very small values for D. Region $R2$: (**a**) $b = 21.5$ and $D = 0.001$ and, Region $R3$: (**b**) $b = 23.24$ and $D = 0.001$.

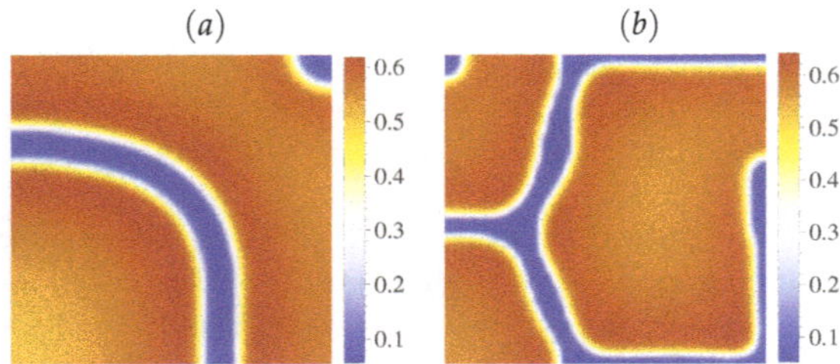

Figure 10. Examples of Turing patterns obtained for equilibrium E_1 in different regions of the parameters b with greater values for D. Region $R3$: (**a**) $b = 23.27$ and $D = 0.012$ and, Region $R4$: (**b**) $b = 23.5$ and $D = 0.008$.

An interesting phenomenon is observed in region $R2$. For b close to region R_3, initial perturbations of equilibrium E_1 lead to heterogeneous cold spots as those illustrated in Figure 5a. However, for small values of b in region R_2 (approximately $20.92 < b < 22$) and $D \geq 0.005$ initial perturbations around E_1 grow and the prey is lead to a homogeneous distribution at the equilibrium E_2 as represented in Figure 11. The blue points indicate the prey and predator densities in all the domain. For $t = 0$ (Figure 11a), all the points are around the equilibrium E_1. Without diffusion, all the values would converge to E_1. However, with diffusion, the solution in all the domain migrate during the transients (see Figure 11b–e) and finally tend to equilibrium E_2 (Figure 11f), even though the parameters belong to the Turing region. Nevertheless, there is no contradiction with Turing's theory since it does not state that when diffusion is present, heterogeneous patterns will necessarily takes place. It says that the homogeneous distribution is unstable to small perturbations, what is the case here: perturbations around E_1 grow and the system is led to a uniform distribution at the equilibrium value E_2. This phenomenon promotes a rupture of the effect of hysteresis observed in the local dynamics; in the spatially distributed model, the system state jumps from E_1 to E_2 due to perturbations caused by movement even though the initial distribution is inside the basin of attraction of E_1. We also observe that the unstable limit-cycles observed for small values b in R_2 (see Figure 7) do not have any influence on the patterns obtained (for $D < 0.005$) and on the solution migration to E_2 (for greater values of D).

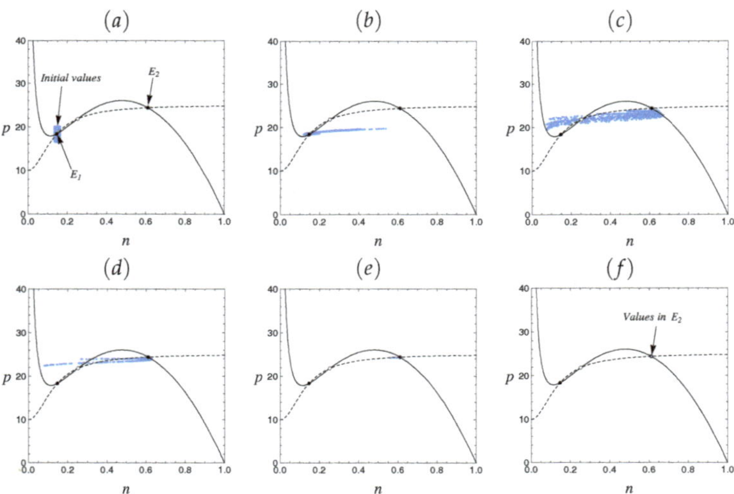

Figure 11. $n-$ and $p-$nullclines along with the values of prey and predator densities in the whole domain (indicated in blue), for $b = 21.5$ and $D = 0.05$, at time: (**a**) $t = 0$, (**b**) $t = 60$, (**c**) $t = 100$, (**d**) $t = 120$, (**e**) $t = 140$ and (**f**) $t = 200$.

5.3. Set III

Parameters' set III presents a much more complex local dynamics than the previous two cases. Differently from Set II, no bistability is observed. However, there is the coexistence of a stable equilibrium with a stable limit-cycle. We detected six intervals of b with different behaviour related to pattern formation which will be explained in what follows.

Figure 12 illustrates the bifurcation diagram for the prey with parameter b as controlling parameter. We will call E_1 the prey equilibrium with lower density and E_2 the higher prey equilibrium value. Figures 13 and 14 show the combinations of parameters b and D for which diffusive instability is obtained, respectively, for equilibrium E_1 and E_2 in different intervals for b. The dots in these two figures indicate the combinations tested and the type of pattern obtained.

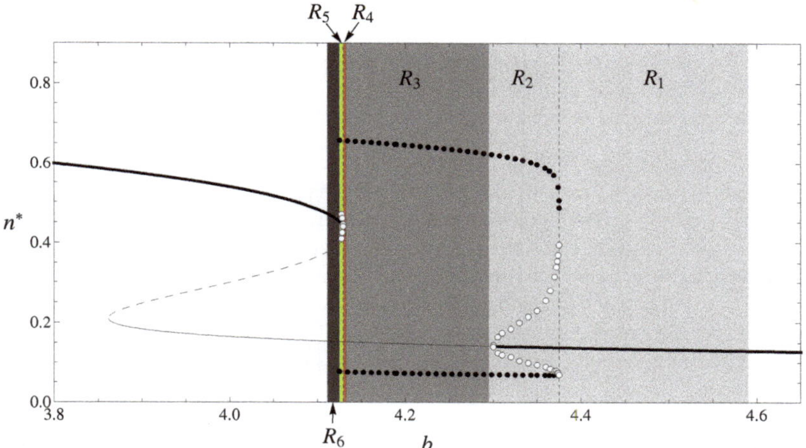

Figure 12. Bifurcation diagram for the prey equilibrium with respect to b.

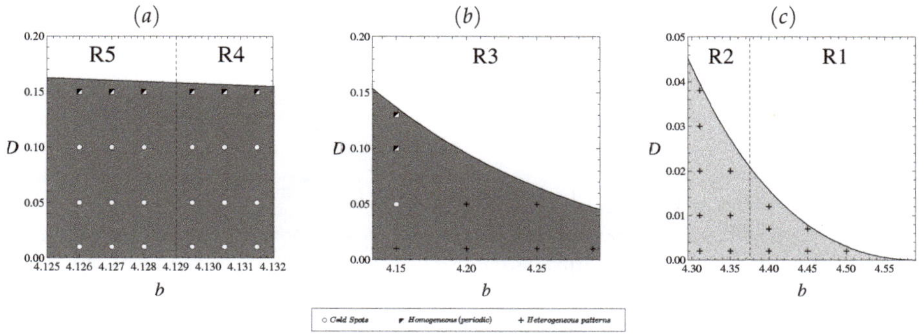

Figure 13. Regions in the $b - D$ parameters' space where Turing or Turing-Hopf instabilities are obtained for equilibrium E_1, illustrated in (**a–c**) for better visualization. Dots indicate the combinations of parameters used for tests.

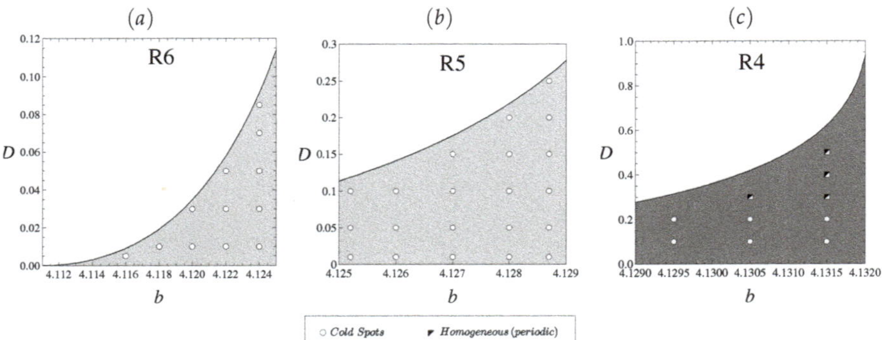

Figure 14. Regions in the $b - D$ parameters' space where Turing or Turing-Hopf instabilities are obtained for equilibrium E_2, illustrated in (**a–c**) for better visualization. Dots indicate the spatial pattern obtained for the corresponding combination of parameters.

In the intervals corresponding to R_1 of Figure 12, E_1 is the only stable equilibrium. All the diffusive instability conditions are satisfied and hence, Turing patterns are possible for E_1 in R_1. For D in gray region of Figure 13c, different heterogeneous patterns and stripes were obtained.

In R_2 of Figure 12, E_1 is still stable but now it is surrounded by two limit-cycles: one unstable and the other stable. Turing patterns are possible for E_1 in R_2 since all the conditions for diffusive instability can be satisfied for suitable combinations of b and D. The resulting patterns for D taken inside the corresponding gray region R_2 of Figure 13c are heterogeneous patterns. Since for b in R_2 there is an unstable limit-cycle surrounding E_1, we initially perturbed this equilibrium up to 1% of its value in each point.

For b in region R_3, E_1 becomes unstable but the stable limit cycle is still surrounding it (see Figure 12). The condition (23) can still be satisfied what characterize region R_3 as a region where Turing-Hopf patterns are possible. For small values of D inside the gray region of Figure 13b, heterogeneous patterns and cold spots are obtained. However, with larger values of D, still inside the Turing region, the system is driven to a homogeneous periodic distribution which assumes the corresponding values of the stable limit cycle that surrounds E_1.

Regions R_4 and R_5 are Turing-Hopf regions for E_1 since it is an unstable equilibrium surrounded by a stable limit-cycle. The combinations of b and D for which diffusive instability is observed can be seen in Figure 13a. Small perturbations of the homogeneous equilibrium distributions lead to heterogeneous as well as homogeneous periodic solutions for b and D in these regions. Cold spots patterns appear for low values of D in both regions R_4 and R_5 while homogeneous periodic solutions can be seen for larger values of D.

A saddle-node bifurcation takes place as b cross the border of region R_3 towards R_4 and the equilibrium E_2, at high prey levels, appears. E_2, which is also unstable for all values of b in R_4, is surrounded by a stable limit-cycle . However, condition (23) holds for b and D in a certain range of values as illustrated in Figure 14c, which characterizes it as a Turing-Hopf region for equilibrium E_2. We tested several combinations of parameters in this region and observed *cold spots* patterns for low values of D. For higher values of D, still inside the gray region, homogeneous periodic solutions are obtained. We can observe an unstable limit-cycle around E_2; with high values of D, the population values in all the domain jump outside this cycle and the system is lead to the stable outer cycle. It is worth noting that R_4 is also a Turing-Hopf region for equilibrium E_1 as mentioned above.

Moving b to region R_5, equilibrium E_2 becomes stable. However, it is surrounded by an unstable limit-cycle and by the outer stable limit-cycle. All the conditions (21) to (23) can be satisfied for E_2 in this region. For b and D the region illustrated in Figure 14b, the Turing patterns obtained are *cold spots*.

Finally, for b in region R_6, both limit-cycles disappear and E_2 is still stable. In R_6 the diffusive instability conditions can be satisfied only for E_2 if D is taken in the appropriate range illustrated in Figure 14a. *Cold spots* are the kind of distribution we obtained in this case.

6. Discussion

In this paper, we have analysed the effects of spatial distribution and species movement on the generalist predator-prey dynamics studied by Erbach et al. (2013) [27]. In particular, we have investigated the scenarios of pattern formation in the rich dynamics that the above mentioned authors proposed. Our main goal was to elucidate how is pattern formation influenced by limit-cycles and bistability present in the rich local dynamics. We then derived the conditions for Turing instability and developed numerical simulations for different sets of relevant parameters for the local kinetics. We found the regions in the parameter space b (related to the predation intensity) and D (the relation between prey and predator motility) for which Turing and Turing-Hopf patterns can be obtained, in most cases, for parameters far way from the bifurcation. By means of numerical simulations, we observed that hot spots, cold spots, stripes and a mixture of them can emerge depending on the predation intensity. With small values of b, predation on the focal prey is low and we observe isolated regions of lower prey (and predator) density; that is, cold spots are the typical patterns for low values of b. On the other hand, when predation is high (great values of b), the prey remains confined to limited regions of habitat in the so called hot spots patterns if predator movement rate is great compared to the prey. As consequence, the type of pattern can be associated with the effectiveness of the predator on controlling the focal prey. If hot spots are observed, prey density is around its low equilibrium level.

6.1. Spatiotemporal Chaos

In the Turing-Hopf instability domain, we obtained either stationary patterns (for small values of D, including hot/cold spots and stripes depending on b) or oscillating homogeneous distributions (for larger values of D). Spatiotemporal chaos, frequently observed in specialist [20,21] as well as generalist [25] predator-prey systems for parameters, close or inside, the Turing-Hopf instability regions, was not observed for any combinations of parameters, neither close the bifurcation nor into the region where chaotic behaviour would be expected. We also simulated Equation (2), for the parameters set III with $b = 4.2$ and $D = 0.05$; that is, inside the Turing-Hopf domain, from the initial perturbations $n(x,y,0) = n^* + 0.01\sin(\varepsilon(y-20))$ and $p(x,y,0) = p^* + 0.01\sin(\varepsilon(y-20))$, for $\varepsilon = 0.006$, $\varepsilon = 0.025$ and $\varepsilon = 0.25$. In all these simulations the system also converged to either stationary heterogeneous patterns or homogeneous oscillating distributions. None of the initial perturbation tested, produced spatiotemporal chaos.

Contrary to [20,31] who found spatially irregular non-stationary patterns and spatiotemporal chaos for parameters in the vicinity of a Turing-Hopf bifurcation, Banerjee and Petrovskii [21] obtained heterogeneous stationary patterns for parameters close to this type of bifurcation. Spatiotemporal chaos, nonetheless, was obtained in [21] for parameters into the Turing-Hopf range. Furthermore, there are evidences that Turing instability is not necessary to trigger chaos; Banerjee and Petrovskii [21] showed that chaotic spatial patterns can appear in a specialist predator-prey system, outside the Turing domain, as a result of non-homogeneous perturbations in a region of parameters where limit-cycles are present in the local dynamics. We then tested the above mentioned parameters in set III with $D = 1$ (that is, outside the Turing domain) and again no irregular spatiotemporal pattern was observed. Chakraborty [25] have obtained chaotic spatiotemporal oscillations in a generalist predator-prey interaction, however in his model, predation occurs according to the Holling type II functional response. Our results, on the other hand, indicate that neither Turing-Hopf nor Hopf bifurcations can always be associated with spatiotemporal chaos in continuous reaction-diffusion predator-prey systems. A similar behaviour has already been found by Rodrigues and colleagues [32] in a discrete Coupled Map Lattice for predator-prey system, in which stationary patterns are observed even when the local dynamics is oscillatory. Although more investigation is needed, our findings suggest that the Holling type III function response such as we have used in the present work, may have a stabilizing effect on the spatiotemporal behaviour of the generalist predator-prey dynamics.

6.2. Bistability and Regime Shifts

The model studied here presents bistability that emerge from saddle-node bifurcations, which is the scenario for jumps and hysteresis to occur [33].

When the prey is an agricultural pest controlled by the generalist predator, the hysteresis effect observed in the local dynamics means that once the prey density falls from the high equilibrium density E_2 to the desired low equilibrium E_1, as a result of increasing b, predation efforts can be relaxed to keep the control. That is, the pest density would only grow back to the high level E_2 if the predation decreases below the leftmost saddle-node bifurcation point, called now b_{c_1} (see Figure 7). Our results, nonetheless, show that spatial movement of the species (as it occurs with many species in real world plantations) may cause a shift back of the prey density to the upper levels if predation intensity is slightly relaxed below the level that causes the prey density decay, the rightmost saddle-node bifurcation point b_{c_2}. Close to this parameter value, the basin of attraction of the lower density equilibrium E_1 is considerable large (see Figure 7) so that, small perturbations caused by movement do not promote a shift back to high density E_2. However, as the parameter b decreases from b_{c_2}, the basin of attraction of E_1 and consequently, its resilience, diminishes. Now, in the reaction-diffusion model, for values of $b > b_{c1}$, perturbations due to movement can promote the shift back to high prey densities, in what we call an hysteresis break due to movement. In terms of pests control, our results suggest that in order to keep the prey at low densities, predation efforts should be kept as close as possible to b_{c_2}.

From a different perspective, if the prey is an endangered specie to be conserved, the regime shift and hysteresis described above may be undesirable since a regime shift can be hard to revert. From the species conservation point of view, it is important to detect regime shifts prior to its occurrence and there has been considerable effort of modelers to find hints that help identifying that the system is close to a regime shift and designing strategies of control to preventing catastrophes [28,34–36]. In the present study, we observed that for the upper prey equilibrium E_2 (see regions R_3 and R_6 in Figures 7 and 12, respectively), pattern formation is only observed close to the catastrophic transition value. The typical patterns observed in this case are cold spots what suggests that when the prey species under predation of a generalist predator, starts being scarce in some geographical regions, it may be a signal that some dramatic change is approaching if predation pressure increases. Since the avoidance of regime shifts demands the development and implementation of strategies of control, the indication of a regime shift close to its occurrence may, unfortunately, represent another difficulty of impending it. Our results, thus suggest that monitoring ecological systems is essential for impeding regime shifts.

We hope that our results can, from one hand side, address part of the questions raised by Erbach and colleagues and, on the other side, instigate researchers interested in population dynamics to further investigate generalist predator-prey interactions.

Author Contributions: The authors have contributed equally to this work. All authors have read and agreed to the published version of the manuscript.

Funding: VWR was funded by Coordenação de Aperfeiçoamento de Pessoal de Nível Superior.

Conflicts of Interest: The authors declare no conflict of interest.

References

1. Gibert, J.P.; Yeakel, J.D. Laplacian matrices and Turing bifurcations: Revisiting Levin 1974 and the consequences of spatial structure and movement for ecological dynamics. *Theor. Ecol.* **2019**, *12*, 265–281. [CrossRef]
2. Nicholson, A.J.; Bailey, V.A. The balance of animal population. Part I. *Proc. Zool. Soc. Lond.* **1935**, *3*, 551–598. [CrossRef]

3. Comins, H.N.; Hassell, M.P.; May, R.M. The spatial dynamics of host-parasitoid systems. *J. Anim. Ecol.* **1992**, *61*, 735–748. [CrossRef]
4. Hassell, M.P.; Comins, H.N.; May, R.M. Spatial structure and chaos in insect population dynamics. *Nature* **1991**, *353*, 255–258. [CrossRef]
5. Cantrell, R.S.; Cosner, C. *Spatial Ecology via Reaction-Diffusion Equations*; Wiley: London, UK, 2003.
6. Edelstein-Keshet, L. *Mathematical Models in Biology*; Random House: New York, NY, USA, 1988.
7. Holmes, E.E.; Lewis, M.A.; Banks, J.E.; Veit, R.R. Partial differential equations in ecology: Spatial interactions and populations dynamics. *Ecology* **1994**, *75*, 17–29. [CrossRef]
8. Okubo, A.O.; Levin, S.A. *Diffusion and Ecological Problems. Modern Perspectives*; Springer: New York, NY, USA, 2001.
9. Lewis, M.A.; Kareiva, P. Allee dynamics and the spread of invading organisms. *Theor. Pop. Biol.* **1993**, *43*, 141–158. [CrossRef]
10. Owen, M.; Lewis, M.A. How predation can stop, slow or reverse a prey invasion. *Bull. Math. Biol.* **2001**, *63*, 655–689. [CrossRef]
11. Sherratt, J.A.; Lewis, M.A.; Fowler, A.C. Ecological chaos in invasion. *Proc. Natl. Acad. Sci. USA* **1995**, *92*, 2524–2528. [CrossRef]
12. Shigesada, N.; Kawasaki, K. *Biological Invasions: Theory and Practice*; Oxford University Press: Oxford, UK, 1997.
13. Kierstead, H.; Slobodkin, L.B. The size of water masses containing plankton blooms. *J. Mar. Res.* **1953**, *12*, 141–147.
14. Malchow, H.; Petrovskii, S.V.; Venturino, E. *Spatiotemporal Patterns in Ecology and Ecoepidemiology: Theory, Models, and Simulations*; Champman & and Hall: London, UK, 2008.
15. Murray, J.D. *Mathematical Biology II: Spatial Models and Biomedical Applications*, 3rd ed.; Springer: Berlin, Germany, 2003.
16. Segel, L.A.; Jackson, J.L. Dissipative structure: an explanation and an ecological example. *J. Theor. Biol.* **1972**, *37*, 545–559. [CrossRef]
17. Turing, A.M. The chemical basis of morphogenesis. *Phil. Trans. R. Soc. Lond. B* **1952**, *237*, 37–72.
18. Gierer, A.; Meinhardt, H. A theory for biological pattern formation. *Kybernetik* **1972**, *12*, 30–39. [CrossRef] [PubMed]
19. Meinhardt, H. *Models of Biological Pattern Formation*; Academic Press: London, UK, 1982.
20. Baurmann, M.; Gross, T.; Feudel, U. Instabilities in spatially extended predator-prey systems: Spatio-temporal patterns in the neighborhood of Turing-Hopf bifurcations. *J. Theor. Biol.* **2007**, *245*, 220–229. [CrossRef] [PubMed]
21. Banerjee, M.; Petrovskii, S. Self-organized spatial patterns and chaos in a ratio-dependent predator-prey system. *Theor. Ecol.* **2011**, *4*, 37–53. [CrossRef]
22. Morozov, A.; Petrovskii, S. Excitable population dynamics, biological control failure, and spatiotemporal pattern formation in a model ecosystem. *Bull. Math. Biol.* **2009**, *71*, 863–887. [CrossRef]
23. Kumari, N. Pattern formation in spatially extended tritrophic food chain model systems: Generalist versus specialist top predator. *ISRN Biomath.* **2013**, *2013*, 198185. [CrossRef]
24. Magal, C.; Cosner, C.G.; Ruan, S.; Casas, J. Control of invasive hosts by generalist parasitoids. *Math. Med. Biol.* **2008**, *25*, 1–20. [CrossRef]
25. Chakraborty, S. The influence of generalist predators in spatially extended predator-prey systems. *Ecol. Complex.* **2015**, *23*, 50–60. [CrossRef]
26. Holling, C.S. Some characteristics of simple types of predation and parasitism. *Can. Entomol.* **1959**, *91*, 385–398. [CrossRef]
27. Erbach, A.; Lutscher, F.; Seo, G. Bistability and limit cycles in generalist predator-prey dynamics. *Ecol. Complex.* **2013**, *14*, 48–55. [CrossRef]
28. Biggs, R.; Carpenter, S.R.; Brock, W.A. Turning back from the brink: Detecting an impeding regime shift in time to avert it. *Proc. Natl. Acad. Sci. USA* **2009**, *106*, 826–831. [CrossRef] [PubMed]
29. Segel, L.A. *Modeling Dynamic Phenomena in Molecular and Cellular Biology*; Cambridge University Press: Cambridge, UK, 1984.
30. Jones, D.S.; Sleeman, B.D. *Differential Equations and Mathematical Biology*; Chapman & Hall/CRC: London, UK, 2003.

31. Meixner, M.; De Wit, A.; Bose, S.; Scholl, E. Generic spatiotemporal dynamics near codimension-two Turing-Hopf bifurcations. *Phys. Rev. E* **1997**, *55*, 6690–6697. [CrossRef]
32. Rodrigues, L.A.D.; Mistro, D.C.; Petrovskii, S. Pattern Formation, Long-Term Transients, and the Turing–Hopf Bifurcation in a Space- and Time-Discrete Predator–Prey System. *Bull. Math. Biol.* **2011**, *73*, 1812–1840. [CrossRef] [PubMed]
33. Strogatz, S.H. *Nonlinear Dynamics and Chaos. With Applications to Physics, Biology, Chemistry, and Engineering*, 2nd ed.; Westview Press: Philadelphia, PA, USA, 2015.
34. Kéfi, S.; Maarten, B.E.; de Ruiter, P.C.; Rietkerk, M. Bistability and regular patterns in arid systems. *Theor. Ecol.* **2010**, *3*, 257–269. [CrossRef]
35. Rietkerk, M.; Dekker, S.C.; de Ruiter, P.C.; van de Koppel, J. Self-Organized Patchiness and Catastrophic Shifts in Ecosystems. *Science* **2004**, *305*, 1926–1929. [CrossRef]
36. Scheffer, M.; Carpenter, S.; Foley, J. A.; Walker, B. Catastrophic shifts in ecosystems. *Nature* **2001**, *413*, 591–596. [CrossRef] [PubMed]

Article

Carrying Capacity of a Population Diffusing in a Heterogeneous Environment

D.L. DeAngelis [1,*], Bo Zhang [2], Wei-Ming Ni [3,4] and Yuanshi Wang [5]

1 U.S. Geological Survey, Wetland and Aquatic Research Center, Gainesville, Florida, FL 32653, USA
2 Department of Environmental Science and Policy, University of California at Davis, Davis, CA 95616, USA; bozhangophelia@gmail.com
3 School of Mathematics, University of Minnesota, Minneapolis, MN 55455, USA; weiming.ni@gmail.com
4 School of Science and Engineering, Chinese University of Hong Kong-Shenzhen, Shenzhen 518000, China
5 School of Mathematics, Sun Yat-sen University, Guangzhou 510275, China; mcswys@mail.sysu.edu.cn
* Correspondence: don_deangelis@usgs.gov; Tel.: +1-305-284-1690

Received: 7 November 2019; Accepted: 23 December 2019; Published: 1 January 2020

Abstract: The carrying capacity of the environment for a population is one of the key concepts in ecology and it is incorporated in the growth term of reaction-diffusion equations describing populations in space. Analysis of reaction-diffusion models of populations in heterogeneous space have shown that, when the maximum growth rate and carrying capacity in a logistic growth function vary in space, conditions exist for which the total population size at equilibrium (i) exceeds the total population that which would occur in the absence of diffusion and (ii) exceeds that which would occur if the system were homogeneous and the total carrying capacity, computed as the integral over the local carrying capacities, was the same in the heterogeneous and homogeneous cases. We review here work over the past few years that has explained these apparently counter-intuitive results in terms of the way input of energy or another limiting resource (e.g., a nutrient) varies across the system. We report on both mathematical analysis and laboratory experiments confirming that total population size in a heterogeneous system with diffusion can exceed that in the system without diffusion. We further report, however, that when the resource of the population in question is explicitly modeled as a coupled variable, as in a reaction-diffusion chemostat model rather than a model with logistic growth, the total population in the heterogeneous system with diffusion cannot exceed the total population size in the corresponding homogeneous system in which the total carrying capacities are the same.

Keywords: carrying capacity; spatial heterogeneity; Pearl-Verhulst logistic model; reaction-diffusion model; energy constraints; total realized asymptotic population abundance; chemostat model

1. Introduction

Partial differential equations have long been used in spatial ecology, usually in the form of reaction-diffusion equations, to describe such phenomena as spatial pattern formation (e.g., references [1–3]), the spread of populations in space (e.g., references [4,5]), and the effects of spatial heterogeneity on populations (e.g., reference [6]).

The last of these mentioned, the effects of spatial heterogeneity on populations, has relevance to the general question of what population size can be supported on a landscape or region that is heterogeneous in habitat quality for various species; i.e., resource availability. The question arises in relation to a population that is being exploited for economic needs or that is an object of conservation. The term 'carrying capacity' is usually used to mean the maximum population size that could be sustained in a given habitat (e.g., references [7,8]) The term was originally used in livestock management to describe the number of domestic animals that can be sustained in a given managed area, but was extended by ecologists in the 1930s to describe the number of wild animals that could be sustained in a natural area [9], such as predators in a nature reserve [10]. Further, carrying capacity is used to describe the sustained ecological services that can be provided for human and ecological populations by terrestrial or aquatic ecosystems [11–14]. Therefore, carrying capacity now covers a range of uses. Because of its importance, the effects of various factors, such as climate change [15], environmental stressors [16], and human impact [17] on carrying capacity in various contexts have been studied.

Carrying capacity has usually been assumed to reflect the amount of resources available to the population. With this interpretation, carrying capacity can be formulated in a mathematical way, by equating carrying capacity with the constant K in the Pearl-Verhulst version of the logistic equation [18]:

$$\frac{du}{dt} = r\left(1 - \frac{u}{K}\right)u \tag{1}$$

where u is population size or abundance, r is the maximum grow rate of the population, and t is time. This formulation focuses the problem on measuring K. However, such measurement is often not easy when dealing with populations in the wild, which are affected by many environmental factors, not just resources, and are observed to fluctuate strongly through time. Therefore, ways to include time dependence have been developed in various directions [19,20].

Despite some vagueness in their meaning and measurement, both the Pearl-Verhulst logistic equation and carrying capacity have remained major features in population ecology textbooks and theoretical models. Studying equations such as (1), even in the absence of precise data, can give insights into ecological questions, such as how the population size might respond when subject to perturbations. The need to extend the logistic population model to space and to include capacity for organism movement was realized [21,22]. To extend this model to populations over a spatial area such as a landscape or region partial differential equations are useful tools. If the spatial movement of the population can be approximated as diffusion, then the logistic equation may look like:

$$\frac{du}{dt} = D\Delta u + r\left(1 - \frac{u}{K}\right)u \tag{2}$$

where Δu is the Laplacian and D is the diffusion coefficient, where now u is a function of both time and space, x, defined on region Ω. If the environmental conditions of the landscape or region of interest are homogeneous, then the parameters r and K will be constants. If, however, that is not true, then $r(x)$ and $K(x)$, can be defined as the local growth rates and carrying capacities at locations x.

Although carrying capacity has largely been used with the assumption of a relatively uniform or homogeneous environment, it might at first seem that the extension of carrying capacity to the whole spatial region would be a fairly straightforward adding up of the local carrying capacities, through integrating $K(x)$ over region Ω; i.e., $\int_\Omega K(x)$. However, extending the logistic equation to space leads to results that show that such an extension is not a simple matter. Using a simplified form of Equation (2), in which maximum growth rate and carrying capacity are combined into one spatially varying parameter, $g(x)$, Lou [23] (see also references [24,25]) solved for the total population size of a population diffusing in a heterogeneous spatial region described by the equation:

$$\frac{du}{dt} = D\Delta u + [g(x) - u]u \tag{3}$$

where $g(x)$ varies with the spatial distance x but is constant temporally. The population diffuses at a constant rate D, with Neumann (no-flux) boundary conditions on u. Lou (2006) [23] showed, dividing the terms of the right-hand side by u and integrating over space Ω, that the following equation holds at equilibrium:

$$D\int_\Omega \frac{\lfloor \nabla u \rfloor^2}{u^2} + \int_\Omega [g(x) - u(x)]dx = 0. \tag{4}$$

Because the first term is necessarily positive; i.e.,

$$D\int_{\Omega}\frac{\lfloor\nabla u\rfloor^{2}}{u^{2}} > 0 \tag{5}$$

it implies that

$$\int_{\Omega}[g(x) - u(x)]dx < 0. \tag{6}$$

The total population size can be defined as $\int_{\Omega} u(x)$. We will follow reference [26] to use the term 'total realized asymptotic population abundance' for $\int_{\Omega} u(x)$. (*TRAPA*). It is clear from (6) that:

$$TRAPA \equiv \int_{\Omega} u(x) > \int_{\Omega} g(x) \tag{7}$$

that is, it is greater than the integral over all local values of the carrying capacity, $g(x)$.

The results of reference [23] are for a special case of the logistic equation in which one spatially varying parameter, $g(x)$, replaces both $r(x)$ and $K(x)$ of the Pearl-Verhulst growth function. To examine the separate effects of growth and carrying capacity, DeAngelis, et al. [27] extended the analysis to Equation (2). The equation:

$$\frac{\partial u}{\partial t} = D\Delta u + r(x)\left[1 - \frac{u}{K(x)}\right]u \tag{8}$$

is difficult to analyze for all values of diffusion, but can be evaluated in the limits $D \to 0$ *and* $D \to \infty$. When the diffusion rate is small, it was established in [27,28] that it is possible for $\int_{\Omega} u(x) > \int_{\Omega} K(x)$, although this becomes an equality in the limit as $D \to 0$. For the case D approaches infinity (i.e., diffusion rate becomes much larger than growth rate), the authors show that $\int_{\Omega} u(x) > \int_{\Omega} K(x)$ can hold, if there is a positive correlation between $K(x)$ and $r(x)$. Specifically, they show that for $D \to \infty, u$ approaches the asymptotic value u_d, where:

$$u_d = \frac{\int_{\Omega} r(x)}{\int_{\Omega} \frac{r(x)}{K(x)}}. \tag{9}$$

From (9) it can be shown that, in the case that $r(x)/K(x)$ is strictly decreasing with x:

$$u_d < \overline{K} \equiv \frac{1}{\lfloor\Omega\rfloor}\int_{\Omega} K(x). \tag{10}$$

In the case that $r(x)/K(x)$ is strictly increasing with x:

$$u_d > \overline{K}. \tag{11}$$

Therefore, the situation in which TRAPA exceeds the spatial integral over $K(x)$ is only a special case of a positive correlation of $r(x)$ and $K(x)$, and the reverse is also possible. The result of reference [23] in Equation (3) is due to the fact that $g(x)$ is both the maximum growth rate and carrying capacity, which results in these being positively correlated.

The question can be asked: What is the reason for the puzzling result that TRAPA can exceed the integral of $K(x)$ over space? A qualitative interpretation of both inequalities (7) and (11) is that diffusion of individuals away from areas of locally high growth rate, $r(x)$, and high local carrying capacity, $K(x)$, keeps the population levels in those areas below $K(x)$, so that new production exceeds losses at steady state in those areas. At the same time, diffusion from these areas of higher production to areas of lower $r(x)$ and $K(x)$ allows population levels in those low production areas to exceed the local $K(x)$ in those areas. Sometimes it causes the total population over all space to exceed that which would occur in a homogeneous space with the same total carrying capacity. But this situation is limited to positive correlation between $r(x)$ and $K(x)$. In reality, the parameters $r(x)$ and $K(x)$ may not necessarily be positively correlated, so the increase in TRAPA with diffusion in heterogeneous space is not a general phenomenon, although we will see later that a positive correlation is likely.

2. Analogy and Insight from Spatially Discretized Model

The puzzling effects shown by the reaction-diffusion equation for population growth and diffusion in heterogeneous space can perhaps be better understood by referring to historically earlier analysis of a simplified system of a population diffusing between two different discrete patches with different values of r and K. As above, assume that growth is described by the Pearl-Verhulst logistic equation, and that symmetric transfer rates, D, thus random or diffusive movement, are assumed to exist between the patches. The equations for the two patches then have the form:

$$\frac{dU_1}{dt} = r_1\left(1 - \frac{U_1}{K_1}\right)U_1 - DU_1 + DU_2 \tag{12a}$$

$$\frac{dU_2}{dt} = r_2\left(1 - \frac{U_2}{K_2}\right)U_2 - DU_2 + DU_1 \tag{12b}$$

where r_i is the maximum growth rate and K_i is henceforth defined as the local carrying capacity on patch i. It can be shown that for $D \to \infty$; that is, when the movement rates between the compartments are large compared to population growth rates, so that perfect mixing of the population between the patches occurs, the following expression for population size holds at equilibrium:

$$TRAPA = U_1^* + U_2^* = K_1 + K_2 + (K_1 - K_2)\frac{r_1K_2 - r_2K_1}{r_1K_2 + r_2K_1}, \tag{13}$$

This expression was found by reference [29], and corrected for typos by reference [30] and reference [26]. From Equation (13) it follows that, if $K_1 > K_2$ and $r_1/K_1 > r_2/K_2$, the total realized asymptotic population abundance at equilibrium ($TRAPA$) exceeds the sum of the local carrying capacities of the two patches; that is, $U_1^* + U_2^* > K_1 + K_2$. These conditions on the values of r_i and K_i are the discrete space version of the positive correlation between $r(x)$ and $K(x)$. This result was extended to n patches by Zhang et al. [31], where it was shown that this inequality applied for all $D > 0$. In addition, in reference [31], experimental manipulations simulating diffusion of a floating aquatic plant, duckweed, among patches lent support to the theoretical results.

Equation (13) also implies that in the heterogeneous case, i.e., $K_1 \neq K_2$, the total equilibrium population size with diffusion, $U_1^* + U_2^*$ ($TRAPA$), for $K_1 > K_2$ and $r_1/K_1 > r_2/K_2$, with $D \to \infty$, exceeds the total populations on patches for the case in which the carrying capacities are homogeneous, which occurs where $K_1 = K_2 = (K_1 + K_2)/2$. This result also applies to other forms of the logistic equation beyond the Pearl-Verhulst form. When the growth terms are replaced by the original Verhulst form:

$$\frac{dU_1}{dt} = r_1U_1 - \alpha_1U_1^2 - DU_1 + DU_2 \tag{14a}$$

$$\frac{dU_2}{dt} = r_2U_2 - \alpha_2U_2^2 - DU_2 + DU_1 \tag{14b}$$

a variation on (13):

$$U_1^* + U_2^* = \frac{r_1}{\alpha_1} + \frac{r_2}{\alpha_2} + (\alpha_1 - \alpha_2)\frac{1}{\alpha_1\alpha_2}\frac{r_1\alpha_2 - r_2\alpha_1}{r_1\alpha_2 + r_2\alpha_1} \tag{15}$$

can be derived, where again the sum of the equilibrium populations, $U_1^* + U_2^*$, can exceed the sum of the individual carrying capacities, $\frac{r_1}{\alpha_1} + \frac{r_2}{\alpha_2}$ [28].

The result that TRAPA for a landscape in which the total carrying capacity is heterogeneous (local carrying capacities K_1 plus K_2 in a two-patch system) is greater than that of the homogeneous landscape ($\frac{1}{2}$ ($K_1 + K_2$) in each patch) when $K_1 > K_2$ and $r_1/K_1 > r_2/K_2$, has a simple explanation. Consider the two-patch system. The total input of energy or limiting resources in the heterogeneous case is r_1K_1 + r_2K_2, while the the equivalent in the homogeneous case is $[r_1(K_1 + K_2)/2 + r_2(K_1 + K_2)/2]$. These are not the same, so the total population sizes supported should also not be the same. It is the input of limiting energy (or other limiting resource) that is fundamental in nature. Carrying capacity is simply a resultant quantity. Therefore, there is a hidden assumption, when one is simply varying the local carrying capacities, of changing the total energy (or other limiting resource) input. These considerations apply not only to the two-patch system, but to all spatial systems with a potentially heterogeneous growth rate and carrying capacity.

To see this, consider the Pearl-Verhulst equation:

$$\frac{dU_i}{dt} = r_i\left(1 - \frac{U_i}{K_i}\right)U_i \tag{16}$$

where U_i is in terms of energy or, alternatively, a limiting resource such as a nutrient. At steady state equilibrium, $U_i^* = K_i$, there is an influx of energy into the population, $r_iU_i^* = r_iK_i$, producing births and growth of individuals per unit time, and a loss flux of energy $-r_iU_i^{*2}/K_i = -r_iK_i$, due to metabolic maintenance and mortality resulting from interference with other members of the population [20,32]. Thus, energy through-flow maintains the population. Therefore, when comparing systems with different values of r_i and K_i in the patch system, or $r(x)$ and $K(x)$ in the continuous system, it is also necessary to consider if the total energy (or other limiting resource) input is the same in the two systems being compared.

It can be shown that the apparent paradox of greater TRAPA for cases of a heterogeneous region, with or without diffusion, versus TRAPA in the homogeneous case with the same summed local carrying capacities, arises in cases in which greater energy influx has implicitly been assumed for the heterogeneous case. This can be examined mathematically. To show that is the case, we follow reference [33] to consider the Pearl-Verhulst equations with the energy input requirement for the population implemented. Energetic (or other limiting resource) requirements on the population growth will be shown to explain the paradoxical behavior for $D \to \infty$.

For the Pearl-Verhulst model for the two-patch system, we have:

$$TRAPA_{homogeneous} = \frac{K_1 + K_2}{2} + \frac{K_1 + K_2}{2} = K_1 + K_2 \tag{17a}$$

$$TRAPA_{heterogeneous,\ no\ diffusion} = K_1 + K_2 \tag{17b}$$

$$TRAPA_{heterogenuous,diffusion} = K_1 + K_2 + (K_1 - K_2)\frac{r_1K_2 - r_2K_1}{r_1K_2 + r_2K_1}. \tag{17c}$$

This implies that, for $K_1 > K_2$ and $r_1/K_1 > r_2/K_2$, the relationships

$$TRAPA_{heterogeneous,\ diffusion} \geq TRAPA_{heterogeneous,\ no\ diffusion} = TRAPA_{homogeneous} \tag{18}$$

hold, where $K_1 + K_2$ has the same value for each of these three cases and the values of r_1 and r_2 are also the same in all cases. The values of all of the TRAPAs are shown as functions of r_1 in Figure 1 for comparison. We can also calculate the influx of energy (or other limiting resource) that is needed by each of the *TRAPAs* above to be maintained at those levels. We call these 'ENERGY TRAPAs':

$$ENERGY\ TRAPA_{homogeneous} = r_1\frac{K_1 + K_2}{2} + r_2\frac{K_1 + K_2}{2} = \frac{1}{2}(r_1 + r_2)(K_1 + K_2) \tag{19a}$$

$$ENERGY\ TRAPA_{heterogeneous,\ no\ diffusion} = r_1K_1 + r_2K_2 \tag{19b}$$

$$\begin{aligned} &ENERGY\ TRAPA_{heterogeneous,\ diffusion} \\ &= \tfrac{1}{2}(r_1 + r_2)\left(K_1 + K_2 + (K_1 - K_2)\,\tfrac{r_1K_2 - r_2K_1}{r_1K_2 + r_2K_1}\right) \end{aligned} \tag{19c}$$

Authors [33] demonstrate that, for the following set of inequalities, $K_1 > K_2$ and $r_1/K_1 > r_2/K_2$:

$$ENERGY\ TRAPA_{heterogeneous,no\ diffusion} \geq ENERGY\ TRAPA_{heterogeneous,diffusion} \geq ENERGY\ TRAPA_{homogeneous}\ . \quad (20)$$

These inequalities imply first, that for $r_1 > 7$ in Figure 2, a greater influx of energy (or other limiting resource) is required to support the heterogeneous system without diffusion (20) than to support the homogeneous system, even though they have the same total population (18). The values of each of the ENERGY TRAPAs as a function of r_1 are shown in Figure 2 for comparison. The extra energy input of the heterogeneous system is can be considered to be wasted through greater density-dependent mortality and respiration of the $-r_1 U_1^{*2}/K_1$ term. Second, when diffusion is allowed in the heterogeneous case, less energy is needed to support the heterogeneous system with diffusion than the heterogeneous system without diffusion, although the population increases when diffusion occurs. Therefore, it is true that $ENERGY\ TRAPA_{heterogeneous,no\ diffusion} \geq ENERGY\ TRAPA_{heterogeneous,diffusion}$, even though $TRAPA_{heterogeneous,diffusion} \geq TRAPA_{heterogeneous,\ no\ diffusion}$. This implies that the heterogeneous system with diffusion is more efficient than the heterogeneous system without diffusion in that it requires less maintenance energy for a higher population. But if only a limited amount of energy (or other limiting resource) is available, then to achieve the largest population, it is not possible to do better than to spread it homogeneously between the patches. Reference [33] further shows that when the energy constraints are applied to limit the way r_i and K_i can be assigned among the patches, such that the same amount of energy is input in both the homogeneous and heterogeneous systems, then it is impossible for $TRAPA_{heterogeneous,\ diffusion} > TRAPA_{homogeneous}$.

The results presented in this section are for either two-or multi-patch, systems that are spatially discrete. Extensions to partial differential equations should be straightforward, although we do not know of such analysis for the above results of reference [33].

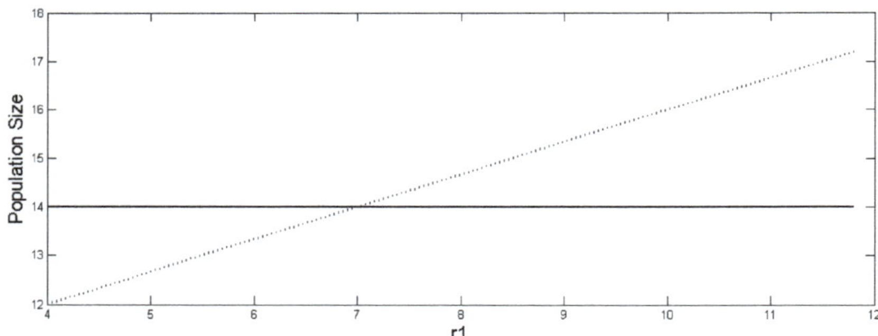

Figure 1. Size of $TRAPA_{heterogeneous,diffusion}$ (dotted) and $TRAPA_{heterogeneous,no\ diffusion}$ and $TRAPA_{homogeneous}$, (solid, overlapping) as function of r_1. The ratio $r_1/K_1 > r_2/K_2$ holds after $r_1 > 7$. The other parameter values are $r_2 = 3.5$, $K_1 = 14$, and $K_2 = 7$.

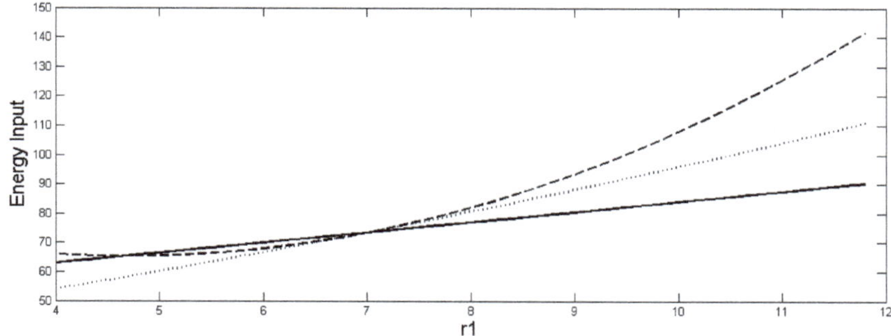

Figure 2. Size of $ENERGY\ TRAPA_{heterogeneous,diffusion}$ (dotted) and $ENERGY\ TRAPA_{heterogeneous,no\ diffusion}$ (dashed) and $ENERGY\ TRAPA_{homogeneous}$, (solid) as function of r_1. $r_1/K_1 > r_2/K_2$ holds after $r_1 > 7$. The parameter values are the same as in Figure 1.

3. Analysis and Experiment: Consumer-Resource Chemostat Model

With the above constraints on energy, the reaction-diffusion models with logistic growth can be used to compare effects of different parameterizations of the model on population dynamics. However, the use of such logistic equation models for patches connected by diffusion has been criticized in that they are limited to the assumption of fixed resource levels [26,31]. Resources such as energy and nutrients are exploitable and behave as variables. This includes study of microbial populations in laboratory settings, which can be used for testing theory. Therefore, population growth is best modeled using a mechanistic, bottom-up approach with a population of consumers utilizing variable resources [34,35]. This approach is suitable for describing a population in space. It will be shown there that this approach has the advantage of avoiding some of the counter-intuitive results mentioned above for spatial models involving the logistic model to describe growth [36,37].

A general pair of equations for the consumer-resource system is:

$$\frac{\partial u(x,t)}{\partial t} = D\frac{\partial^2 u(x,t)}{\partial x^2} + \frac{r_{max}n(x,t)u(x,t)}{k+n(x,t)} - m(x)u(x,t) - g(x)u(x,t)^2 \quad (21a)$$

$$\frac{dn(x,t)}{dt} = N_{input}(x) - \eta n(x,t) - \frac{r_{max}n(x,t)u(x,t)}{\gamma(k+n(x,t))} \quad (21b)$$

where $u(x,t)$ is the consumer population abundance and $n(x,t)$ is the nutrient concentration. Only the consumer is assumed to be diffusing. As before, D is the diffusion rate, while r_{max} is the asymptotic growth rate under infinite resources. Now k is the half-saturation coefficient, $m(x)$ is the density-independent mortality rate, $g(x)$ is the density-dependent loss rate, $N_{input}(x)$ is the nutrient input, η is the loss rate of nutrient from the system, and γ is the yield, representing consumer individuals per unit nutrient.

The consumer-resource chemostat model has the advantage that input of the limiting resource, nutrient in (21b), is explicitly denoted by a single driving function, $N_{input}(x)$. Therefore, it can be specified across a spatial region with no ambiguity in what the total energy input is at steady state (or any time); $\int_\Omega N_{input}(x)$. This differs from the spatial logistic model studied earlier, in which the energy input at steady state is the product $r(x)K(x)$, where $r(x)$ and $K(x)$ are separate variables. The model (21a,b), discretized as 12 patches in a one-dimensional row to facilitate analysis, was used by reference [36] in parallel with an experimental setup on yeast in which a growing yeast population could be manipulated to disperse. The purpose of the model and experiment was to examine the relationship between the size of a population and diffusing in a discretized version of this consumer-resource model with n patches in a one-dimensional line. All patches, except for no-flux conditions for the two end patches, were connected only to their two neighbors. The analysis was designed to evaluate two results that have come out of earlier analyses of the reaction-diffusion model with logistic growth. Zhang et al. [36] did the analysis for two versions of (21a,b); Model 1 with $\eta = 0$ and $m_i = 0$, and Model 2 with $g_i = 0$. Two important results came out of the analysis of the consumer-resource model compared with the logistic model and were supported by the experiment.

The first result of the logistic reaction-diffusion model is that if a consumer population exists in an environment in which an exploited renewable resource input is heterogeneously distributed, and there is a positive relationship between growth rate and carrying capacity, then the total steady state of a diffusing population (TRAPA) can attain a greater abundance than the non-diffusing population. It was proven in both Models 1 and 2 of the consumer-resource model over a range of parameter values relevant to the experiment, that:

$$TRAPA_{heterogeneous,diffusion} > TRAPA_{heterogeneous,no\ diffusion} \quad (22)$$

for the consumer-resource model [36]. Thus, the logistic reaction-diffusion and consumer-resource models are in agreement.

The second result of the logistic reaction-diffusion is that a population, $TRAPA_{heterogeneous,diffusion}$, that is diffusing in an environmental space in which there is a heterogeneously distributed input of exploited limiting resource, can reach a greater steady state abundance (TRAPA) than a population either diffusing ($TRAPA_{homogeneous,diffusion}$) or not diffusing ($TRAPA_{homogeneous,no\ diffusion}$) in an environmental space in which the same total input of resources are spread homogeneously in the space. It was shown in reference [36] that the result of the consumer-resource model disagrees with that result, meaning it is shown that:

$$TRAPA_{homogeneous,diffusion\ or\ not} \geq TRAPA_{heterogeneous,diffusion} \geq TRAPA_{heterogeneous,no\ diffusion} \quad (23)$$

These inequalities are shown in Figure 3 over a range of values of $g_{low\ nutrient}$, which is the density-dependent mortality rate, g, in odd-numbered patches in the 12-patch model (which receive low nutrient input), and where $g_{high\ nutrient}$ is the density-dependent mortality rate in the even-numbered patches (with high nutrient input).

The details of the mathematical analysis and the confirmation by experiments on yeast are given in reference [36] and will not be detailed here, except to say that the experiments used patches, so the discretized version of equations (21a,b) was appropriate for describing those experiments. Separate experiments showed that conditions $r_iK_i > r_j/K_j$ held for $K_i > K_j$, which agrees with the general relationships between r and K in microbial populations found by reference [38].

The Model 1 for the continuous PDE version was studied by reference [39], with equivalent results.

4. Discussion

We reviewed two important and related results that have been proved for the reaction-diffusion equation of a population of the type (3) or (8) in a spatially heterogeneous environment when the growth rate, $r(x)$ and carrying capacity $K(x)$ are positively related (which is automatically true for (3)). The first result is that if the population diffuses in space, the TRAPA can be greater than if it does not. The second result is that when the total carrying capacity $\int_\Omega K(x)$ is spread heterogeneously across space and the population diffuses, the TRAPA can be greater than if the same total carrying capacity were spread homogeneously. It was also shown that the same results hold for the analogous system of two discrete patches differing in growth rate and carrying capacity initially studied by references [27,29] showed that models for discrete n-patch systems and other growth functions than the Pearl-Verhulst equation produced similar results. So it appears to be a general property of models based on the general type of model represented by Equation (8) or its discretized variant (12a,b). These results would seem to imply that carrying capacities can be manipulated by creating heterogeneity in a landscape to support higher populations when population movements occur through diffusion.

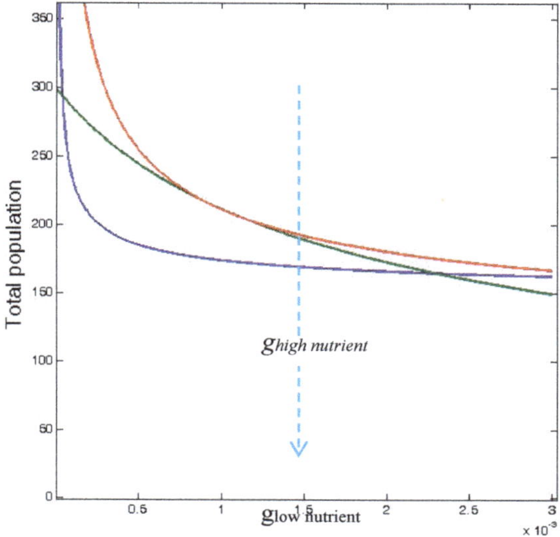

Figure 3. Total carrying capacity $TRAPA_{hetero,no\ diffusion}$ (blue curve), total population abundance $TRAPA_{hetero,diffusion}$ (green curve), and total population for homogeneously distributed inputs (red curve), $TRAPA_{homogeneous,no\ diffusion}$, as functions of the g_i value for the low nutrient input wells, $g_{low\ nutrient}$, (odd patches) for fixed value of the $g_i = 0.001$ for the high nutrient input wells, $g_{high\ nutrient}$ (even patches). $TRAPA_{homogeneous,diffusion}$ coincides with $TRAPA_{heterogeneous,diffusiion}$ for $D \to \infty$. The other parameters have been set to $r_{max} = 0.1$, $k = 0.1$, and $\gamma = 1.0$. $TRAPA_{homogeneous,diffusion}$ and $TRAPA_{homogeneous,no\ diffusion}$ coincide.

The analysis of models of the type of Equations (8) and (12a,b), however, leaves out consideration that an energy or other limiting resource flux is necessary to sustain a population, and that the local energy or limiting resource flux is $r(x)K(x)$ in the continuous case and r_iK_i in the discrete patch case. Not taking this into account can lead to an inadvertent changing in the amount of energy (or limiting resource) input to the system when the local carrying capacities are changed, even when total carrying capacity, $\int_\Omega K(s)$, remains the same. Reference [33] showed that when these energetic constraints are taken into account, the second counter-intuitive result disappears and a homogeneous distribution of energy (or other limiting resource) input becomes most efficient for supporting a population. It was also shown by reference [36] that when a resource is explicitly modeled, as in consumer-resource chemostat models, so that the energy input at each spatial point is clearly expressed in the equations, then some earlier results are better understood. In particular, the counter-intuitive result from logistic reaction-diffusion models that heterogeneously distributed input of exploited renewable limiting resource can reach a greater steady state abundance (TRAPA) than a population either diffusing or not diffusing in an environmental space with the same total input of resources spread homogeneously is now shown to be an artifact of the logistic population equation, which does not model the population's resources as a variable. However, the consumer-resource model is consistent with the logistic population model in showing that a population diffusing in a heterogeneous environment can reach greater abundance at equilibrium than the population in the same heterogeneous environment that is not diffusing.

The results shown here for both the logistic model and the consumer-resource model apply to situations in which the population is assumed supported by an influx of energy or other resource (e.g., nutrient) that are limited, such that carrying capacity and per capita growth rate are positively related. The do not apply in cases in which growth rates have no correlation with carrying capacity. For example, when local carrying capacities are determined in terms of such things as number of nesting sites, refuges, or similar types of factors, rather than an exploitable resource, per capita growth rate is independent of carrying capacity. In that case, there are no paradoxical results from the the logistic equation, and it may be a good description of the population. The modeling by partial differential equations has great importance in understanding ecological populations dispersing in heterogeneous space [40]. For example reference [41], used a reaction-advection diffusion model to study the dispersal of fish moving into newly flooded areas in search for food. Like that model, many other models of population include advection in addition to diffusion, because organisms may have directional movement in many cases. However, reaction diffusion models without any directional movement are also common in ecology, as diffusion is a good approximation of movement in many cases [42].

Reaction-diffusion models of the type studied here play a role in conservation ecology, as populations exist on landscape that are heterogeneous in resources. Some areas may be rich enough in landscapes for a population to grow rapidly, while others may only have sufficient resources for the population to barely survive. Still other areas may be 'sinks' in which a population would decrease in size through time. The dynamics of a population is complicated by movement in the landscape, approximated as diffusion in reaction-diffusion models. It is possible for individuals to be attracted to sink areas, where they die or fail to reproduce, threatening the whole population with extinction [43]. Numerous papers have expanded on the effects of landscape heterogeneity on populations [44–46]. The experiments and model suggest that different rates of dispersal among sources and pseudo-sinks in nature can affect regional population size and certain rates of diffusion could maximize the size of the total population in a heterogeneous region. For example, in an experiment described by references [45,47,48], densities of seeds of the plant *Cakile edentula* were manipulated along a gradient through sand dunes. In this case the seaward or beach end of the gradient was a source of seeds, while the middle and landward sites were net sinks in which mortality was higher than reproduction. A model by reference [49] showed that the plants were most abundant in the sink sites because of the high seed migration from the source. This could be an example of a case where diffusion in a heterogeneous environment results in total population size greater than the total of individual sites along the gradient in the absence of diffusion, although this is difficult to prove. In any case, conservation ecologists must take into account possible effects of this sort on animal and plant populations when designing nature reserves.

5. Conclusions

The reaction-diffusion equation with a logistic growth term is a fundamental equation describing an ecological population. A counterintuitive result of such a system is that the total realized asymptotic population achieved in a heterogeneous system is not the same as the sum of local carrying capacities of homogeneous subareas. This sometimes may lead to misleading results if it is not kept in mind that it is the input of energy or a limiting resource that maintains a population. The carrying capacity is simply a consequence of the spatially varying energy or resource inputs and may not add in a linear way. The use of consumer-resource models is often a better way to relate the input of energy or limiting nutrients to the total size of a population in space.

Author Contributions: D.L.D. and B.Z. designed the research and wrote the first draft. W.-M.N. and Y.W. contributed to the theory. All authors contributed to revisions of the manuscript. All authors have read and agreed to the published version of the manuscript.

Funding: This research was funded by U.S. Geological Survey's Greater Everglades Priority Ecosystem Science Grant GEPES ATLSS T5.17, Project Title KA30CTF.

Acknowledgments: D.L.D. and B.Z. were supported by the USGS Greater Everglades Priority Ecosystems Science. B.Z. was also supported by the UC Davis Chancellor's postdoc fellowship. The research of W.-M.N. and Y.W. were partially supported by NSF and Chinese NSF. We appreciate comments of Adrian Lam and Donald Schoolmaster on an earlier version of this manuscript.

Conflicts of Interest: The authors declare no conflict of interest.

References

1. Holmes, E.E.; Lewis, M.A.; Banks, J.E.; Veit, R.R. Partial-differential dquations in dcology—Spatial interactions and population-dynamics. *Ecology* **1994**, *75*, 17–29. [CrossRef]
2. Levin, S.A.; Segel, L.A. Hypothesis for origin of planktonic patchiness. *Nature* **1976**, *259*, 659. [CrossRef]
3. Okubo, A. *Diffusion and Ecological Problems: Mathematical Models*; Springer: Berlin, Germany, 1980.
4. Cantrell, R.S.; Cosner, C.; Lou, Y. Evolution of dispersal in heterogeneous landscapes. In *Spatial Ecology*; CRC Press: Boca Raton, FL, USA, 2009; pp. 213–229. [CrossRef]
5. Shigesada, N.; Kawasaki, K. *Biological Invasions: Theory and Practice*; Oxford University Press: Oxford, UK, 1997.
6. Cantrell, R.S.; Cosner, C. The effects of spatial heterogeneity in population-dynamics. *J. Math. Biol.* **1991**, *29*, 315–338. [CrossRef]
7. Errington, P.L. Vulnerability of Bob-White populations to predation. *Ecology* **1934**, *15*, 110–127. [CrossRef]
8. Leopold, A. The conservation ethic. *J. For.* **1933**, *31*, 634–643.
9. Dhondt, A.A. Carrying-capacity—A confusing concept. *Acta Oecol.-Oecol. Gener.* **1988**, *9*, 337–346.
10. Hayward, M.W.; O'Brien, J.; Kerley, G.I.H. Carrying capacity of large African predators: Predictions and tests. *Biol. Conserv.* **2007**, *139*, 219–229. [CrossRef]
11. Qin, G.H.; Li, H.X.; Wang, X.; Ding, J. Research on water resources design carrying capacity. *Water Res.* **2016**, *8*, 157. [CrossRef]
12. Wang, H.; Zhou, Y.Y.; Tang, Y.; Wu, M.A.; Deng, Y.Q. Fluctuation of the water environmental carrying capacity in a huge river-connected lake. *Int. J. Environ. Res. Public Health* **2015**, *12*, 3564–3578. [CrossRef]
13. Zeng, C.; Liu, Y.L.; Liu, Y.F.; Hu, J.M.; Bai, X.G.; Yang, X.Y. An integrated approach for assessing aquatic ecological carrying capacity: A case study of Wujin District in the Tai Lake Basin, China. *Int. J. Environ. Res. Public Health* **2011**, *8*, 264–280. [CrossRef]
14. Zhou, R.X.; Pan, Z.W.; Jin, J.L.; Li, C.H.; Ning, S.W. Forewarning model of regional water resources carrying capacity based on combination weights and entropy Principles. *Entropy* **2017**, *19*, 574. [CrossRef]
15. Schell, D.M. Declining carrying capacity in the Bering Sea: Isotopic evidence from whale baleen. *Limnol. Oceanogr.* **2000**, *45*, 459–462. [CrossRef]
16. Sibly, R.M.; Williams, T.D.; Jones, M.B. How environmental stress affects density dependence and carrying capacity in a marine copepod. *J. Appl. Ecol.* **2000**, *37*, 388–397. [CrossRef]
17. Vasconcellos, M.; Gasalla, M.A. Fisheries catches and the carrying capacity of marine ecosystems in southern Brazil. *Fish Res.* **2001**, *50*, 279–295. [CrossRef]
18. Tuckwell, H. A study of some diffusion models of population growth. *Theor. Popul. Biol.* **1974**, *5*, 345–357. [CrossRef]
19. Ang, T.K.; Safuan, H.M. Harvesting in a toxicated intraguild predator-prey fishery model with variable carrying capacity. *Chaos Solitons Fractals* **2019**, *126*, 158–168. [CrossRef]
20. Rapport, D.J.; Turner, J.E. Feeding rates and population-growth. *Ecology* **1975**, *56*, 942–949. [CrossRef]
21. Goss-Custard, J.D.; Stillman, R.A.; Caldow, R.W.G.; West, A.D.; Guillemain, M. Carrying capacity in overwintering birds: When are spatial models needed? *J. Appl. Ecol.* **2003**, *40*, 176–187. [CrossRef]
22. Van Gils, J.A.; Edelaar, P.; Escudero, G.; Piersma, T. Carrying capacity models should not use fixed prey density thresholds: A plea for using more tools of behavioural ecology. *Oikos* **2004**, *104*, 197–204. [CrossRef]
23. Lou, Y. On the effects of migration and spatial heterogeneity on single and multiple species. *J Differ. Equ.* **2006**, *223*, 400–426. [CrossRef]
24. He, X.Q.; Ni, W.M. The effects of diffusion and spatial variation in Lotka-Volterra competition-diffusion system I: Heterogeneity vs. homogeneity. *J. Differ. Equ.* **2013**, *254*, 528–546. [CrossRef]
25. He, X.Q.; Ni, W.M. The effects of diffusion and spatial variation in Lotka-Volterra competition-diffusion system II: The general case. *J. Differ. Equ.* **2013**, *254*, 4088–4108. [CrossRef]
26. Arditi, R.; Lobry, C.; Sari, T. Is dispersal always beneficial to carrying capacity? New insights from the multi-patch logistic equation. *Theor. Popul. Biol.* **2015**, *106*, 45–59. [CrossRef] [PubMed]
27. DeAngelis, D.L.; Ni, W.M.; Zhang, B. Dispersal and spatial heterogeneity: Single species. *J. Math. Biol.* **2016**, *72*, 239–254. [CrossRef]
28. DeAngelis, D.L.; Ni, W.M.; Zhang, B. Effects of diffusion on total biomass in heterogeneous continuous and discrete-patch systems. *Theor. Ecol.* **2016**, *9*, 443–453. [CrossRef]

29. Freedman, H.I.; So, J.W.H.; Waltman, P. Predator influence on the growth of a population with three Genotypes. III. Persistence and Extinction. *J. Math. Anal. Appl.* **1987**, *128*, 287–304. [CrossRef]
30. Holt, R.D.; Gomulkiewicz, R.; Barfield, M. The phenomenology of niche evolution via quantitative traits in a 'black-hole' sink. *Proc. R. Soc. Lond. Ser. B Biol. Sci.* **2003**, *270*, 215–224. [CrossRef]
31. Zhang, B.; Liu, X.; DeAngelis, D.L.; Ni, W.M.; Wang, G.G. Effects of dispersal on total biomass in a patchy, heterogeneous system: Analysis and experiment. *Math. Biosci.* **2015**, *264*, 54–62. [CrossRef]
32. Jensen, A.L. Comparison of Logistic Equations for Population-Growth. *Biometrics* **1975**, *31*, 853–862. [CrossRef]
33. Wang, Y.S.; DeAngelis, D.L. Energetic constraints and the paradox of a diffusing population in a heterogeneous environment. *Theor. Popul. Biol.* **2019**, *125*, 30–37. [CrossRef]
34. MacArthur, R.H. *Geographical Ecology*; Harper & Row: New York, NY, USA, 1972.
35. Tilman, D. *Resource Competition and Community Structure*; Princeton University Press: Princeton, NJ, USA, 1982.
36. Zhang, B.; Kula, A.; Mack, K.M.L.; Zhai, L.; Ryce, A.L.; Ni, W.M.; DeAngelis, D.L.; Van Dyken, J.D. Carrying capacity in a heterogeneous environment with habitat connectivity. *Ecol. Lett.* **2017**, *20*, 1118–1128. [CrossRef] [PubMed]
37. Van Dyken, J.D.; Zhang, B. Carrying capacity of a spatially-structured population: Disentangling the effects of dispersal, growth parameters, habitat heterogeneity and habitat clustering. *J. Theor. Biol.* **2018**. [CrossRef] [PubMed]
38. Hendriks, A.J.; Maas-Diepeveen, J.L.M.; Heugens, E.H.W.; Van Straalen, N.M. Meta-analysis of intrinsic rates of increase and carrying capacity of populations affected by toxic and other stressors. *Environ. Toxicol. Chem.* **2005**, *24*, 2267–2277. [CrossRef] [PubMed]
39. He, X.Q.; Lam, K.Y.; Lou, Y.; Ni, W.M. Dynamics of a consumer-resource reaction-diffusion model: Homogeneous versus heterogeneous environments. *J. Math. Biol.* **2019**, *78*, 1605–1636. [CrossRef] [PubMed]
40. Cantrell, R.S.; Cosner, C.; Lou, Y. Advection-mediated coexistence of competing species. *Proc. R. Soc. Edinb. Sect. A* **2007**, *137*, 497–518. [CrossRef]
41. DeAngelis, D.L.; Trexler, J.C.; Cosner, C.; Obaza, A.; Jopp, F. Fish population dynamics in a seasonally varying wetland. *Ecol. Model.* **2010**, *221*, 1131–1137. [CrossRef]
42. Levin, S.A.; Pacala, S.W.; Tilman, D. Theories of simplification and scaling of spatially distributed processes. In *Ecology: Achievement and Challenge*; Princeton University Press: Princeton, NJ, USA, 1997; pp. 271–296.
43. Pulliam, H.R. Sources, sinks, and population regulation. *Am. Nat.* **1988**, *132*, 652–661. [CrossRef]
44. Amarasekare, P.; Hoopes, M.F.; Mouquet, N.; Holyoak, M. Mechanisms of coexistence in competitive metacommunities. *Am. Nat.* **2004**, *164*, 310–326. [CrossRef]
45. Watkinson, A.R.; Sutherland, W.J. Sources, sinks and pseudo-sinks. *J. Anim. Ecol.* **1995**, *64*, 126–130. [CrossRef]
46. Wilson, D.S. Complex interactions in metacommunities, with implications for biodiversity and higher levels of selection. *Ecology* **1992**, *73*, 1984–2000. [CrossRef]
47. Keddy, P.A. Experimental demography of the sand-dune annual, *Cakile-Edentula*, growing along an environmental gradient in Nova-Scotia. *J. Ecol.* **1981**, *69*, 615–630. [CrossRef]
48. Keddy, P.A. Population ecology on an environmental gradient—*Cakile-Edentula* on a sand dune. *Oecologia* **1982**, *52*, 348–355. [CrossRef] [PubMed]
49. Watkinson, A.R. On the abundance of plants along an environmental gradient. *J. Ecol.* **1985**, *73*, 569–578. [CrossRef]

Article

Quantitatively Inferring Three Mechanisms from the Spatiotemporal Patterns

Kang Zhang [1] and Wen-Si Hu [2] and Quan-Xing Liu [1,2,3,*]

1. School of Ecological and Environmental Sciences, East China Normal University, Shanghai 200241, China; zero360@foxmail.com
2. State Key Laboratory of Estuarine and Coastal Research, School of Ecological and Environmental Sciences, East China Normal University, Shanghai 200241, China; wensiwho@126.com
3. Shanghai Key Lab for Urban Ecological Processes and Eco-Restoration & Center for Global Change and Ecological Forecasting, East China Normal University, Shanghai 200241, China

* Correspondence: qxliu@sklec.ecnu.edu.cn

Received: 18 November 2019; Accepted: 7 January 2020; Published: 10 January 2020

Abstract: Although the diversity of spatial patterns has gained extensive attention on ecosystems, it is still a challenge to discern the underlying ecological processes and mechanisms. Dynamical system models, such partial differential equations (PDEs), are some of the most widely used frameworks to unravel the spatial pattern formation, and to explore the potential ecological processes and mechanisms. Here, comparing the similarity of patterned dynamics among Allen–Cahn (AC) model, Cahn–Hilliard (CH) model, and Cahn–Hilliard with population demographics (CHPD) model, we show that integrated spatiotemporal behaviors of the structure factors, the density-fluctuation scaling, the Lifshitz–Slyozov (LS) scaling, and the saturation status are useful indicators to infer the underlying ecological processes, even though they display the indistinguishable spatial patterns. First, there is a remarkable peak of structure factors of the CH model and CHPD model, but absent in AC model. Second, both CH and CHPD models reveal a hyperuniform behavior with scaling of -2.90 and -2.60, respectively, but AC model displays a random distribution with scaling of -1.91. Third, both AC and CH display uniform LS behaviors with slightly different scaling of 0.37 and 0.32, respectively, but CHPD model has scaling of 0.19 at short-time scales and saturation at long-time scales. In sum, we provide insights into the dynamical indicators/behaviors of spatial patterns, obtained from pure spatial data and spatiotemporal related data, and a potential application to infer ecological processes.

Keywords: Allen–Cahn model; Cahn–Hilliard model; spatial patterns; spatial fluctuation; dynamic behaviors

---

1. Introduction

The reaction–diffusion systems have been posed to address the spatially extended interactions among species or chemical substances since 1937 [1], when the traveling waves behaviors of the reaction–diffusion equations were discovered and studied. To date, the spatial systems are widely used to explore the spatiotemporal behaviors going beyond the initial frameworks such as existence and stability of spatial solutions in biology, geology, physics, and ecology [2–9]. The solutions to reaction–diffusion equations display diverse behaviors including the formation of traveling waves [10], wave-like phenomena [11], spatial self-organized patterns [12], and coarsening behaviors [13].

In mathematics, the spatial extend models can be classified into the following three categories with the simplest version. Firstly, the model describes the Brownian dispersion and growth-mortality processes on single species [14,15]. Herein, the species abundance can dynamically fluctuate around their equilibria that is limited by carrying capacity. Secondly, we can describe the mass-conserving systems with the Cahn–Hilliard model if species have a biased movement behavior such as a local density-dependent relationship [13,16–18]. In fact, both of these processes may exist in an ecological system that lead to a third theoretical model, the Cahn–Hilliard model with demographical processes of species [19–21]. These models display similar patterns for the reasonable parameters on the same systems. How to infer the underlying mechanisms from the spatial behaviors are crucial themes for ecologists. Many spatial hallmarks were proposed to characterize their spatiotemporal behaviors including spatial correlation functions [22,23], spatial structure factor [24], LS law [25–27], the exponent of cluster size [28], and the exponent of spatial fluctuations [29], but few studies are hitherto proposed to comprehensively compare the hallmarks on the above-mentioned models.

In this paper, we present a comprehensive comparison of three typical PDE systems in two-dimensional space: the Allen–Cahn model, the mass-conserving Cahn–Hilliard model, and the Cahn–Hilliard model with demographical processes. The central theme is the characterization of the spatiotemporal dynamics of such systems displaying a spatially patterned behavior. A key insight from our spatial analysis, provided in Section 4, is how spatial patterns and spatial fluctuations can be used to infer underlying ecological processes and mechanisms from the indicators of the spatial fluctuation, spatial structure factor, spatial correlation functions, and LS growth behavior. Importantly, different ecological processes may lead a similar spatial patterns in some systems even for strikingly different ecosystems. Here, our proposed integrated indicators have potential for distinguishing the underpinning patterned behaviors.

2. Theoretical Models

2.1. Allen–Cahn Model

The original Allen–Cahn model is a reaction–diffusion equation describing the phase separation process in a multi-component alloy system [30], including from disordered to ordered transitions. Recently, McNally et al. [18] showed that Allen–Cahn model can well describe the spatial pattern formation of microbial community, where microbes use T6SS to kill neighbors to separate the initially mixed flora, forming cloned plaques that grow over time. Theoretically, the model is given as follows assuming the species density as variable u at the time t within a spatial domain of Ω ($\Omega \subset R^2$), i.e.,

$$\begin{aligned} u_t &= -f(u) + \varepsilon \Delta u, u \in \Omega, \\ \boldsymbol{n} \cdot \nabla u &= 0, u \in \partial\Omega, \end{aligned} \tag{1}$$

where u_t is the partial derivative of u over time, ε is the diffusion coefficient, Δ is the Laplacian operator, $\boldsymbol{n}$ is the normal vector of the outer boundary, and ∇ is the gradient operator. Here, we take $f(u) = u(u-\alpha)(u-\beta)$, and $0 < \alpha < \beta$, then the zero points of $f(u)$ is $u_0 = 0$, α, and β. Here, u_0 are the equilibrium points of the model in Equation (1).

It can be easily proved that 0 and β are stable equilibria of the model in Equation (1), whereas α is an unstable equilibrium. Over time, the solutions of the model in Equation (1) would be around 0 or β. In the spatial model in Equation (1), the solutions of phase separation will occur if $\varepsilon > 0$.

2.2. Cahn–Hilliard Model

The Cahn–Hilliard model describes the spontaneous separation of a mixed fluid into two pure phases [31], resulting in the formation of spatial patterns. This phase separation principle was used to elucidate the mechanism of pattern formation on small scales in a mussel bed ecosystem [13]. This self-organizing process results from the relationship between movement speed and local mussel density, which differs from the existing Turing principle. The general form of the Cahn–Hilliard model is given by

$$\begin{aligned} u_t &= \varepsilon_B \Delta(f(u) - \varepsilon_A \Delta u), u \in \Omega, \\ \boldsymbol{n} \cdot \nabla u &= 0, u \in \partial\Omega, \end{aligned} \tag{2}$$

where u_t is the partial derivative of u over time, ε_A and ε_B are the local and nonlocal diffusion coefficients, Δ is the Laplacian operator, $\boldsymbol{n}$ is the normal vector of the outer boundary, and ∇ is the gradient operator. We take $f(u) = u(u-\alpha)(u-\beta)$ and $0 < a < \beta$, then the zero points of $f(u)$ are $u_0 = 0$, α, and β.

Mathematically, the stable equilibria of the Cahn–Hilliard model in Equation (2) are determined by the minimum values of the potential function $F(u)$ with $F'(u) = f(u)$. Now, the potential function is expressed as $F(u) = \frac{1}{4}u^4 - \frac{\alpha+\beta}{3}u^3 + \frac{\alpha\beta}{2}u^2 + C$ with constant C. α is the local maximum of $F(u)$, 0 and β are the global minimum of $F(u)$. Therefore, 0 and β are stable equilibria, and α is an unstable equilibrium. Note that Cahn–Hilliard model describes a mass-conservation ecosystem without mortality and birth processes.

2.3. Cahn–Hilliard with Population Demographics Model

It is natural to ask that what will happen when the population demographics are involved in the Cahn–Hilliard model. In general, we address the Cahn–Hilliard model with a population demographics (hereafter, called CHPD), which describes the growth and aggregation of a single species [16]. A similar model was used to describe patterns formation result from density-dependent motility rather than traditional chemotaxis behavior [20]. The general model is given by

$$\begin{aligned} &u_t = \varepsilon_B \Delta(f(u) - \varepsilon_A \Delta u) + ru(1 - \frac{u}{K}), u \in \Omega, \\ &\boldsymbol{n} \cdot \nabla u = 0, u \in \partial\Omega, \end{aligned} \tag{3}$$

where u_t is the partial derivative of u over time, ε_A and ε_B are the local and nonlocal diffusion coefficients, and Δ is the Laplacian operator. The last term is the logistic growth law, r $(r > 0)$, is the net growth rate population, K is the environmental carrying capacity $(K > 0)$, $\boldsymbol{n}$ is the normal vector of the outer boundary, and ∇ is the gradient operator. Now, the equilibria of the model in Equation (3) are 0 and K.

Linearizing the nonspatial model in Equation (3) using a first-order Taylor expansion at $u_0 = 0$ and K, respectively, yields

$$u_t = g(u_0) + g'(u_0)(u - u_0), \tag{4}$$

where $g'(u) = r - \frac{2ur}{K}$. One sees that $u_0 = 0$ is an unstable equilibrium since $g'(0) = r > 0$, whereas $g'(K) = -r < 0$.

However, the spatial model in Equation (3) is linearized around $u(\boldsymbol{x}) = u_0$ in Fourier space, i.e., $u(\boldsymbol{x}) = u_0 + \sum_k \delta_{u_k} e^{i\boldsymbol{k}\cdot\boldsymbol{x} + \Lambda_k t}$ yields

$$\begin{aligned} &\dot{\delta} u_{\boldsymbol{k}} = \Lambda_{\boldsymbol{k}} \delta_{\boldsymbol{k}}, \\ &\Lambda_{\boldsymbol{k}} = -r - \varepsilon_B f'(u_0)\boldsymbol{k}^2 - \varepsilon_A \varepsilon_B \boldsymbol{k}^4, \end{aligned} \tag{5}$$

where $i = \sqrt{-1}$, $\boldsymbol{x}$ is position, and $\boldsymbol{k}$ is wave vector. If $\Lambda_{\boldsymbol{k}} \le 0$ is satisfied for all $\boldsymbol{k}$, the system is stable. If $\Lambda_{\boldsymbol{k}} > 0$ is established, the system is unstable. Since $r > 0$, when $k = 0$, $\Lambda_{\boldsymbol{k}} < 0$. Hence, instability occurs if

$$\begin{aligned} &\varepsilon_A f'(u_0) < 0, \\ &\varepsilon_B f'(u_0) f'(u_0) > 4\varepsilon_A r. \end{aligned} \tag{6}$$

From this instability condition in Equation (6), one can see that the instability of the system is determined by the population growth rate (r) and the diffusion coefficients (ε_A and ε_B). If the logistic growth term is ignored, then 0 and β are the stable equilibrium points and α is the unstable equilibrium point. The pattern of the model in Equation (3) will be phase separated at 0 and β. If both the diffusion term and the logistic term are considered, the phase separation is suppressed by the logistic growth term. Over time, the values of the patterned solution will be around 0 and K, which ascribes the Turing patterns.

3. Materials and Methods

3.1. Numerical Simulations

We use the finite difference methods [32] to solve the equation numerically in two dimensions. The Laplacian terms of the model in Equations (1)–(3) Care approximated by the following formula

$$\Delta u(x,y) = \frac{u(x+\Delta x, y) + u(x-\Delta x, y) - 2u(x,y)}{(\Delta x)^2} + \frac{u(x, y+\Delta y) + u(x, y-\Delta y) - 2u(x,y)}{(\Delta y)^2}, \quad (7)$$

with step length $\Delta x = \frac{L_x}{m} = 1.0$ and $\Delta y = \frac{L_y}{m} = 1.0$. Here, x and y indicate the spatial coordinate. L_x and L_y are the physical length of the simulation in x and y directions and m is the grid number. The numerical simulations were implemented in two-dimensional space with grid numbers from $L_x = L_y = 2048$ to $L_x = L_y = 20,000$, and periodic boundary conditions. For AC and CH models, the initial values are random numbers with a mean of 0.5. For CHPD model, the initial values are random numbers with a mean of 0.33.

The iteration over time uses the first-order Euler method, i.e.,

$$u(x,y,t+\Delta t) = u(x,y,t) + \Delta t f(u), \quad (8)$$

with $\Delta t = 0.0025$ that satisfied the stability condition [33].

All numerical simulations were implemented in Python 3.7 (Python Software Foundation, CWI, Amsterdam, Netherlands) and MATLAB 2019a (The MathWorks Inc., Natick, MA, USA). PyOpencl (https://documen.tician.de/pyopencl/) was used to accelerate the simulations; the code is available from Github (https://github.com/Kang-Zhang).

3.2. Spatial Correlation Functions

Spatial correlation function is widely applied to characterized spatial structure and its correlations biophysics [34]. It is usually expressed as

$$g(\boldsymbol{r}_i, \boldsymbol{r}_j) = \langle (\varphi(\boldsymbol{r}_i) - \langle \varphi(\boldsymbol{r}_j) \rangle) \times (\varphi(\boldsymbol{r}_j) - \langle \varphi(\boldsymbol{r}_j) \rangle) \rangle = \langle \varphi(\boldsymbol{r}_i)\varphi(\boldsymbol{r}_j) \rangle - \langle \varphi(\boldsymbol{r}_i) \rangle \langle \varphi(\boldsymbol{r}_j) \rangle, \quad (9)$$

where $\boldsymbol{r}_i$ and $\boldsymbol{r}_j$ are two different locations; $\varphi(\boldsymbol{r}_i)$, $\varphi(\boldsymbol{r}_i)$ are numerical solutions at $\boldsymbol{r}_i$ and $\boldsymbol{r}_j$, respectively; and $\langle\ \rangle$ is the mathematical expectation. Thus, $g(\boldsymbol{r}_i, \boldsymbol{r}_j)$ depends only on the displacement $\boldsymbol{r}$ between the two locations. That is,

$$g(\boldsymbol{r}_i, \boldsymbol{r}_j) = g(\boldsymbol{r}_i, \boldsymbol{r}_i + \boldsymbol{r}) = g(\boldsymbol{r}). \quad (10)$$

Thus, the average of the correlation functions of all two points with a displacement of $\boldsymbol{r}$ is obtained as follows

$$g(r) = \langle \sum_{|\boldsymbol{r}_i - \boldsymbol{r}_j| = r} [\langle \varphi(\boldsymbol{r}_i)\varphi(\boldsymbol{r}_j) \rangle - \langle \varphi(\boldsymbol{r}_i) \rangle \langle \varphi(\boldsymbol{r}_j) \rangle] \rangle. \quad (11)$$

The spatial correlation function is a measure of the correlation of a variable between two spatial positions. If the value of the variable fluctuates in one direction at the same time, a positive value is taken; otherwise, a negative value is taken. If their fluctuations are completely unrelated, then the value is zero.

3.3. Structure Factors

Structure factor is an important indicator for analyzing long-range spatial correlations and scattering that is a function of scattering vector ($\boldsymbol{k}$) and time (t). The normalized structural factor [35] is written as

$$s(\boldsymbol{k}, t) = \frac{\langle \frac{1}{N} | \sum_r \exp^{-i\boldsymbol{k}\cdot\boldsymbol{r}} [\varphi(\boldsymbol{r}, t) - \langle \varphi \rangle] |^2 \rangle}{\langle \varphi^2(t) \rangle - \langle \varphi \rangle^2}, \quad (12)$$

where $\varphi(\boldsymbol{r}, t)$ is the numerical solution at time t and position $\boldsymbol{r}$ and $\langle\varphi\rangle$ is the mean of all $\varphi(\boldsymbol{r}, t)$. Here, the structural factors $s(\boldsymbol{k}, t)$ are obtained by the Fourier transform on the solutions, and then the square of the modulus is taken and averaged in all directions.

3.4. Density Fluctuation

Density fluctuation is a measure to directly quantify observations of local crowd density, the rules that predict mass behaviors under new circumstances [36,37]. In a spherical window of radius R, the local variance $\sigma_{\varphi}^{2}(R)$ associated with field fluctuations can be expressed in terms of the autocovariance function or the spectral function [29]. It is given by

$$\sigma_{\varphi}^{2}(R) = \frac{1}{v_1(R)} \int_{R^d} \psi(\boldsymbol{r}) \alpha_2(\boldsymbol{r}; R) d\boldsymbol{r} = \frac{1}{v_1(R)(2\pi)^d} \int_{R^d} \widetilde{\psi}(\boldsymbol{k}) \widetilde{\alpha}_2(\boldsymbol{k}; R) d\boldsymbol{k}, \tag{13}$$

where $v_1(R)$ is the volume of a d-dimensional sphere of radius R; $\boldsymbol{r} = \boldsymbol{r}_2 - \boldsymbol{r}_1$. $\psi(\boldsymbol{r})$ describes the autocorrelation functions, i.e., $\langle(\varphi(\boldsymbol{r}_1) - \langle\varphi(\boldsymbol{r}_1)\rangle)(\varphi(\boldsymbol{r}_2) - \langle\varphi(\boldsymbol{r}_2)\rangle)\rangle$; and $\widetilde{\psi}(\boldsymbol{k})$ depicts the Fourier transform of $\psi(\boldsymbol{r})$. $\alpha_2(\boldsymbol{r}; R)$ is the scaled intersection volume [37] of two spherical windows with radius r and R and $\widetilde{\alpha}_2(\boldsymbol{k}; R)$ is its Fourier transform. One can call that φ is a hyperuniform state if its autocorrelation function subjects to the condition

$$\int_{R^d} \psi(\boldsymbol{r}) d\boldsymbol{r} = 0. \tag{14}$$

4. Results and Discussion

4.1. Spatial Patterns

Allen–Cahn model reveals a continual coarsening behavior, where cluster size become bigger and bigger and converge to one of its stable states, as shown in Figure 1. The system only displays a characteristic spatial scale at certain stages. A scale-free behavior appears at early stages. On the contrary, Cahn–Hilliard model displays a clearly characteristic spatial scale (Figure 2), and this scale follows a power law relationship with time. They are indistinguishable spatial patterns from Turing patterns (cf. Figures 2 and 3). It is notable that CH model describes the mass-conservation ecological processes that remarkably differ from the Allen–Cahn model despite both models reveal a coarsening behavior. The CHPD model displays classical Turing patterns with a stationary spatial scale when we consider a species birth–mortality process (Figure 3). In sum, it is difficult to infer the underlying ecological processes and potential mechanism from the merely observed spatial patterns.

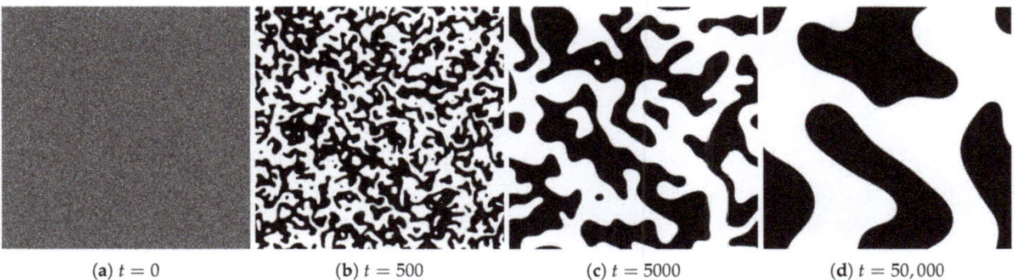

(a) $t = 0$ **(b)** $t = 500$ **(c)** $t = 5000$ **(d)** $t = 50,000$

Figure 1. The patterned dynamics of Allen–Cahn model in Equation (1) solved with the simple Euler algorithm starting a random initial condition: **(a–d)** four typical patterns at $t = 0$, $t = 500$, $t = 5000$, and $t = 50,000$. Numerical simulation was implemented on the discrete 2048×2048 lattices with $\Delta x = \Delta y = 1.0$ and $\Delta t = 0.0025$. It can be observed that the AC model has a relatively scattered wavelength on the spatial scales. Parameter values for the simulation: $\alpha = 0.5$, $\beta = 1.0$, and $\varepsilon = 0.5$.

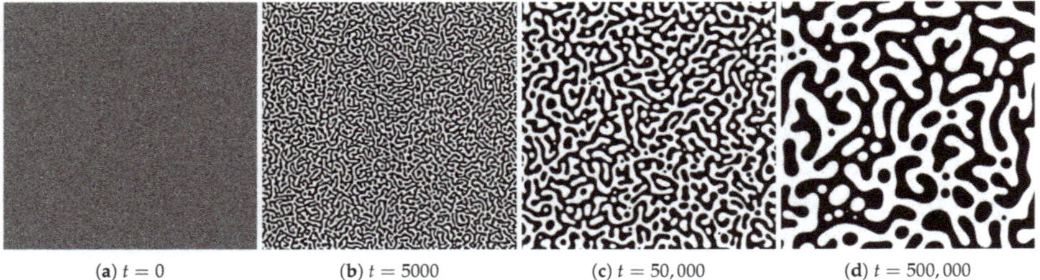

(**a**) $t = 0$ (**b**) $t = 5000$ (**c**) $t = 50,000$ (**d**) $t = 500,000$

Figure 2. The patterned dynamics of Cahn–Hilliard model in Equation (2) solved with the simple Euler algorithm starting a random initial condition: (**a–d**) four typical patterns at $t = 0$, $t = 5000$, $t = 50{,}000$, and $t = 500{,}000$. Numerical simulation was implemented on the discrete 2048×2048 lattices with $\Delta x = \Delta y = 1.0$ and $\Delta t = 0.0025$. It can be observed that the CH model has a more concentrated wavelength on the spatial scale than the AC model, which seems to be more regular on the time scales. Parameter values for the simulation: $\alpha = 0.5$, $\beta = 1.0$, $\varepsilon_A = 0.5$, and $\varepsilon_B = 1.0$.

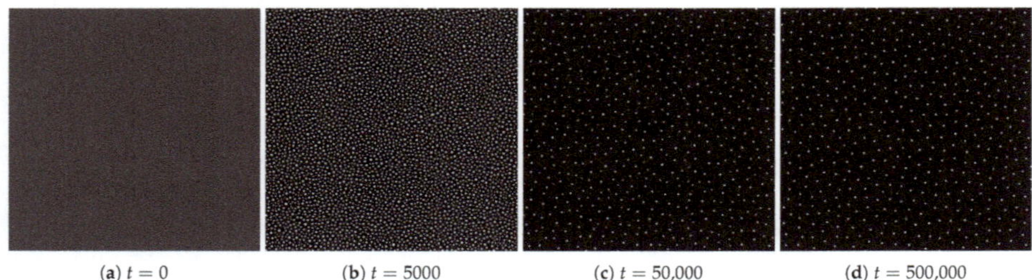

(**a**) $t = 0$ (**b**) $t = 5000$ (**c**) $t = 50{,}000$ (**d**) $t = 500{,}000$

Figure 3. The patterned dynamics of Cahn–Hilliard with population demographics model solved with the simple Euler algorithm starting a random initial condition: (**a–d**) four typical patterns at $t = 0$, $t = 5000$, $t = 50{,}000$, and $t = 500{,}000$. Numerical simulation was implemented on the discrete 2048×2048 lattices with $\Delta x = \Delta y = 1.0$ and $\Delta t = 0.0025$. It can be observed that the pattern of the Cahn–Hilliard with population demographics model is dense at the initial moment, and the dots gradually become spatial regular patterns. The spatial patterns approximate a stable state after the critical time scales. At this time, the coarsening processes is suppressed by the population mortality and birth processes. Parameter values for the simulation: $\alpha = 0.5$, $\beta = 1.0$, $\varepsilon_A = 0.5$, $\varepsilon_B = 1.0$, $r = 0.0001$, and $K = 0.3$.

4.2. Spatial Correlation Functions

Figure 4 illustrates the spatial correlation of the three models from $t = 1$ to $t = 10{,}000$. The spatial correlation functions reveal differently dampened oscillation behaviors at long time scales, where the location of minimal values increases with time (Figure 4). It is consistent with the coarsening processes of the spatial patterns. However, this behavior is not obvious in AC model, which means that the pattern has no characteristic scale. In sum, the spatial correlation function is not a significant indicator to distinguish three models.

4.3. Structure Factors

To investigate the characteristic wavelength of the patterns, we plot the structural factor $s(\boldsymbol{k}, t)$ versus wave numbers $\boldsymbol{k}$ at different times in Figure 5. A larger value of $s(\boldsymbol{k}, t)$ means a stronger dominant effect of its wave numbers.

For AC model, there is no dominant wave numbers of $s(\boldsymbol{k})$, implying absence of spatial characteristic length on this system. It displays a power law decline with exponent of -3.12, predicting scale-free patterns (Figure 5d). However, $s(\boldsymbol{k})$ has obvious dominant peak in the CH and CHPD models, indicating the regular spatial patterns. Furthermore, with time, the peak shifts to a smaller wave number, indicating that the spatial characteristic length is gradually increasing. For CHPD model, there is no significant change in the structure factor after a critical time scale (here 50,000), indicating that the spatial patterns do not change over time. This saturated behavior indicates the logistic growth term is playing a leading role (the phase separation is suppressed). In sum, the structure factors can serve as an excellent indicator to distinguish AC and CH/CHPD model.

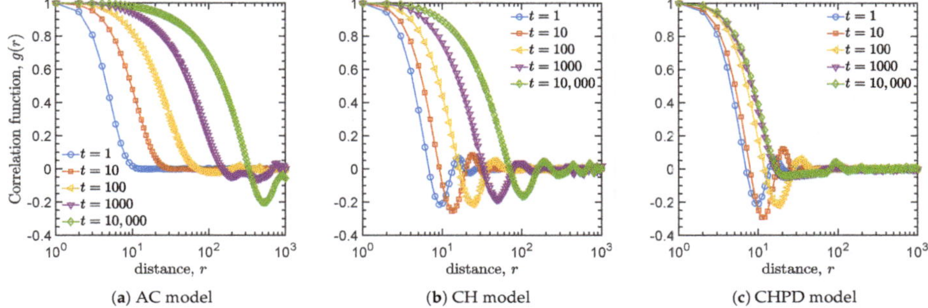

Figure 4. Spatial correlation functions for the self-organized patterns at different times: (**a**) Allen–Cahn model; (**b**) Cahn–Hilliard model; and (**c**) Cahn–Hilliard with population demographics model. Different colors indicate the different times.

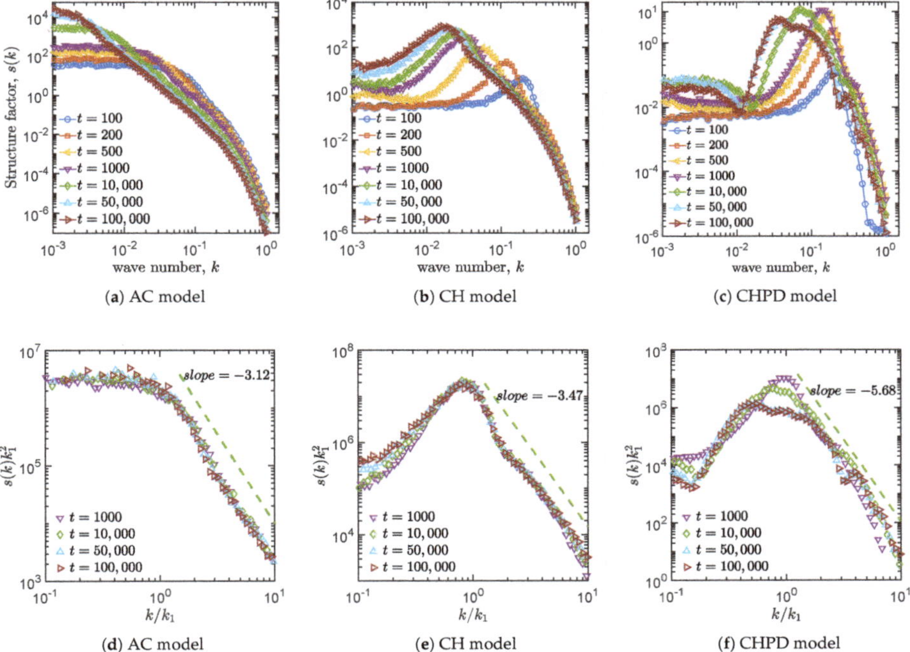

Figure 5. (**a–c**) The structure factors $s(k)$ of the spatial patterns at different times. (**d–f**) The scaled structure factors $s(k)k_1^2$ versus k/k_1 at different times. Different colors indicate the different times.

Moreover, all systems display a universal scaling behavior at high wave numbers (i.e., small wavelength). They collapse to master curves when the structure factors were normalized by k/k_1 for different times [38], as shown in Figure 5d–f. Unfortunately, to date we do not know the ecological significance of those slope values.

4.4. Density Fluctuation

The exponent of density fluctuation can be used to quantitatively measure the spatial distribution of the pattern [36,37]. As mentioned in the Method Section, we plot the variable-field fluctuations as a function of the window size L in Figure 6a. As expected, the exponent of density fluctuations predicts remarkable differences between AC model and CH/CHPD model despite it being impossible to directly distinguish from spatial patterns (cf. Figures 1–3). The exponent of the AC model is around -1.91, implying a random distribution of patterns at large-spatial scales. Interestingly, CH and CHPD models have the exponents of -2.90 and -2.60, respectively. Generally, this exponential value of the scaling law is used to describe the spatial-ordered features. It is called "hyperuniformity" when the exponent of the density fluctuation is less than -2.0, where the long-wavelength density fluctuations are suppressed [39]. In sum, the exponent of density fluctuation is also useful to quantify the spatial patterns between AC model and CH/CHPD models.

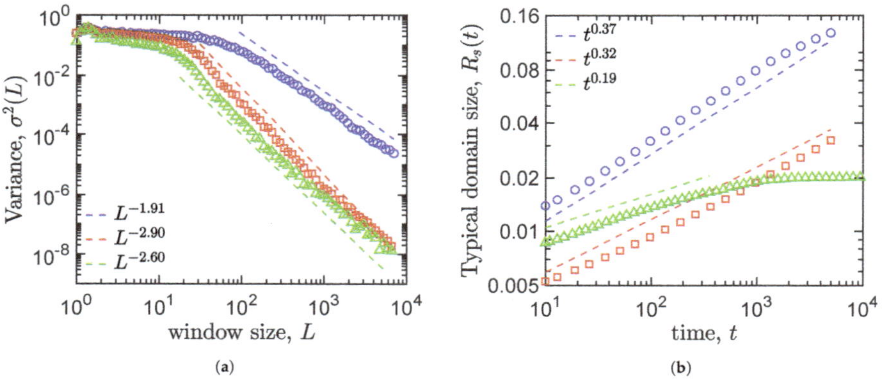

Figure 6. Spatiotemporal hallmarks of three different dynamics models (Allen–Cahn model ○, Cahn–Hilliard model □, and Cahn–Hilliard with population demographics model △). (**a**) The variable-field fluctuations as a function of the window size length L at $t = 500,000$ associated with $20,000 \times 20,000$ systems. (**b**) The scaling behavior of the spatial wavelength $R_s(t)$ versus increased times.

4.5. Growth Law

To measure how the spatial structure of the patterns evolves over time, we calculate the characteristic length, $R_s(t)$, with location of the maximum of $s(\boldsymbol{k}, t)$ and its first moment at time t, i.e.,

$$k_1(t) = \frac{\sum_k k\, s(\boldsymbol{k}, t)}{\sum_k s(\boldsymbol{k}, t)}. \tag{15}$$

Then, the spatial characteristic length (also called wavelength) change over time can be obtained as

$$R_s(t) = \frac{2\pi}{k_1(t)}. \tag{16}$$

The scaling behaviors of the spatial wavelength $R_s(t)$ are shown in Figure 6b. The AC and CH models reveal power law relationship with exponents around 0.37 and 0.32, respectively. It should be noted that this scaling property is called the Lifshitz–Slyozov–Wagner law [40]. However, CHPD model has phase separation patterns at an early stage with scaling of 0.19, but it saturates at long-time scales. These patterns are equivalent to the Turing principle. In sum, this saturated behavior of growth rate can well distinguish CH model and CHPD model, which are impossible to be separated by structure factor, density fluctuation, and spatial correlation function (see Table 1 for summary).

Table 1. Summary of the different indicators on three mechanism models.

Model Types	Structure Factors	Density Fluctuation	Growth Law	Saturation
Allen–Cahn	−3.12	−1.91	0.37	No
Cahn–Hillard	−3.47	−2.90	0.32	No
Cahn–Hillard and logistic term	−5.68	−2.60	0.19	Yes
Validity of the indicators	Yes	Yes	Yes	Yes

5. Conclusions

Although PDE models have been widely used to describe the spatial patterns in many ecological and biological systems, few studies focus on inferring the linkage between these visual patterned characters and their potential mechanisms [41–43]. To do this, one of the effective ways is to build the quantitative indicators of these macrocosm patterns, which can serve as the hallmarks of the underlying ecological processes. Thus, it is a key issue to explain the ecological mechanisms by comparing the temporal and spatial behavior of the spatial patterns [44–46].

Here, we use structure factors, density fluctuation, growth law, and saturation status (Table 1) to quantify the spatiotemporal dynamics of AC model, CH model, and CHPD model. There is a remarkable peak of structure factors in the CH model and CHPD model, but absent in AC model. The differences of spatial structures between the AC model and CH/CHPD model can be well distinguished by the exponent of density fluctuations. Both CH and CHPD models reveal a hyperuniform behavior with scaling of −2.90 and −2.60, respectively, but AC model displays a random distribution with scaling of −1.91. The differences in time scales among the patterns of the three models can be well distinguished by LS behaviors. Both AC and CH model display the LS behaviors with scaling of 0.37 and 0.32, respectively, but CHPD model with scaling of 0.19 at short-time scales and saturation at long-time scales.

We showed that the underlying mechanisms can well be identified by the spatial information, i.e., the structure factors, the density fluctuation, the growth law, and the saturation; however, seeking real ecological cases is still a challenge (see Table 2 for summary). To the best of our knowledge, the first indicator (structure factors) can be applied to calculate the spatial wavelength, with only spatial data. The second indicator (density fluctuation) can distinguish the random distribution and the hyperuniform state at large spatial scales. It also reveals the intensity of long wave suppression. The third indicator (growth law) can be obtained from the spatiotemporal observed sequences. It describes the change and the stability of spatial structure with time. There are still many other potential mechanisms to generate spatially self-organized patterns in the ecosystems [47,48] such as animal aggregation due to taxis and density-dependence [49], but they are beyond the scope of the discussion here. It is interesting for further comparison in the future research.

To sum, we find that the density-fluctuation exponent and LS behaviors can quantitatively infer the three underlying mechanisms from the spatiotemporal patterns. We need to integrate multiple indicators to distinguish the underlying ecological processes and mechanisms, such as Turing principle and phase separation principle.

Table 2. A summary of the three mechanisms on spatial self-organization in ecosystems.

Model	Ecosystem	Ecological Processes	Refs.
AC	Bacteria (*Vibrio cholerae*)	Phase separation caused by competition	[18]
	Plant (*Centaurea maculosa*)	Allelopathy and exotic plant invasion	[50]
	Coral reef	Allelopathy and spatial competition	[51]
CH	Mussels	Density-dependent movement behavior	[8,13]
	Ants	Density-dependent movement behavior	[52]
	Bacteria (*Escherichia coli*)	Density-dependent chemo-taxis behavior	[53,54]
	Birds	Resource-dependent movement behavior	[55]
	Elk (*Cervus canadensis*)	Socially inform	[56]
	Zebrafish	Run-and-chase behavior movement	[57]
	Sperm	Integrated geometry with minima drag	[58]
CHPD	Bacteria (*Escherichia coli, Bacillus subtilis*)	Density-dependent motility and birth-death	[19,20]
	Stones and soil	Freeze–thaw cycles	[59]
	Insect	General theory	[16]

Author Contributions: K.Z. performed research; W.-S.H. conceived the numerical analysis; Q.-X.L. constructed research; and K.Z., W.-S.H., and Q.-X.L. wrote the manuscript. All authors have read and agreed to the published version of the manuscript.

Funding: This work was supported by the National Natural Science Foundation of China (41676084) and the National Key R&D Program of China (2017YFC0506001).

Conflicts of Interest: The authors declare no conflict of interest.

Abbreviations

The following abbreviations are used in this manuscript:

AC	Allen–Cahn
CH	Cahn–Hilliard
CHPD	Cahn–Hilliard with population demographics
LS	Lifshitz–Slyozov scaling
PDE	partial differential equation

References

1. Fisher, R.A. The wave of advance of advntageous genes. *Ann. Eugen.* **1937**, *7*, 355–369. [CrossRef]
2. Murray, J.D. *Mathematical Biology*; Springer: Berlin, Germany, 2002.
3. Woodward, D.E.; Tyson, R.; Myerscough, M.R.; Murray, J.D.; Budrene, E.O.; Berg, H.C. Spatio-temporal patterns generated by Salmonella typhimurium. *Biophys. J.* **1995**, *68*, 2181–2189. [CrossRef]
4. Meinhardt, H. *The Algorithmic Beauty of Sea Shells*; Springer Science & Business Media: Berlin/Heidelberg Germany, 2009.
5. Klausmeier, C.A. Regular and irregular patterns in semiarid vegetation. *Science* **1999**, *285*, 838–838. [CrossRef]
6. Ouyang, Q.; Swinney, H.L. Transition from a Uniform State to Hexagonal and Striped Turing Patterns. *Nature* **1991**, *352*, 610–612. [CrossRef]
7. Kondo, S.; Miura, T. Reaction-Diffusion Model as a Framework for Understanding Biological Pattern Formation. *Science* **2010**, *329*, 1616–1620. [CrossRef]
8. Van de Koppel, J.; Rietkerk, M.; Dankers, N.; Herman, P.M.J. Scale-Dependent Feedback and Regular Spatial Patterns in Young Mussel Beds. *Am. Nat.* **2005**, *165*, 66–77. [CrossRef]
9. Segel, L.A.; Jackson, J.L. Dissipative structure: An explanation and an ecological example. *J. Theor. Biol.* **1972**, *37*, 545–559. [CrossRef]
10. Cabrera, C.V. Lateral inhibition and cell fate during neurogenesis in Drosophila: The interactions between scute, Notch and Delta. *Development* **1990**, *110*, 733–742.
11. Dormann, D.; Weijer, C.J. Propagating chemoattractant waves coordinate periodic cell movement in Dictyostelium slugs. *Development* **2001**, *128*, 4535–4543.

12. Rietkerk, M.; van de Koppel, J. Regular pattern formation in real ecosystems. *Trends Ecol. Evol.* **2008**, *23*, 169–175. [CrossRef]
13. Liu, Q.X.; Doelman, A.; Rottschäfer, V.; de Jager, M.; Herman, P.M.; Rietkerk, M.; van de Koppel, J. Phase separation explains a new class of self-organized spatial patterns in ecological systems. *Proc. Natl. Acad. Sci. USA* **2013**, *110*, 11905–11910. [CrossRef]
14. Aihara, S.; Takaki, T.; Takada, N. Multi-phase-field modeling using a conservative Allen–Cahn equation for multiphase flow. *Comput. Fluids* **2019**, *178*, 141–151. [CrossRef]
15. Allen, S.M.; Cahn, J.W. Ground state structures in ordered binary alloys with second neighbor interactions. *Acta Metall.* **1972**, *20*, 423–433. [CrossRef]
16. Cohen, D.S.; Murray, J.D. A Generalized Diffusion-Model for Growth and Dispersal in a Population. *J. Math. Biol.* **1981**, *12*, 237–249. [CrossRef]
17. Liu, Q.X.; Rietkerk, M.; Herman, P.M.; Piersma, T.; Fryxell, J.M.; van de Koppel, J. Phase separation driven by density-dependent movement: A novel mechanism for ecological patterns. *Phys. Life Rev.* **2016**, *19*, 107–121. [CrossRef]
18. McNally, L.; Bernardy, E.; Thomas, J.; Kalziqi, A.; Pentz, J.; Brown, S.P.; Hammer, B.K.; Yunker, P.J.; Ratcliff, W.C. Killing by Type VI secretion drives genetic phase separation and correlates with increased cooperation. *Nat. Commun.* **2017**, *8*, 14371. [CrossRef]
19. Vicsek, T. Ecological patterns emerging as a result of the density distribution of organisms: Comment on "Phase separation driven by density-dependent movement: A novel mechanism for ecological patterns" by Quan-Xing Liu et al. *Phys. Life Rev.* **2016**, *19*, 139–141. [CrossRef]
20. Cates, M.E.; Marenduzzo, D.; Pagonabarraga, I.; Tailleur, J. Arrested phase separation in reproducing bacteria creates a generic route to pattern formation. *Proc. Natl. Acad. Sci. USA* **2010**, *107*, 11715–11720. [CrossRef]
21. Wittkowski, R.; Tiribocchi, A.; Stenhammar, J.; Allen, R.J.; Marenduzzo, D.; Cates, M.E. Scalar $\phi 4$ field theory for active-particle phase separation. *Nat. Commun.* **2014**, *5*, 4351. [CrossRef]
22. Toral, R.; Chakrabarti, A.; Gunton, J.D. Droplet distribution for the two-dimensional Cahn-Hilliard model: Comparison of theory with large-scale simulations. *Phys. Rev. A* **1992**, *45*, 2147–2150. [CrossRef]
23. Chen, X.; Dong, X.; Be'er, A.; Swinney, H.L.; Zhang, H.P. Scale-Invariant Correlations in Dynamic Bacterial Clusters. *Phys. Rev. Lett.* **2012**, *108*, 148101. [CrossRef]
24. Zhu, J.; Chen, L.-Q.; Shen, J.; Tikare, V. Coarsening kinetics from a variable-mobility Cahn-Hilliard equation: Application of a semi-implicit Fourier spectral method. *Phys. Rev. E* **1999**, *60*, 3564–3572. [CrossRef]
25. Collet, J.F.; Goudon, T. On solutions of the Lifshitz-Slyozov model. *Nonlinearity* **2000**, *13*, 1239–1262. [CrossRef]
26. Kukushkin, S.A.; Slyozov, V.V. Theory of the Ostwald ripening in multicomponent systems under non-isothermal conditions. *J. Phys. Chem. Solids* **1996**, *57*, 195–204. [CrossRef]
27. Svoboda, J.; Fischer, F.D. Generalization of the Lifshitz-Slyozov-Wagner coarsening theory to non-dilute multi-component systems. *Acta Mater* **2014**, *79*, 304–314. [CrossRef]
28. Kéfi, S.; Rietkerk, M.; Alados, C.L.; Pueyo, Y.; Papanastasis, V.P.; ElAich, A.; De Ruiter, P.C. Spatial vegetation patterns and imminent desertification in Mediterranean arid ecosystems. *Nature* **2007**, *449*, 213–215. [CrossRef]
29. Torquato, S. Hyperuniform states of matter. *Phys. Rep.* **2018**, *745*, 1–95. [CrossRef]
30. Allen, S.M.; Cahn, J.W. A microscopic theory for antiphase boundary motion and its application to antiphase domain coarsening. *Acta Metall.* **1979**, *27*, 1085–1095. [CrossRef]
31. Cahn, J.W.; Hilliard, J.E. Free energy of a nonuniform system. I. Interfacial free energy. *J. Chem. Phys.* **1958**, *28*, 258–267. [CrossRef]
32. Chow, S.-N.; Conti, R.; Johnson, R.; Mallet-Paret, J.; Nussbaum, R. *Dynamical systems: Lectures Given at the CIME Summer School Held in Cetraro, Italy, June 19–26, 2000*; Springer: Berlin, Germany, 2003.
33. Richtmyer, R.D.; Dill, E. Difference methods for initial-value problems. *Phys. Today* **1959**, *12*, 50. [CrossRef]
34. Newman, M.; Barkema, G. *Monte Carlo Methods in Statistical Physics*; Oxford University Press: New York, NY, USA, 1999.
35. Toral, R.; Chakrabarti, A.; Gunton, J.D. Large scale simulations of the two-dimensional Cahn-Hilliard model. *Phys. A Stat. Mech. Its Appl.* **1995**, *213*, 41–49. [CrossRef]
36. Zhang, H.P.; Be'er, A.; Florin, E.L.; Swinney, H.L. Collective motion and density fluctuations in bacterial colonies. *Proc. Natl. Acad. Sci. USA* **2010**, *107*, 13626–13630. [CrossRef] [PubMed]

37. Klatt, A.M.; Torquato, S. Characterization of maximally random jammed sphere packings. II. Correlation functions and density fluctuations. *Phys. Rev. E* **2016**, *94*, 022152. [CrossRef]
38. Brown, G.; Rikvold, P.A.; Grant, M. Universality and scaling for the structure factor in dynamic order-disorder transitions. *Phys. Rev. E* **1998**, *58*, 5501–5507. [CrossRef]
39. Lei, Q.L.; Ni, R. Hydrodynamics of random-organizing hyperuniform fluids. *Proc. Natl. Acad. Sci. USA* **2019**, *116*, 22983–22989. [CrossRef]
40. Cheng, S.Z. *Phase Transitions in Polymers: The Role of Metastable States*; Elsevier: Amsterdam, The Netherlands, 2008.
41. Hammond, M.P.; Kolasa, J. Spatial variation as a tool for inferring temporal variation and diagnosing types of mechanisms in ecosystems. *PLoS ONE* **2014**, *9*, 89245–89245. [CrossRef]
42. Borgogno, F.; D'Odorico, P.; Laio, F.; Ridolfi, L. Mathematical models of vegetation pattern formation in ecohydrology. *Rev. Geophys.* **2009**, *47*. [CrossRef]
43. Liu, Q.X.; Weerman, E.J.; Herman, P.M.; Olff, H.; van de Koppel, J. Alternative mechanisms alter the emergent properties of self-organization in mussel beds. *Proc. R. Soc. B Biol. Sci.* **2012**, *279*, 2744–2753. [CrossRef]
44. Ball, P. *Patterns in Nature: Why the Natural World Looks the Way it Does*; University of Chicago Press: Chicago, IL, USA, 2006.
45. Mueller, E.N.; Wainwright, J.; Parsons, A.J.; Turnbull, L. *Patterns of Land Degradation in Drylands*; Springer: Berlin, Germany, 2014.
46. Thompson, S.E.; Daniels, K.E. A Porous Convection Model for Grass Patterns. *Am. Nat.* **2010**, *175*, 10–15. [CrossRef]
47. Petrovskii, S. Pattern, process, scale, and model's sensitivity Comment on "Phase separation driven by density-dependent movement: Anovel mechanism for ecological patterns". *Phys. Life Rev.* **2016**, *19*, 131–134. [CrossRef]
48. Ellis, J.; Petrovskaya, N.; Petrovskii, S. Effect of density-dependent individual movement on emerging spatial population distribution: Brownian motion vs Levy flights. *J. Theor. Biol.* **2019**, *464*, 159–178. [CrossRef]
49. Tyutyunov, Y.; Senina, I.; Arditi, R. Clustering due to acceleration in the response to population gradient: A simple self-organization model. *Am. Nat.* **2004**, *164*, 722–735. [CrossRef]
50. Bais, H.P.; Vepachedu, R.; Gilroy, S.; Callaway, R.M.; Vivanco, J.M. Allelopathy and exotic plant invasion: From molecules and genes to species interactions. *Science* **2003**, *301*, 1377–1380. [CrossRef]
51. Jackson, J.B.C.; Buss, L.E.O. Alleopathy and spatial competition among coral reef invertebrates. *Proc. Natl. Acad. Sci. USA* **1975**, *72*, 5160–5163. [CrossRef]
52. Theraulaz, G.; Bonabeau, E.; Nicolis, S.C.; Solé, R.V.; Fourcassié, V.; Blanco, S.; Fournier, R.; Joly, J.L.; Fernández, P.; Grimal, A.; et al. Spatial patterns in ant colonies. *Proc. Natl. Acad. Sci. USA* **2002**, *99*, 9645–9649. [CrossRef]
53. Liu, C.; Fu, X.; Liu, L.; Ren, X.; Chau, C.K.; Li, S.; Xiang, L.; Zeng, H.; Chen, G.; Tang, L.H.; et al. Sequential establishment of stripe patterns in an expanding cell population. *Science* **2011**, *334*, 238–241. [CrossRef]
54. Fu, X.; Tang, L.H.; Liu, C.; Huang, J.D.; Hwa, T.; Lenz, P. Stripe formation in bacterial systems with density-suppressed motility. *Phys. Rev. Lett.* **2012**, *108*, 198102. [CrossRef]
55. Folmer, E.O.; Olff, H.; Piersma, T. The spatial distribution of flocking foragers: Disentangling the effects of food availability, interference and conspecific attraction by means of spatial autoregressive modeling. *Oikos* **2012**, *121*, 551–561. [CrossRef]
56. Haydon, D.T.; Morales, J.M.; Yott, A.; Jenkins, D.A.; Rosatte, R.; Fryxell, J.M. Socially informed random walks: Incorporating group dynamics into models of population spread and growth. *Proc. R. Soc. B Biol. Sci.* **2008**, *275*, 1101–1109. [CrossRef]
57. Yamanaka, H.; Kondo, S. In vitro analysis suggests that difference in cell movement during direct interaction can generate various pigment patterns in vivo. *Proc. Natl. Acad. Sci. USA* **2014**, *111*, 1867–1872. [CrossRef]
58. Fisher, H.S.; Giomi, L.; Hoekstra, H.E.; Mahadevan, L. The dynamics of sperm cooperation in a competitive environment. *Proc. R. Soc. B Biol. Sci.* **2014**, *281*, 20140296. [CrossRef] [PubMed]
59. Kessler, M.A.; Werner, B.T. Werner. Self-organization of sorted patterned ground. *Science* **2003**, *299*, 380–383. [CrossRef] [PubMed]

Article

Cross Diffusion Induced Turing Patterns in a Tritrophic Food Chain Model with Crowley-Martin Functional Response

Nitu Kumari * and Nishith Mohan

School of Basic Sciences, Indian Institute of Technology Mandi, Mandi, Himachal Pradesh 175001, India; nishithmohan888@gmail.com

* Correspondence: nitu@iitmandi.ac.in

Received: 20 January 2019; Accepted: 23 February 2019; Published: 1 March 2019

Abstract: Diffusion has long been known to induce pattern formation in predator prey systems. For certain prey-predator interaction systems, self diffusion conditions ceases to induce patterns, i.e., a non-constant positive solution does not exist, as seen from the literature. We investigate the effect of cross diffusion on the pattern formation in a tritrophic food chain model. In the formulated model, the prey interacts with the mid level predator in accordance with Holling Type II functional response and the mid and top level predator interact via Crowley-Martin functional response. We prove that the stationary uniform solution of the system is stable in the presence of diffusion when cross diffusion is absent. However, this solution is unstable in the presence of both self diffusion and cross diffusion. Using a priori analysis, we show the existence of a inhomogeneous steady state. We prove that no non-constant positive solution exists in the presence of diffusion under certain conditions, i.e., no pattern formation occurs. However, pattern formation is induced by cross diffusion because of the existence of non-constant positive solution, which is proven analytically as well as numerically. We performed extensive numerical simulations to understand Turing pattern formation for different values of self and cross diffusivity coefficients of the top level predator to validate our results. We obtained a wide range of Turing patterns induced by cross diffusion in the top population, including floral, labyrinth, hot spots, pentagonal and hexagonal Turing patterns.

Keywords: cross diffusion; Turing patterns; non-constant positive solution

---

1. Introduction

The two primary objectives of studying systems ecology are to get an understanding of the dynamics of the ecological systems and the nature of the forces, which determine the community structure. Earlier, the classical ecological models focused on two species interactions. Continuous time models of two interacting species have been studied extensively in the literature (see [1] and references therein). Mathematically, only two basic patterns are exhibited in these models—approach to a steady state or to a limit cycle [2]—but the ecological models of nature exhibit very complex dynamical behavior. Price et al. [3] proposed that community behavior must be based on three or more trophic levels. Three species continuous time models have observed more complex dynamical behavior [4–7]. All these studies depend upon the classical types of functional responses, which include Holling Types I, II and III and Holling-Tanner ratio dependent functional response.

The Crowley-Martin [8] functional response is one of the predator dependent functional responses, i.e., the functional response is function of both prey and predator abundance because of predator interference. The basic assumption behind the formulation of Crowley-Martin functional response is that there is a decrease in predator feeding rate because of high predator density even if there is a high prey density. Hence, the effect of predator interference on the feeding rate is important, even when an individual predator is handling or searching for a prey at a given instant of time. Here, v and r are population density of prey and predator, respectively. The parameters w_2, d, b are positive parameters that describe the effects of capture rate, handling time and magnitude of interference among predators on the feeding rate, respectively. The per capita rate in this formulation is given by:

$$f(v,r) = \frac{w_2 v}{1 + dv + br + bdvr}.$$

The Crowley-Martin functional response is used for datasets that indicate an asymptotic feeding rate that is affected by predator density. A continuous time model to analyze the dynamics of a three species food chain with Crowley-Martin functional response was studied by Upadhyay et al. [9]. The model system in [9] is based on the assumption that the species under consideration are spatially homogeneously distributed; however, in nature, the species distribution is always inhomogeneous. Therefore, to model a realistic food chain scenario, reaction diffusion mechanism should be considered, as employed in [10–12]. A qualitative analysis of a two species model with Crowley-Martin functional response and diffusion is carried out in [13]. In the above model, the movement of a species is determined solely by their own characteristics, i.e., the movements of the species in these models are physically affected by the population pressures due to the mutual interference among the individuals of the same species. Therefore, such models take into account only the self diffusion of the species in concern. However, in the case of prey-predator interacting systems, the movement of one species may affect the movement or motility of the other. The above models fail to describe such situation, which is more realistic. In many systems, the predators develop migratory strategies to take advantage over the prey. In the case of a three species prey-predator model, such a migratory behavior depends on the concentration of both predators (i.e., the mid level predator and the top level predator). This leads to *cross diffusive system*, in addition to each species natural tendency to diffuse i.e., self diffusion; such models are studied in [14–18]. The cross diffusive coefficient may have positive or negative values. The positive cross diffusivity coefficient represents that one species tends to move in the direction of lower concentration of another species. The negative cross diffusion term represents that the population flux of one species is in the direction of higher concentration of the other species.

In the present paper, we consider a tritrophic food chain model in which the species interact via Crowley-Martin and Holling type II functional response at different levels. We introduce a nonlinear cross diffusion among top level and mid level predators. The objective is to understand the effect of both types of diffusion self and cross on the stability of the system. An analytic approach is adopted to understand the formation of inhomogeneous steady state solutions. Numerical simulation depicting spatial pattern formation was performed for a wide range of parameters to understand the predation behavior due to the effect of cross and self diffusion.

In Section 2, we formulate the temporal model using various assumptions and spatially extend it to formulate a nonlinear reaction diffusion system and with cross diffusivity (a strongly coupled parabolic system). In Section 3, we perform a linear stability analysis of the temporal model and obtain the conditions under which the stationary uniform solution is locally asymptotically stable. In Section 4, we provide analytical proof of stability of the stationary uniform solution in the presence of diffusion and in the absence of cross diffusion but unstable in the presence of cross diffusion. In Section 5, we provide an analytical proof of the existence of inhomogeneous steady states. Section 6 deals with the numerical simulation of the cross diffusive system and explains Turing pattern formation for different values of cross diffusive and self diffusive coefficients of the top level predator.

2. Model System

We first consider a temporal model governing the dynamics of tritrophic food chain, consisting of Holling type-II and Crowley Martin type functional responses, defined by a system of differential equations [9]. Here, $U(T)$ is the population density of the lowest trophic level species (prey), $V(T)$ is the population density of the middle trophic level species (intermediate predator) and $R(T)$ is the population density of the highest trophic level species (top predator), at any time T. We made the following basic assumptions for formulation of the proposed model system.

Assumption 1. *The prey U grows with an intrinsic growth rate a_1, and has a carrying capacity of K in the absence of predator V. D is a measure of the extent to which the environment can provide protection to U and w is the maximum value which per capita reduction rate of U can attain. In addition, the mid level predator V predates on U in accordance with Holling Type II functional response, $\frac{wUV}{U+D}$. Therefore, considering the above assumptions, we formulate the first equation of the model system as:*

$$\frac{dU}{dT} = a_1U\left(1 - \frac{U}{K}\right) - \frac{wUV}{U+D} \tag{1}$$

Assumption 2. *The intermediate or mid level predator V has a natural death rate of a_2. w_1 describes the maximum value which per capita reduction rate of V can attain, D_1 is similar to D of U and p represents the internal competition coefficient among the members of V. The predation of V over U is governed by Holling Type II functional response. The top level predator R predates on V in accordance with Crowley-Martin functional response. The constants w_2, b and d are the saturating Crowley-Martin type functional response parameters, where w_2 describes capture rate effect, b represents handling time of prey and d is the magnitude of interference among predators. The second equation of the model system is given as:*

$$\frac{dV}{dT} = -a_2V - pV^2 + \frac{w_1UV}{U+D_1} - \frac{w_2VR}{1+dV+bR+bdVR} \tag{2}$$

Assumption 3. *The top level predator R dies at a natural death rate of c. As the predating behavior of R over V is described by the Crowley-Martin functional response. w_3 is the saturating Crowley-Martin type functional response parameter similar to w_2. The third equation of the model system takes the following form:*

$$\frac{dR}{dT} = -cR + \frac{w_3VR}{1+dV+bR+bdVR} \tag{3}$$

All parameters described above take only positive values and the model system is a three species food chain model involving a hybrid type of prey dependent and predator dependent functional response. Therefore, using the above assumptions, the model system takes the following form:

$$\begin{aligned}
\frac{dU}{dT} &= a_1U\left(1 - \frac{U}{K}\right) - \frac{wUV}{U+D} \\
\frac{dV}{dT} &= -a_2V - pV^2 + \frac{w_1UV}{U+D_1} - \frac{w_2VR}{1+dV+bR+bdVR} \\
\frac{dR}{dT} &= -cR + \frac{w_3VR}{1+dV+bR+bdVR}
\end{aligned} \tag{4}$$

The model system described by Equation (4) consists of thirteen parameters, which makes the analysis quite cumbersome, therefore the number of parameters are reduced by rescaling the model system. The model is rescaled, using the following new variables and parameters:

$$\begin{aligned}
&t = a_1T, \quad u = \frac{U}{K}, \quad v = \frac{wV}{a_1K}, \quad r = \frac{ww_2R}{a_1^2dK}, \quad w_4 = \frac{D}{K}, \quad w_5 = \frac{a_2}{a_1}, \quad w_6 = \frac{w_1}{a_1}, \quad w_7 = \frac{D_1}{K}, \\
&w_8 = \frac{a_1b}{w_2}, \quad w_9 = \frac{a_1^2bdK}{ww_2}, \quad w_{10} = \frac{w}{a_1dK}, \quad w_{11} = \frac{c}{a_1}, \quad w_{12} = \frac{w_3}{a_1d}, \quad e = \frac{pw}{a_1^2K}
\end{aligned} \tag{5}$$

The rescaled system is as follows:

$$\begin{aligned}
\frac{du}{dt} &= u\left(1-u-\frac{v}{u+w_4}\right)\\
\frac{dv}{dt} &= v\left(-w_5+\frac{w_6u}{u+w_7}-ev-\frac{r}{v+(w_8+w_9v)\,r+w_{10}}\right)\\
\frac{dr}{dt} &= r\left(-w_{11}+\frac{w_{12}v}{v+(w_8+w_9v)\,r+w_{10}}\right)
\end{aligned} \tag{6}$$

The model system proposed in Equation (4) and the rescaled form of the model given in Equation (6) are based on the assumption that the species under consideration are homogeneously distributed but in nature the species distribution is always inhomogeneous. Therefore, to model a realistic food chain scenario, we consider the model system with diffusion. Consider the spatially explicit three species predator prey food chain model system. At any location (x, y) and time t, the interaction of three species populations $u(x, y, t)$, $v(x, y, t)$ and $r(x, y, t)$ can be modeled with the reaction diffusion equations given in Equation (7). Here, Δ denotes two dimensional Laplacian operator given by $\left(\Delta \equiv \frac{\partial^2}{\partial x^2}+\frac{\partial^2}{\partial y^2}\right)$, where, $x, y \in \Omega$ and $t > 0$. Ω is a bounded domain in $\mathbb{R}^2$ with a smooth boundary $\partial\Omega$.

$$\begin{aligned}
&\frac{\partial u}{\partial t}-d_1\Delta u = u\left(1-u-\frac{v}{u+w_4}\right)\\
&\frac{\partial v}{\partial t}-d_2\Delta v = v\left(-w_5+\frac{w_6u}{u+w_7}-ev-\frac{r}{v+(w_8+w_9v)\,r+w_{10}}\right)\\
&\frac{\partial r}{\partial t}-d_3\Delta r = r\left(-w_{11}+\frac{w_{12}v}{v+(w_8+w_9v)\,r+w_{10}}\right)\\
&\frac{\partial u}{\partial n}=\frac{\partial v}{\partial n}=\frac{\partial r}{\partial n}=0.
\end{aligned} \tag{7}$$

The boundary conditions $\frac{\partial u}{\partial n}=\frac{\partial v}{\partial n}=\frac{\partial r}{\partial n}=0$ are directional derivative normal to the boundary and n is the outward normal to Ω.

In the above system, the movement of a species is determined solely by their own characteristics, i.e., the movements of the species can be physically affected by the population pressures due to the mutual interference between the individuals of the same species. Therefore, we refer to the constants d_1, d_2 and d_3 as the *self diffusion rates* of species u, v and r, respectively. However, in the case of a prey-predator interacting system, the movement of one species may affect the movement or motility of the other. Hence, the above model cannot be used to describe such a situation. There are chances of development of migratory strategies among the predators to take advantage over the prey. This behavior can be described by taking into account the effect of *cross diffusion*. Along with the self tendency to move, i.e., self diffusion, the species also migrate or move under the influence of other. Such behavior takes into account concentration of both predators—the mid and the top level predators—constituting a cross diffusive system. The coefficient of cross diffusion may be positive or negative. The positive cross diffusion coefficient represents that one species tends to move in the direction of lower concentration of another species. The negative cross diffusion term represents that the population flux of one species is in the direction of higher concentration of the other. Therefore, we propose the interaction of the above three species population model with cross diffusion as follows:

$$\begin{aligned}
&\frac{\partial u}{\partial t} - d_1 \Delta u = u\left(1 - u - \frac{v}{u + w_4}\right) \\
&\frac{\partial v}{\partial t} - d_2 \Delta v = v\left(-w_5 + \frac{w_6 u}{u + w_7} - ev - \frac{r}{v + (w_8 + w_9 v)\, r + w_{10}}\right) \\
&\frac{\partial r}{\partial t} - d_3 \Delta\left(r + d_4 v r\right) = r\left(-w_{11} + \frac{w_{12} v}{v + (w_8 + w_9 v)\, r + w_{10}}\right) \\
&\frac{\partial u}{\partial n} = \frac{\partial v}{\partial n} = \frac{\partial r}{\partial n} = 0.
\end{aligned} \tag{8}$$

where Δ denotes two dimensional Laplacian operator given by $\left(\Delta \equiv \frac{\partial^2}{\partial x^2} + \frac{\partial^2}{\partial y^2}\right)$, $(x, y) \in \Omega$.

The nonlinear diffusion terms in the equation governing the dynamics of top level predator r implies that the direction of dispersion of top predator contains a self diffusion term by which it moves from a region of its higher concentration to a region of its lower concentration and a cross diffusive term. The top predator r diffuses with the flux

$$\mathbf{J} = -\nabla\left(d_3 r + d_3 d_4 v r\right) = -d_3 d_4 r \nabla v - \left(d_3 + d_3 d_4 v\right) \nabla r.$$

where the $-d_3 d_4 \nabla r < 0$, the $-d_3 d_4 r \nabla v$ part of the flux $\mathbf{J}$ is directed towards the decreasing population density of mid level predator v. This is because, in certain prey-predator systems, several prey form a huge group to protect themselves from the attacking predators. In addition, as in the model system the predation of top predator is impossible, to extensively study the role of cross diffusion on the model system, we induce it at r, i.e., the top predator.

3. Linear Stability Analysis of Temporal Model

Under the assumption that the model system in Equation in Equation (6) has the unique positive stationary uniform solution, which we denote by $\mathbf{u}^* = (u^*, v^*, r^*)$ (where $\mathbf{u}^*$ is a three coordinate vector), we derive the conditions under which it is locally asymptotically stable.

Theorem 1. *The positive equilibrium solution given by $\mathbf{u}^* = (u^*, v^*, r^*)$ of the model system in Equation (6) is locally asymptotically stable if the parameters satisfy,*

$$\frac{r^* v^* (1 + w_9 r^*)}{\left[v^* + (w_8 + w_9 v^*)\, r^* + w_{10}\right]^2} < e \quad and \quad \frac{v^*}{(u^* + w_4)^2} < 1. \tag{9}$$

Proof. We consider the following notation given in Equation (10):

$$\mathbf{G}(\mathbf{u}) = \begin{pmatrix} G_1(\mathbf{u}) \\ G_2(\mathbf{u}) \\ G_3(\mathbf{u}) \end{pmatrix} = \begin{pmatrix} u g_1(\mathbf{u}) & \cong u\left(1 - u - \frac{v}{u + w_4}\right) \\ v g_2(\mathbf{u}) & \cong v\left(-w_5 + \frac{w_6 u}{u + w_7} - ev - \frac{r}{v + (w_8 + w_9 v)\, r + w_{10}}\right) \\ r g_3(\mathbf{u}) & \cong r\left(-w_{11} + \frac{w_{12} v}{v + (w_8 + w_9 v)\, r + w_{10}}\right) \end{pmatrix} \tag{10}$$

Calculating $\mathbf{G}_{\mathbf{u}}(\mathbf{u}^*)$, and putting, $\mathbf{G}_{\mathbf{u}}(\mathbf{u}^*) = 0$.

$$\mathbf{G}_{\mathbf{u}}(\mathbf{u}^*) = \begin{pmatrix} \frac{u^* v^*}{(u^* + w_4)^2} - u^* & \frac{-u^*}{u^* + w_4} & 0 \\ \frac{w_6 w_7 v^*}{(u^* + w_7)^2} & -ev^* + \frac{r^* v^* (1 + w_9 r^*)}{\left[v^* + (w_8 + w_9 v^*)\, r^* + w_{10}\right]^2} & -\frac{v^* (v^* + w_{10})}{\left[v^* + (w_8 + w_9 v^*)\, r^* + w_{10}\right]^2} \\ 0 & \frac{r^* w_{12} (w_8 r^* + w_{10})}{\left[v^* + (w_8 + w_9 v^*)\, r^* + w_{10}\right]^2} & \frac{-w_{12} v^* r^* (w_8 + w_9 v^*)}{\left[v^* + (w_8 + w_9 v^*)\, r^* + w_{10}\right]^2} \end{pmatrix} \tag{11}$$

Here, $\rho(\lambda)$ denotes the characteristic polynomial of $\mathbf{G_u}(\mathbf{u}^*)$ given by:

$$\rho(\lambda) = \lambda^3 + H_1\lambda^2 + H_2\lambda + H_3$$

where

$$H_1 = \frac{w_{12}v^*r^*\left(w_8 + w_9v^*\right)}{\left[v^* + \left(w_8 + w_9v^*\right)r^* + w_{10}\right]^2} + v^*\left(e - \frac{r^*v^*(1 + w_9r^*)}{\left[v^* + \left(w_8 + w_9v^*\right)r^* + w_{10}\right]^2}\right) + u^*\left(1 - \frac{v^*}{(u^* + w_4)^2}\right)$$

$$\begin{aligned}
H_2 =& u^*\left(1 - \frac{v^*}{(u + w_4)^2}\right)\left\{\frac{w_{12}v^*r^*(w_8 + w_9v^*)}{\left[v^* + \left(w_8 + w_9v^*\right)r^* + w_{10}\right]^2} + v^*\left(e - \frac{r^*v^*(1 + w_9r^*)}{\left[v^* + \left(w_8 + w_9v^*\right)r^* + w_{10}\right]^2}\right)\right\} \\
&+ \frac{w_6w_7u^*v^*}{(u^* + w_4)(u^* + w_7)^2} + \frac{v^*r^*w_{12}(v^* + w_{10})(w_8r^* + w_{10})}{\left[v^* + \left(w_8 + w_9v^*\right)r^* + w_{10}\right]^2} + \frac{w_{12}v^{*2}r^*(w_8 + w_9)}{\left[v^* + \left(w_8 + w_9v^*\right)r^* + w_{10}\right]^2} \\
&\times\left(e - \frac{r^*v^*(1 + w_9r^*)}{\left[v^* + \left(w_8 + w_9v^*\right)r^* + w_{10}\right]^2}\right)
\end{aligned}$$

$$\begin{aligned}
H_3 =& \frac{w_6w_7u^*v^{*2}r^*(w_8 + w_9r^*)}{(u + w_4)(u + w_7)^2\left[v^* + \left(w_8 + w_9v^*\right)r^* + w_{10}\right]^2} + \left\{\frac{v^*r^*w_{12}(v^* + w_{10})(w_8r^* + w_{10})}{\left[v^* + \left(w_8 + w_9v^*\right)r^* + w_{10}\right]^2}\right. \\
&\left. + \left(\frac{w_{12}v^{*2}r^*(w_8 + w_9)}{\left[v^* + \left(w_8 + w_9v^*\right)r^* + w_{10}\right]^2}\right)\left(e - \frac{r^*v^*(1 + w_9r^*)}{\left[v^* + \left(w_8 + w_9v^*\right)r^* + w_{10}\right]^2}\right)\right\} \\
&\times\left\{u^*\left(1 - \frac{v^*}{(u^* + w_4)^2}\right)\right\}
\end{aligned}$$

By using the criteria stated in Theorem 1, it is easy to verify that $H_1, H_2, H_3 > 0$ and $H_1H_2 - H_3 > 0$. Therefore, from the Routh-Hurwitz criteria that the three roots $\rho(\lambda) = 0$ have negative real parts, the positive equilibrium solution $\mathbf{u}^*$ of the model system in Equation (6) is locally asymptotically stable under the condition in Equation (9). □

4. Analysis of Spatially Extended Model

In this section, we focus on the spatiotemporal models in Equation (7) in presence of diffusion but absence of cross diffusion and Equation (8) in presence of both self and cross diffusion. Our objective is to derive the conditions under which the positive equilibrium solution is stable with diffusion but without cross diffusion but unstable with diffusion and cross diffusion both. This phenomenon is called *cross diffusion driven instability* [16,18].

4.1. Without Cross Diffusion

We now consider the system in Equation (7). To discuss the local stability analysis of the system of parabolic Equations (7) and (8), we use the notation used in [16].

Notation 1. *Let $0 = \mu_1 < \mu_2 < \ldots$ be the eigenvalues of $-\Delta$ on Ω under no-flux boundary conditions, and $E(\mu_i)$ be the space of eigenfunctions corresponding to μ_i. We define the following space decomposition*

(i) $\mathbf{X_{ij}} := \left\{\mathbf{c}\ .\ \phi_{ij} : \mathbf{c} \in \mathbf{R^3}\right\}$ *where $\{\phi_{ij}\}$ are orthonormal basis of $E(\mu_i)$ for $j = 1, \ldots, dimE(\mu_i)$.*

(ii) $\mathbf{X} := \left\{\mathbf{u} \in \left[C^1(\bar{\Omega})\right]^3 : \partial_n u = \partial_n v = \partial_n r = 0 \text{ on } \partial\Omega\right\}$, *and so* $\mathbf{X} = \bigoplus_{i=1}^{\infty}\mathbf{X_i}$, *where* $\mathbf{X_i} = \bigoplus_{j=1}^{dimE(\mu_i)}\mathbf{X_{ij}}$.

Theorem 2. *The positive equilibrium solution* $\mathbf{u}^*$ *of the model system in Equation (7) is locally asymptotically stable if*

$$\frac{r^*v^*(1+w_9r^*)}{\left[v^*+(w_8+w_9v^*)\,r^*+w_{10}\right]^2}<e \quad and \quad \frac{v^*}{(u^*+w_4)^2}<1. \tag{12}$$

Proof. The spatially extended model system without cross diffusion in Equation (7) has been linearized at $\mathbf{u}^*$ and expressed as [16]:

$$\mathbf{u_t}=(D\Delta+\mathbf{G_u}(\mathbf{u}^*))\,\mathbf{u}, \tag{13}$$

where,

$$D=\begin{bmatrix} d_1 & 0 & 0\\ 0 & d_2 & 0\\ 0 & 0 & d_3\end{bmatrix}$$

Notation 1 suggests that $\mathbf{X_i}$ is invariant under the operator $D\Delta+\mathbf{G_u}(\mathbf{u}^*)$, and λ is an eigenvalue of this operator on $\mathbf{X_i}$ if and only if it is an eigenvalue of the matrix in $D\Delta+\mathbf{G_u}(\mathbf{u}^*)$. The characteristic polynomial of the matrix $\mu_i D+\mathbf{G_u}(\mathbf{u}^*)$ is given by

$$\psi_\lambda=\lambda^3+A_1\lambda^2+A_2\lambda+A_3$$

where

$$\begin{aligned}
A_1 =&\mu_i\,(d_1+d_2+d_3)+\frac{w_{12}v^*r^*\,(w_8+w_9v^*)}{\left[v^*+(w_8+w_9v^*)\,r^*+w_{10}\right]^2}+v^*\left(e-\frac{r^*v^*(1+w_9r^*)}{\left[v^*+(w_8+w_9v^*)\,r^*+w_{10}\right]^2}\right)\\
&+u^*\left(1-\frac{v^*}{(u^*+w_4)^2}\right)
\end{aligned}$$

$$\begin{aligned}
A_2 =&\mu_i d_1\left\{\mu_i(d_2+d_3)+\frac{w_{12}v^*r^*(w_8+w_9v^*)}{\left[v^*+(w_8+w_9v^*)\,r^*+w_{10}\right]^2}+v^*\left(e-\frac{r^*v^*(1+w_9r^*)}{\left[v^*+(w_8+w_9v^*)\,r^*+w_{10}\right]^2}\right)\right\}\\
&+u^*\left(1-\frac{v^*}{(u^*+w_4)^2}\right)\left\{\mu_i(d_2+d_3)+\frac{w_{12}v^*r^*(w_8+w_9v^*)}{\left[v^*+(w_8+w_9v^*)\,r^*+w_{10}\right]^2}\right.\\
&\left.+v^*\left(e-\frac{r^*v^*(1+w_9r^*)}{\left[v^*+(w_8+w_9v^*)\,r^*+w_{10}\right]^2}\right)\right\}+\frac{w_6w_7u^*v^*}{(u^*+w_4)(u^*+w_7)^2}+\mu_i^2d_2d_3\\
&+\frac{\mu_i d_2w_{12}v^*r^*(w_8+w_9v^*)}{\left[v^*+(w_8+w_9v^*)\,r^*+w_{10}\right]^2}+\frac{v^*r^*w_{12}(v^*+w_{10})(w_8r^*+w_{10})}{\left[v^*+(w_8+w_9v^*)\,r^*+w_{10}\right]^2}\\
&+\left(\mu_i d_3v^*+\frac{w_{12}v^{*2}r^*(w_8+w_9)}{\left[v^*+(w_8+w_9v^*)\,r^*+w_{10}\right]^2}\right)\left(e-\frac{r^*v^*(1+w_9r^*)}{\left[v^*+(w_8+w_9v^*)\,r^*+w_{10}\right]^2}\right)
\end{aligned}$$

$$\begin{aligned}
A_3 =&\mu_i d_3\frac{w_6w_7u^*v^*}{(u^*+w_4)(u^*+w_7)^2}+\frac{w_6w_7u^*v^{*2}r^*(w_8+w_9r^*)}{(u^*+w_4)(u+w_7)^2\left[v^*+(w_8+w_9v^*)\,r^*+w_{10}\right]^2}\\
&+\left\{\mu_i^2d_2d_3+\frac{\mu_i d_2w_{12}v^*r^*(w_8+w_9v^*)}{\left[v^*+(w_8+w_9v^*)\,r^*+w_{10}\right]^2}+\frac{v^*r^*w_{12}(v^*+w_{10})(w_8r^*+w_{10})}{\left[v^*+(w_8+w_9v^*)\,r^*+w_{10}\right]^2}\right.\\
&\left.+\left(\mu_i d_3v^*+\frac{w_{12}v^{*2}r^*(w_8+w_9)}{\left[v^*+(w_8+w_9v^*)\,r^*+w_{10}\right]^2}\right)\left(e-\frac{r^*v^*(1+w_9r^*)}{\left[v^*+(w_8+w_9v^*)\,r^*+w_{10}\right]^2}\right)\right\}\\
&\times\left\{\mu_i d_3+u^*\left(1-\frac{v^*}{(u^*+w_4)^2}\right)\right\}
\end{aligned}$$

□

By using the stated criteria in Theorem 2, it seems clear that $A_1, A_2, A_3 > 0$ and $A_1A_2 - A_3 > 0$. Therefore, it follows from the Routh-Hurwitz criteria that, for each $i \geq 1$, all three roots $\lambda_{i,1}, \lambda_{i,2}, \lambda_{i,3}$ of $\psi_i(\lambda) = 0$ have negative real parts. Hence, the positive equilibrium solution $\mathbf{u}^*$ of Equation (7) is locally asymptotically stable under the stated condition in Equation (12).

From Theorem 2, it is clear that addition of self diffusion terms to the temporal model system in Equation (6) results in stable positive equilibrium solution under the stated condition. Therefore, diffusion driven instability has not yet occurred. Now, we now consider the effect of cross diffusion on the model system.

4.2. With Cross Diffusion

In this section, we study in detail the system given in Equation (8).

Theorem 3. *Provided that Theorem 2 holds, assuming $d_4 > 0$ and considering the following inequality holds,*

$$\begin{aligned}&\left\{e - \frac{r^*(1+w_9r^*)}{[v^* + (w_8 + w_9v^*)\,r^* + w_{10}]^2}\right\}\left\{1 - \frac{v^*}{(u^*+w_4)^2}\right\} + \frac{w_6w_7}{(u^*+w_4)(u^*+w_7)^2} \\ &< \frac{d_4r^*(r^*+w_{10})}{(1+d_4v^*)\,[v^* + (w_8 + w_9v^*)\,r^* + w_{10}]^2}\left(1 - \frac{v^*}{(u^*+w_4)^2}\right)\end{aligned} \tag{14}$$

if $\mu_2 < \tilde{\mu}$, where μ_2 is as explained in Notation 1 and $\tilde{\mu}$ is given by Equation (24), then there exists $d_3^* > 0$, such that, when $d_3 \geq d_3^*$, the positive equilibrium solution $\mathbf{u}^*$ of the cross diffusive system in Equation (8) becomes unstable.

We denote $\Phi(\mathbf{u}) = (d_1u, d_2v, d_3r(1+d_4v))^T$. Linearizing the system in Equation (8) at $\mathbf{u}^*$, we have

$$\mathbf{u_t} = (\Phi_u\Delta + \mathbf{G_u}(\mathbf{u}^*))\,\mathbf{u}, \tag{15}$$

where,

$$\Phi_u = \begin{bmatrix} d_1 & 0 & 0 \\ 0 & d_2 & 0 \\ 0 & d_3d_4r^* & d_3 + d_3d_4v^* \end{bmatrix}$$

The characteristic polynomial of $-\mu_i\Phi_u + \mathbf{G_u}(\mathbf{u}^*)$ is given by,

$$\psi_\lambda = \lambda^3 + B_1\lambda^2 + B_2\lambda + B_3 \tag{16}$$

where

$$\begin{aligned}B_1 =& \mu_i(d_1+d_2+d_3) + \mu_id_3d_4v^* + \frac{w_{12}v^*r^*(w_8+w_9v^*)}{[v^* + (w_8 + w_9v^*)\,r^* + w_{10}]^2} \\ &+ v^*\left(e - \frac{r^*v^*(1+w_9r^*)}{[v^* + (w_8 + w_9v^*)\,r^* + w_{10}]^2}\right) + u^*\left(1 - \frac{v^*}{(u^*+w_4)^2}\right)\end{aligned}$$

$$B_2 = \Bigg\{ \mu_i(d_2+d_3) + \mu_i d_3 d_4 v^* + \frac{w_{12}v^*r^*(w_8+w_9v^*)}{[v^*+(w_8+w_9v^*)\,r^*+w_{10}]^2} + v^*\left(e - \frac{r^*v^*(1+w_9r^*)}{[v^*+(w_8+w_9v^*)\,r^*+w_{10}]^2}\right)\Bigg\} \times \left\{\mu_i d_1 + u^*\left(1-\frac{v^*}{(u^*+w_4)^2}\right)\right\}$$
$$+ \mu_i^2 d_2 d_3(1+d_4v^*) + \mu_i d_3 e v^*(1+d_4v^*) + \frac{w_{12}ev^{*2}r^*(w_8+w_9v^*)}{[v^*+(w_8+w_9v^*)\,r^*+w_{10}]^2}$$
$$+ \frac{\mu_i d_2 w_{12} v^* r^*(w_8+w_9v^*)}{[v^*+(w_8+w_9v^*)\,r^*+w_{10}]^2} + \frac{w_{12}v^*r^*(v^*+w_{10})(w_8r^*+w_{10})}{[v^*+(w_8+w_9r^*)\,v^*+w_{10}]^4}$$
$$- \frac{\mu_i d_3 r^* v^*(1+w_9r^*)}{[v^*+(w_8+w_9v^*)\,r^*+w_{10}]^2}(1+d_4v^*) - \frac{\mu_i d_3 d_4 r^* v^*(v^*+w_{10})}{[v^*+(w_8+w_9v^*)\,r^*+w_{10}]^2}$$
$$- \frac{w_{12}v^{*2}r^{*2}(1+w_9r^*)(w_8+w_9v^*)}{[v^*+(w_8+w_9r^*)\,v^*+w_{10}]^4}$$

$$B_3 = \frac{\mu_i d_3 w_6 w_7 u^* v^*}{(u^*+w_4)(u^*+w_7)^2}(1+d_4v^*) + \frac{w_6 w_7 u^* v^{*2} r^*(w_8+w_9v^*)}{(u^*+w_4)(u^*+w_7)^2\,[v^*+(w_8+w_9v^*)\,r^*+w_{10}]^2}$$
$$\left\{\mu_i d_1 + u^*\left(1-\frac{v^*}{(u^*+w_4)^2}\right)\right\} \times \Bigg\{ \mu_i^2 d_2 d_3(1+d_4v^*) + \mu_i d_3 e v^*(1+d_4v^*)$$
$$+ \frac{w_{12}ev^{*2}r^*(w_8+w_9v^*)}{[v^*+(w_8+w_9v^*)\,r^*+w_{10}]^2} + \frac{\mu_i d_2 w_{12} v^* r^*(w_8+w_9v^*)}{[v^*+(w_8+w_9v^*)\,r^*+w_{10}]^2}$$
$$+ \frac{w_{12}v^*r^*(v^*+w_{10})(w_8r^*+w_{10})}{[v^*+(w_8+w_9r^*)\,v^*+w_{10}]^4} - \frac{\mu_i d_3 d_4 r^* v^*(v^*+w_{10})}{[v^*+(w_8+w_9v^*)\,r^*+w_{10}]^2}$$
$$- \frac{\mu_i d_3 r^* v^*(1+w_9r^*)}{[v^*+(w_8+w_9v^*)\,r^*+w_{10}]^2}(1+d_4v^*) - \frac{w_{12}v^{*2}r^{*2}(1+w_9r^*)(w_8+w_9v^*)}{[v^*+(w_8+w_9r^*)\,v^*+w_{10}]^4} \Bigg\}$$

If $\lambda_1(\mu_i), \lambda_2(\mu_i), \lambda_3(\mu_i)$ are the three roots of $\psi(\lambda)=0$, then

$$\lambda_1(\mu_i).\lambda_2(\mu_i).\lambda_3(\mu_i) = -B_3. \tag{17}$$

For at least one $\mathbf{Re}(\lambda_i(\mu_i)) > 0$, it suffices to show that $B_3 < 0$. In addition,

$$B_3 = \det(\mu_i\phi_i - \mathbf{G_u}(\mathbf{u}^*)) \tag{18}$$

Hence, we have,

$$B_3 = b_3\mu_i^3 + b_2\mu_i^2 + b_1\mu_i + b_0 \tag{19}$$

where

$b_3 = d_1d_2d_3(1+d_3v^*)$

$$b_2 = d_1d_3(1+d_4v^*)v^*\left(e - \frac{r^*v^*(1+w_9r^*)}{[v^*+(w_8+w_9v^*)\,r^*+w_{10}]^2}\right) + \frac{d_1d_2w_{12}(w_8+w_9v^*)v^*r^*}{[v^*+(w_8+w_9v^*)\,r^*+w_{10}]^2}$$
$$- \frac{d_1d_3d_4r^*v^*(v^*+w_{10})}{[v^*+(w_8+w_9v^*)\,r^*+w_{10}]^2} + d_2d_3(1+d_4v^*)\left(u^* - \frac{u^*v^*}{(u+w_4)^2}\right)$$

$$
\begin{aligned}
b_1 =& d_1 v^* \left(e - \frac{r^* v^* (1 + w_9 r^*)}{\left[v^* + (w_8 + w_9 v^*)\, r^* + w_{10}\right]^2} \right) \frac{w - 12 v^* r^* (w_8 + w_9 v^*)}{\left[v^* + (w_8 + w_9 v^*)\, r^* + w_{10}\right]^2} \\
& \frac{d_1 w_{12} v^* r^* (v^* + w_{10})(w_8 r^* + w_{10})}{\left[v^* + (w_8 + w_9 r^*)\, v^* + w_{10}\right]^4} + \frac{d_2 w_{12} v^* r^* (w_8 + w_9 v^*)}{\left[v^* + (w_8 + w_9 v^*)\, r^* + w_{10}\right]^2} \left(u^* - \frac{u^* v^*}{(u + w_4)^2} \right) \\
& d_3 (1 + d_4 v^*) \left(e - \frac{r^* v^* (1 + w_9 r^*)}{\left[v^* + (w_8 + w_9 v^*)\, r^* + w_{10}\right]^2} \right) \left(u^* - \frac{u^* v^*}{(u + w_4)^2} \right) \\
& - \frac{d_3 d_4 r^* v^* (v^* + w_{10})}{\left[v^* + (w_8 + w_9 v^*)\, r^* + w_{10}\right]^2} \left(u^* - \frac{u^* v^*}{(u + w_4)^2} \right) + d_3 (1 + d_4 v^*) \frac{w_6 w_7 u^* v^*}{(u + w_4)(u + w_7)^2}
\end{aligned}
$$

$$
\begin{aligned}
b_0 =& \left\{ u^* - \frac{u^* v^*}{(u^* + w_4)^2} \right\} \left(e - \frac{r^* v^* (1 + w_9 r^*)}{\left[v^* + (w_8 + w_9 v^*)\, r^* + w_{10}\right]^2} \right) \frac{w_{12} v^* r^* (w_8 + w_9 v^*)}{\left[v^* + (w_8 + w_9 v^*)\, r^* + w_{10}\right]^2} \\
& + \left\{ u^* - \frac{u^* v^*}{(u^* + w_4)^2} \right\} \frac{v^* r^* w_{12} (v^* + w_{10})(w_8 r^* + w_{10})}{\left[v^* + (w_8 + w_9 r^*)\, v^* + w_{10}\right]^4} \\
& + \frac{w_6 w_7 w_{12} u^* v^{*2} r^* (w_8 + w_9 v^*)}{(u + w_4)(u + w_7)^2 \left[v^* + (w_8 + w_9 v^*)\, r^* + w_{10}\right]^2}
\end{aligned}
$$

From the above expression, we clearly see that $b_0 = -\det(\mathbf{G_u}(\mathbf{u}^*))$.

Let, $\tilde{b}(\mu) = b_3\mu^3 + b_2\mu^2 + b_1\mu - \det(\mathbf{G_u}(\mathbf{u}^*)$ and let $\tilde{\mu}_1, \tilde{\mu}_2, \tilde{\mu}_3$ be the three roots of $b(\tilde{\mu}) = 0$, such that $\mathbf{Re}(\tilde{\mu}_1) \le \mathbf{Re}(\tilde{\mu}_2) \le \mathbf{Re}(\tilde{\mu}_3)$. Now, $\tilde{\mu}_1\, \tilde{\mu}_2\, \tilde{\mu}_3 = \det(\mathbf{G_u}(\mathbf{u}^*))$. As $\det(\mathbf{G_u}(\mathbf{u}^*)) < 0$ due to condition specified in Equation (12), we have,

$$\tilde{\mu}_1 \tilde{\mu}_2 \widetilde{\mu}_3 < 0$$

As $d_1 > 0, d_2 > 0, d_3 > 0$ and $d_4 > 0$, $b_3 > 0$. Using the theory of equations if one of $\widetilde{\mu_1}, \widetilde{\mu_2}, \widetilde{\mu_3}$ is real and negative and the product of the other two is positive.

To obtain the values of d_3^*, we calculate the following limits:

$$\lim_{d_3 \to \infty} \frac{b_3}{d_3} = d_1 d_2 (1 + d_4 v^*) \cong \rho_3 \tag{20}$$

$$
\begin{aligned}
\lim_{d_3 \to \infty} \frac{b_2}{d_3} =& d_1 (1 + d_4 v^*) v^* \left(e - \frac{r^* v^* (1 + w_9 r^*)}{\left[v^* + (w_8 + w_9 v^*)\, r^* + w_{10}\right]^2} \right) - \frac{d_1 d_4 r^* v^* (v^* + w_{10})}{\left[v^* + (w_8 + w_9 v^*)\, r^* + w_{10}\right]^2} \\
& + d_2 (1 + d_4 v^*) \left(u^* - \frac{u^* v^*}{(u + w_4)^2} \right) \cong \rho_2
\end{aligned} \tag{21}
$$

$$
\begin{aligned}
\lim_{d_3 \to \infty} \frac{b_1}{d_3} =& (1 + d_4 v^*) v^* \left(e - \frac{r^* v^* (1 + w_9 r^*)}{\left[v^* + (w_8 + w_9 v^*)\, r^* + w_{10}\right]^2} \right) \left(u^* - \frac{u^* v^*}{(u + w_4)^2} \right) \\
& - \frac{d_4 r^* v^* (v^* + w_{10})}{\left[v^* + (w_8 + w_9 v^*)\, r^* + w_{10}\right]^2} \left(u^* - \frac{u^* v^*}{(u + w_4)^2} \right) + (1 + d_4 v^*) \left(\frac{w_6 w_7 u^* v^*}{(u + w_4)(u + w_7)^2} \right) \cong \rho_1
\end{aligned} \tag{22}
$$

If $\rho_1 < 0$, then the following inequality holds,

$$
\begin{aligned}
& \left\{ e - \frac{r^* (1 + w_9 r^*)}{\left[v^* + (w_8 + w_9 v^*)\, r^* + w_{10}\right]^2} \right\} \left\{ 1 - \frac{v^*}{(u^* + w_4)^2} \right\} + \frac{w_6 w_7}{(u^* + w_4)(u^* + w_7)^2} \\
< & \frac{d_4 r^* (r^* + w_{10})}{(1 + d_4 v^*) \left[v^* + (w_8 + w_9 v^*)\, r^* + w_{10}\right]^2} \left(1 - \frac{v^*}{(u^* + w_4)^2} \right)
\end{aligned}
$$

We have, $\rho_1 < 0 < \rho_3$.

In addition,

$$\lim_{d_3 \to \infty} \frac{\tilde{b}(\mu)}{d_3} = \rho_3\mu^3 + \rho_2\mu^2 + \rho_1\mu = \mu(\rho_3\mu^2 + \rho_2\mu + \rho_1)$$

As $\rho_1 < 0 < \rho_3$, which means that the equation,

$$\lim_{d_3 \to \infty} \frac{\tilde{b}(\mu)}{d_3} = 0$$

has one positive root and one negative root. From continuity, if $d_3 \to \infty$, $\widetilde{\mu}$ is real and negative. Since $\widetilde{\mu_2}\widetilde{\mu_3} > 0$, $\widetilde{\mu_2}$ and $\widetilde{\mu_3}$ are real and positive.

$$\lim_{d_2 \to \infty} \tilde{\mu}_1 = \frac{-\rho_2 - \sqrt{\rho_2^2 - 4\rho_1\rho_3}}{2\rho_3} < 0 \tag{23}$$

$$\lim_{d_2 \to \infty} \tilde{\mu}_3 = \frac{-\rho_2 + \sqrt{\rho_2^2 - 4\rho_1\rho_3}}{2\rho_3} > 0 = \widetilde{\mu} \tag{24}$$

$$\lim_{d_2 \to \infty} \tilde{\mu}_2 = 0 \tag{25}$$

Therefore, there exists $d_3^* > 0$ such that when $d_3 \geq d_3^*$ then the specified criteria holds. In addition, $\tilde{b}(\mu) < 0$ when $\mu \in (-\infty, \tilde{\mu}_1) \cup (\tilde{\mu}_2, \tilde{\mu}_3)$. Therefore, when $0 < \mu_2 < \tilde{\mu}$, then $\mu_2 \in (\tilde{\mu}_2, \tilde{\mu}_3)$, it follows that $\tilde{b}(\mu_2) < 0$. Therefore, $B_3 < 0$, and the proof is complete.

Therefore, from the above theorems, we conclude that cross diffusion destabilizes the stationary uniform solution.

5. Inhomogeneous Steady States

In this section, the justification for the cross diffusion driven instability phenomenon is explained. We now prove that model system in Equation (8) admits the inhomogeneous steady state. We, consider the steady state model system in Equation (8) in the following form,

$$\begin{aligned}
&- d_1\Delta u = u\left(1 - u - \frac{v}{u + w_4}\right) \\
&- d_2\Delta v = v\left(-w_5 + \frac{w_6 u}{u + w_7} - ev - \frac{r}{v + (w_8 + w_9 v)\, r + w_{10}}\right) \\
&- d_3\Delta\left(r + d_4 vr\right) = r\left(-w_{11} + \frac{w_{12} v}{v + (w_8 + w_9 v)\, r + w_{10}}\right) \\
&\frac{\partial u}{\partial n} = \frac{\partial v}{\partial n} = \frac{\partial r}{\partial n} = 0 \qquad x, y \in \Omega.
\end{aligned} \tag{26}$$

The constants $Q^*, \underline{Q}, \bar{Q}$ depend on the domain Ω. As Ω is fixed, this dependance is not mentioned explicitly. The parameters $w_4, w_5, w_6, w_7, e, w_8, w_9, w_{10}, w_{11}, w_{12}$ are collectively denoted by Γ.

5.1. A Priori Estimates

We now attempt to give an a priori upper and lower bounds for the positive solutions to Equation (26). We use Harnack Inequality and Maximum Principle (for details, refer to [16,18]).

Proposition 1. ***Harnack Inequality*** *Let $f \in C^2(\Omega) \cap C^1(\bar{\Omega})$ be a positive solution to $\Delta w(x) + c(x)w(x) = 0$, where $c \in C(\bar{\Omega})$, satisfying the homogeneous Neumann boundary condition. Then, there exists a positive constants C_* that depends only on $||c||_\infty$ such that*

$$\max_{\Omega} w \leq C_* \min_{\Omega} w$$

Proposition 2. ***Maximum principle*** *Let $g \in C(\Omega \times \mathbb{R}^1)$ and $b_{ij} \in C(\bar{\Omega}), j = 1, 2, \ldots, N$*

1. *If $f \in C^2(\Omega) \cap C^1(\bar{\Omega})$ satisfies*

$$\begin{cases} \Delta f(x) + \sum_{j=1}^{N} b_j(x) w_{xj} + g(x, w(x)) \geq 0, & x \in \Omega, \\ \partial_n w(x) \leq 0 & x \in \partial\Omega. \end{cases}$$

and $f(x_0) = \max_{\bar{\Omega}} f$, then $g(x_0, f(x_0)) \geq 0$.

2. *If $f \in C^2(\Omega) \cap C^1(\bar{\Omega})$ satisfies*

$$\begin{cases} \Delta f(x) + \sum_{j=1}^{N} b_j(x) f_{xj} + g(x, f(x)) \leq 0, & x \in \Omega, \\ \partial_n f(x) \geq 0 & x \in \partial\Omega. \end{cases}$$

and $f(x_0) = \min_{\bar{\Omega}} f$, then $g(x_0, f(x_0)) \leq 0$.

Theorem 4. ***(Upper Bound)*** *Let $\delta_1, \delta_2, \delta_3$ and δ_4 be fixed positive constants. Then, there exists positive constants $Q^*(\Gamma, \delta_1, \delta_2, \delta_3, \delta_4)$ and $\bar{Q}(\Gamma, \delta_1, \delta_2, \delta_3, \delta_4)$ such that when $d_1 \geq \delta_1$; $d_2 \geq \delta_2$; $d_3 \geq \delta_3$ and $d_4 \leq \delta_4$, the positive equilibrium solution $\mathbf{u} = (u, v, r)^T$ of Equation (26) satisfies,*

$$\begin{aligned} &\max_{\bar{\Omega}}(u, v, r) \leq \bar{Q}(\Gamma, \delta_1, \delta_2, \delta_3, \delta_4) \\ &\max_{\bar{\Omega}}(u, v, r) \leq Q^*(\Gamma, \delta_1, \delta_2, \delta_3, \delta_4) \min_{\bar{\Omega}}(u, v, r) \end{aligned} \tag{27}$$

Proof. We apply the maximum principle to the first part of Equation (26) and get $\max_{\bar{\Omega}} u \leq 1$. Similarly, applying the maximum principle to the second part of Equation (26) gives $\max_{\bar{\Omega}} v \leq \dfrac{w_6}{e(1 + w_7)}$.

Let $\varphi(x) = d_3(d_4 vr + r)$. Let $x_1 \in \bar{\Omega}$ be such that $\varphi(x_1) \max_{\bar{\Omega}} \varphi$. Now, applying the maximum principle to the third part of Equation (26) gives,$r(x_1) \leq \dfrac{(w_{12} - w_{11})w_6 - e(1 + w_7)w_{10}w_{11}}{e(1 + w_7)w_8 w_{11} + w_6 w_9 w_{11}}$.

Defining $\varphi(x)$ as $d_4 \leq \delta_4$, we have,

$$\begin{aligned} \max_{\bar{\Omega}} r \leq \frac{1}{d_3} \max_{\bar{\Omega}} \varphi(x_1) = \frac{1}{d_3} \varphi(x_1) &= r(x_1) + d_4 v(x_1) r(x_1) \\ &\leq r(x_1) + d_4 \max_{\bar{\Omega}} v(x_1) r(x_1) \\ &\leq r(x_1) \left(1 + d_4 \max_{\bar{\Omega}} v(x_1)\right) \\ \leq \left(1 + \delta_4 \frac{w_6}{e(1 + w_7)}\right) &\left(\frac{(w_{12} - w_{11})w_6 - e(1 + w_7)w_{10}w_{11}}{e(1 + w_7)w_8 w_{11} + w_6 w_9 w_{11}}\right) \end{aligned}$$

Let,

$$\overline{Q}(\Gamma, d) = \max\left\{1, \frac{w_6}{e(1 + w_7)}, \left(1 + \delta_4 \frac{w_6}{e(1 + w_7)}\right) \left(\frac{(w_{12} - w_{11})w_6 - e(1 + w_7)w_{10}w_{11}}{e(1 + w_7)w_8 w_{11} + w_6 w_9 w_{11}}\right)\right\}$$

Therefore, we have,

$$\max_{\bar{\Omega}}(u, v, r) \leq \overline{Q}(\Gamma, \delta_1, \delta_2, \delta_3, \delta_4)$$

□

Now, we show that,

$$\max_{\bar{\Omega}}(u, v, r) \leq Q^*(\Gamma, \delta_1, \delta_2, \delta_3, \delta_4) \min_{\bar{\Omega}}(u, v, r)$$

Using boundedness of u, v and r, we apply the Harnack inequality to the first and second parts of Equation (26), yielding

$$\max_{\bar{\Omega}}(u, v) \leq Q^*(\Gamma, \delta_1, \delta_2, \delta_3, \delta_4) \min_{\bar{\Omega}}(u, v)$$

Let $\vartheta(x) = d_3(d_4 vr + r)$ and we have,

$$\begin{cases} \Delta\vartheta + c(x)\vartheta = 0, & x \in \Omega, \\ \partial_n \vartheta = 0, & x \in \partial\Omega. \end{cases}$$

where $c(x) = \dfrac{(w_{12} - w_{11})v - (w_8 w_{11} + w_9 w_{11} v)r - w_{10} w_{11}}{d_3(1 + d_4 v)(v + (w_8 + w_9 v)r + w_{10})}$ is bounded.

By Harnack inequality,

$$\max_{\bar{\Omega}} \vartheta \leq Q_3^* \min_{\bar{\Omega}}$$

for some positive constant

$$Q_3^* = Q_3^*(\Gamma, \delta_1, \delta_2, \delta_3, \delta_4)$$

Lemma 1. *Let $d_{i,m}$ $i = 1,2,3,4$, be positive constants, $m = 1, 2, \ldots$, and $\mathbf{u_m} = (u_m, v_m, r_m)^T$ be the corresponding positive equilibrium solution of Equation (26) with $d_i = d_{i,m}$. If $\mathbf{u_m} \to \bar{u}$ as $m \to \infty$ and $\bar{\mathbf{u}}$ is a constant vector, then $\bar{\mathbf{u}} = \mathbf{u}^*$, where u^* is the nontrivial solution of $\mathbf{G}(\mathbf{u}) = 0$.*

Proof. For all m, $\int_\Omega u_m \; g_1(\mathbf{u_m})dx = 0$. If $g_1(\bar{\mathbf{u}}) > 0$, then $g_1(\mathbf{u_m}) > 0$ when m is large. However, since u_m is positive, this is not possible. Similarly, $g_1(\bar{\mathbf{u}}) < 0$ is not possible. Therefore, $g_1(\bar{\mathbf{u}}) = 0$. The same argument shows that $g_2(\bar{\mathbf{u}}) = g_3(\bar{\mathbf{u}}) = 0$. □

Theorem 5. ***Lower Bound:*** *Let δ_1, δ_2, δ_3 and δ_4 be fixed positive constants. There exists positive constants $\underline{Q}(\Gamma, \delta_1, \delta_2, \delta_3, \delta_4)$ such that, when $d_1 \geq \delta_1$, $d_2 \geq \delta_2$, $d_3 \geq \delta_3$ and $d_4 \leq \delta_4$, the positive solution $\mathbf{u} = (u, v, r)^T$ of Equation (26) satisfies,*

$$\min_{\Omega}(u, v, r) \geq \underline{Q}(\Gamma, \delta_1, \delta_2, \delta_3, \delta_4) \tag{28}$$

Proof. We integrate the second part of Equation (26) in Ω and consider the inhomogeneous Neumann boundary condition, yielding

$$\int_\Omega v\left(-w_5 + \frac{w_6 u}{u + w_7} - ev - \frac{r}{v + (w_8 + w_9 v)r + w_{10}}\right) = 0$$

therefore, there exists $x_0 \in \Omega$ such that $\dfrac{w_6 u(x_0)}{u(x_0) + w_7} = w_5 + ev(x_0) + \dfrac{r(x_0)}{v(x_0) + (w_8 + w_9 v(x_0))r(x_0) + w_{10}}$. □

As $u \leq 1$, $u(x_0) \geq \dfrac{w_5 w_7}{w_6}$. By using Harnack inequality,

$$\min_{\bar{\Omega}} u(x) \geq \frac{1}{Q^*} \frac{w_5 w_7}{w_6}$$

Now, integrating third part of Equation (26) in Ω

$$\min_{\bar{\Omega}} r(x) \geq \frac{1}{Q^*} w_{10} w_{12}.$$

With reference to Equation (27), to prove Theorem 5, it is sufficient to show that,

$$\max_{\bar{\Omega}} v \geq \underline{Q_1}(\Gamma, \delta_1, \delta_2, \delta_3, \delta_4). \tag{29}$$

Let the inequality in Equation (29) not hold. Then, there exists sequences $\{d_{1,m}, d_{2,m}, d_{3,m}, d_{4,m}\}_{m=1}^{\infty}$ with $d_{1,m} \geq \delta_1, d_{2,m} \geq \delta_2, d_{3,m} \geq \delta_3, d_{4,m} \leq \delta_4$ and the corresponding positive solutions $\mathbf{u_m}$ to Equation (26) such that $\max_{\bar{\Omega}} v_m \to 0$, where, $\mathbf{u_m} = (u_m, v_m, r_m)^T$ satisfies

$$\begin{cases} -d_{1,m}\Delta u_m = u_m\left(1 - u_m - \dfrac{v_m}{u_m + w_4}\right) \\ -d_{2,m}\Delta v_m = v_m\left(-w_5 + \dfrac{w_6 u}{u_m + w_7} - e v_m - \dfrac{r_m}{v_m + (w_8 + w_9 v_m)\, r_m + w_{10}}\right) \\ -d_{3,m}\Delta\left(r_m + d_4 v_m r_m\right) = r_m\left(-w_{11} + \dfrac{w_{12} v_m}{v_m + (w_8 + w_9 v_m)\, r_m + w_{10}}\right) \\ \dfrac{\partial u_m}{\partial n} = \dfrac{\partial v_m}{\partial n} = \dfrac{\partial r_m}{\partial n} = 0. \end{cases} \tag{30}$$

Since $\max_{\bar{\Omega}}(u, r) > 0$, we may assume that $\max_{\bar{\Omega}} u_m \to \bar{u}$ and $\max_{\bar{\Omega}} r_m \to \bar{r}$, where $\bar{u}$ and $\bar{r}$ are positive constants. In addition, we claim that u_m and r_m converge uniformly to positive constants, respectively. In fact, there are two cases of $\{d_{3,m}\}_{m=1}^{\infty}$ to be considered.

Case 1 $\{d_{3,m}\}_{m=1}^{\infty}$ is bounded with respect to m. Set,

$$\eta_m = d_{3,m}(r_m + d_{4,m} v_m r_m).$$

Using upper bound of $\mathbf{u_m}$, we have $\|\eta_m\| \leq C$ for all $m \geq 1$. Since η_m satisfies

$$\Delta\eta_m + \frac{(w_{12} - w_{11})v_m - (w_8 w_{11} + w_9 w_{11} v_m) r_m - w_{10} w_{11}}{d_3(1 + d_4 v_m)(v_m + (w_8 + w_9 v_m) r_m + w_{10})} = 0$$
$$\partial_n \eta_m = 0$$

using L^p estimate and the Sobolev embedding theorems, we have

$$\|\eta_m\|_{Q^{1,\alpha}(\bar{\Omega})} \leq Q\|\eta_m\|_{W_{2,p}(\bar{\Omega})} \leq Q.$$

Similarly, the $C^{2,\alpha}(\bar{\Omega})$ norms of η_m is uniformly bounded with respect to m. Thus, we assume that $\eta_m \to \eta$ in $C^2(\bar{\Omega})$. By the definition of η_m, for sufficiently large m we have,

$$r_m = \frac{\eta_m}{d_3(1 + d_{4,m} v_m)}$$

Since $v_m \to 0$ and $d_{4,m} \leq \delta_4$, we have

$$r_m \to r \equiv \frac{\eta}{d_3}.$$

Hence, η satisfies

$$\begin{cases} d_3^2 \Delta\eta + \eta\left(c d_3 - \dfrac{\eta}{w_{10}}\right) = 0 \\ \partial_n \eta = 0 \end{cases}$$

Hence, $\eta \equiv c w_{10} d_3$; otherwise, $\eta \equiv 0$ on $\bar{\Omega}$ that implies $r = \bar{r} = 0$, which is a contradiction to $\bar{r} > 0$, therefore $r \equiv \dfrac{\eta}{d_3} = c w_{10}$.

Case 2 $\{d_{3,m}\}_{m=1}^{\infty}$ is unbounded with respect to m. We may assume that $d_{3,m} \to \infty$. Set

$$\psi_m = r_m + d_4 v_m r_m.$$

Therefore, proceeding as above

$$-\Delta\psi = 0 \quad \text{in} \quad \Omega \qquad\qquad \partial_n\psi = 0 \quad \text{on} \quad \partial\Omega.$$

Hence, $\psi = r \equiv$ constant > 0. Similarly, we can prove that $u =$ constant ≥ 0. The above argument shows that there are positive constants $c_1, c_3 \geq 0$, such that $(u_m, v_m, r_m) \to (c_1, 0, c_3)$. This is contradiction to Lemma 5.1, thus the proof is completed.

5.2. Existence of Inhomogeneous Positive Steady States

We, now discuss the existence of inhomogeneous positive solutions. Let $\mathbf{X}$ be defined as in Notation 1. Define

$$\mathbf{X}^+ = \{\mathbf{u} \in \mathbf{X} : u > 0, v > 0, r > 0 \quad \text{on} \quad \bar{\Omega}\}$$
$$B(Q) = \{\mathbf{u} \in \mathbf{X} : Q^{-1} < u, v, r < Q \quad \text{on} \quad \bar{\Omega}\}, \quad C > 0.$$

Let $\Phi_u = (d_1 u, d_2 v, d_3 r(1 + d_4 v))^T$. Then, Equation (26) is transformed as follows,

$$\begin{cases} -\Delta\Phi(\mathbf{u}) = G(\mathbf{u}), & x \in \Omega, \\ \partial_n \mathbf{u} = 0, & x \in \Omega. \end{cases} \tag{31}$$

As the determinant of $\Phi_(\mathbf{u})$ is positive for all non-negative $\mathbf{u}$, $\Phi^{-1}(\mathbf{u})$ exists and $\det\Phi_{\mathbf{u}}^{-1}$ is positive. Hence, $\mathbf{u}$ is a positive solution to Equation (31) if and only if

$$F(\mathbf{u}) \cong \mathbf{u} - (I - \Delta)^{-1}\{\Phi_u^{-1}[\mathbf{G}(\mathbf{u}) + \nabla\mathbf{u}.\nabla\mathbf{u}] + \mathbf{u}\} = 0 \quad \text{in} \quad \mathbf{X}^+$$

where $(I - \Delta)^{-1}$ is the inverse of $(I - \Delta)$ in $\mathbf{X}$, where $F(.)$ is a compact perturbation of the identity operator, for any $B = B(Q)$. the Leray-Schauder degree $\deg(F(.), 0, B)$ is well defined if $F(\mathbf{u}) \neq 0$ on ∂B.

Furthermore, we note that

$$D_u F(\mathbf{u}^*) = I - (I - \Delta)^{-1}\left\{\Phi_u^{-1}(\mathbf{u}^*)\mathbf{G}_{\mathbf{u}}(\mathbf{u}^*) + I\right\}$$

and as $D_u F(\mathbf{u}^*)$ is invertible, the index of F at $\mathbf{u}^*$ is defined as index$(F(.), \mathbf{u}^*) = (-1)^\gamma$, where γ is the number of negative eigenvalues of $D_u F(\mathbf{u}^*)$.

For every integer $i \geq 1$ and for each integer $1 \leq j \leq \dim E(\mu_i)$, $\mathbf{X_{ij}}$ is invariant under $D_u F(\mathbf{u}^*)$, and λ is an eigenvalue of $D_u F(\mathbf{u}^*)$ on $\mathbf{X_{ij}}$ if and only if it is an eigenvalue of the matrix,

$$I - \frac{1}{1 + \mu_i}\left[\Phi_u^{-1}\mathbf{u}^*\mathbf{G}_{\mathbf{u}}(\mathbf{u}^*)\right] = \frac{1}{1 + \mu_i}\left[\mu_i I - \Phi_u^{-1}\mathbf{u}^*\mathbf{G}_{\mathbf{u}}(\mathbf{u}^*)\right]$$

Thus, $D_u F(\mathbf{u}^*)$ is invertible if and only if, for all $i \geq 1$, the matrix $I - \dfrac{1}{1 + \mu_i}\left[\Phi_u^{-1}\mathbf{u}^*\mathbf{G}_{\mathbf{u}}(\mathbf{u}^*)\right]$ is non-singular. We have,

$$H(\mu) = H(\mathbf{u}^*; \mu) \cong \det\left\{\mu I - \Phi_u^{-1}\mathbf{u}^*\mathbf{G}_{\mathbf{u}}(\mathbf{u}^*)\right\},$$

Further, if $H(\mu_i) \neq 0$, then for each $1 \leq j \leq \dim E(\mu_i)$, the number of negative eigenvalues of $D_u F(\mathbf{u}^*)$ on $\mathbf{X_{ij}}$ is odd if and only if $H(\mu_i) < 0$. We have results as follows.

Proposition 3. *Suppose that for all, $i \geq 1$, the matrix $\mu_i I - \Phi_i^{-1}(\mathbf{u}^*)G_u(\mathbf{u}^*)$ is nonsingular. Then,*

$$\textit{index}(F(.), \mathbf{u}^*) = (-1)^\tau, \quad \textit{where} \quad \tau = \sum_{\mathbf{i \geq 1, H(\bar{\ }_i) < 0}} \textit{dim}E(\mu_i).$$

To compute $index(F(.), \mathbf{u}^*)$, we consider the sign of $H(\mu_i)$. As we want to study the existence of positive solutions of Equation (31) with respect to d_2, we focus on dependence of $H(\mu_i)$ on d_2, and consider d_1, d_3 and d_4 to be fixed. We denote,

$$H(\mu) = \det\left\{\Phi_u^{-1}(\mathbf{u}^*)\right\}\det\left\{\mu\Phi_u(\mathbf{u}^*) - \mathbf{G_u}(\mathbf{u}^*)\right\}$$

As $\det\left\{\Phi_u^{-1}(\mathbf{u}^*)\right\} > 0$, we only need to consider $\det\left\{\mu\Phi_u(\mathbf{u}^*) - \mathbf{G_u}(\mathbf{u}^*)\right\}$. In fact $B_3 = \det\left\{\mu\Phi_u(\mathbf{u}^*) - \mathbf{G_u}(\mathbf{u}^*)\right\}$, and $B_3 < 0$ by Theorem 3. Therefore, we make the following proposition.

Proposition 4. *Assuming the condition specified in Theorem 3 holds, then there exists a positive number d_3^* such that, for all $d_3 > d_3^*$, the roots $\tilde{\mu}_1, \tilde{\mu}_2, \tilde{\mu}_3$ of $\det\left\{\mu\Phi_u(\mathbf{u}^*) - \mathbf{G_u}(\mathbf{u}^*)\right\} = 0$ are all real and satisfy Equations (23)–(25). Moreover, for all $d_3 \geq d_3^*$.*

$$\begin{cases} -\infty < \tilde{\mu}_1 < 0 < \tilde{\mu}_2 < \tilde{\mu}_3, \\ \det\left\{\mu\Phi_u(\mathbf{u}^*) - \mathbf{G_u}(\mathbf{u}^*)\right\} < 0, \quad \text{when} \quad \mu \in (\infty, \tilde{\mu}_1) \cup (\tilde{\mu}_2, \tilde{\mu}_3), \\ \det\left\{\mu\Phi_u(\mathbf{u}^*) - \mathbf{G_u}(\mathbf{u}^*)\right\} > 0, \quad \text{when} \quad \mu \in (\tilde{\mu}_1, \tilde{\mu}_2) \cup (\tilde{\mu}_3, \infty). \end{cases} \tag{32}$$

To discuss the effect of cross diffusion on the existence of inhomogeneous positive solution to Equation (31), we first deduce a non-existence result when the cross diffusion term is absent.

Theorem 6. *Suppose that*

$$d_1 \geq \frac{1}{\mu_2}, \qquad d_3 \geq \frac{w_{11}}{w_{12}\mu_2}$$

where μ_2 is given previously. Then, there exists a positive constants $\overline{D}_2$ when $d_2 \geq \overline{D}_2$ without cross diffusion term d_4, Equation (26) has no non-constant positive solution. Furthermore,

$$\begin{aligned} &(u, v, r) = (\bar{u}, \bar{v}, \bar{r}), \\ &\textit{where} \quad \bar{u} \cong \frac{1}{\textit{measure}(\Omega)}\int_\Omega u; \quad \bar{v} \cong \frac{1}{\textit{measure}(\Omega)}\int_\Omega v; \quad \bar{r} \cong \frac{1}{\textit{measure}(\Omega)}\int_\Omega r \end{aligned}$$

Proof. Assume $\mathbf{u} = (u, v, r)$ is a positive solution of Equation (26) with $d_4 = 0$.
Let

$$\bar{u} \cong \frac{1}{\text{measure}(\Omega)}\int_\Omega u; \quad \bar{v} \cong \frac{1}{\text{measure}(\Omega)}\int_\Omega v; \quad \bar{r} \cong \frac{1}{\text{measure}(\Omega)}\int_\Omega r$$

Multiplying the first part of Equation (26) by $(u - \bar{u})$, the second part by $(v - \bar{v})$ and the third part by $(r - \bar{r})$ (taking $d_4 = 0$) and integrating by parts over Ω, we have

$$\begin{aligned} d_1\int_\Omega |\nabla u|^2 &= \int_\Omega (u - \bar{u})(ug_1(u, v) - \bar{u}g_1(\bar{u}, \bar{v})) \\ &= \int_\Omega (u - \bar{u})^2\left(1 - u - \bar{u} - \frac{w_4 v}{(u + w_4)(\bar{u} + w_4)}\right) - \int_\Omega (u - \bar{u})(v - \bar{v})(w_4\bar{u} + u\bar{u}) \end{aligned}$$

$$\begin{aligned} d_2\int_\Omega |\nabla v|^2 &= \int_\Omega (v - \bar{v})(vg_2(u, v, r) - \bar{v}g_2(\bar{u}, \bar{v}, \bar{r})) \\ &= \int_\Omega (v - \bar{v})^2 \times \\ &\left[\frac{w_6 u\bar{u}}{(u + w_7)(\bar{u} + w_7)} - w_5 - e(v + \bar{v}) - \frac{w_8 r\bar{r}}{[v + (w_8 + w_9 r)v + w_{10}][\bar{v} + (w_8 + w_9\bar{r})\bar{v} + w_{10}]}\right] \\ &+ \int_\Omega \frac{w_6 w_7(vu - \bar{v}\bar{u})(v - \bar{v})}{(u + w_7)(\bar{u} + w_7)} - \int_\Omega \frac{w_{10}(\bar{v}\bar{r} - rv)(v - \bar{v})}{(u + w_7)(\bar{u} + w_7)} \\ &+ \int_\Omega \frac{v\bar{v}(v - \bar{v})(r - \bar{r})}{[v + (w_8 + w_9 r)v + w_{10}][\bar{v} + (w_8 + w_9\bar{r})\bar{v} + w_{10}]} \end{aligned}$$

$$
\begin{aligned}
d_3 \int_\Omega |\nabla r|^2 &= \int_\Omega (r-\bar{r})(rg_3(v,r) - \bar{r}g_3(\bar{v},\bar{r})) \\
&= \int_\Omega (r-\bar{r})^2 \left[\frac{w_{12} v \bar{v}}{[v + (w_8 + w_9 r)\, v + w_{10}]\, [\bar{v} + (w_8 + w_9 \bar{r})\, \bar{v} + w_{10}]} \right] \\
&+ \int_\Omega \frac{w_{12} v \bar{v} (r+\bar{r})}{[v + (w_8 + w_9 r)\, v + w_{10}]\, [\bar{v} + (w_8 + w_9 \bar{r})\, \bar{v} + w_{10}]} \\
&+ \int_\Omega \frac{w_{10} w_{12} (rv - \bar{r}\bar{v})(r-\bar{r})}{[v + (w_8 + w_9 r)\, v + w_{10}]\, [\bar{v} + (w_8 + w_9 \bar{r})\, \bar{v} + w_{10}]}
\end{aligned}
$$

By using Cauchy inequality, we have $d_1 \int_\Omega |\nabla u|^2 + d_2 \int_\Omega |\nabla v|^2 + d_3 \int_\Omega |\nabla r|^2$

$$
\begin{aligned}
&\leq \int_\Omega (u-\bar{u})^2 \left(1 - u - \bar{u} - \frac{w_4 v}{(u+w_4)(\bar{u}+w_4)} + \epsilon \right) \\
&+ \int_\Omega (v-\bar{v})^2 \times \left[\frac{w_6 u \bar{u}}{(u+w_7)(\bar{u}+w_7)} - w_5 - e(v+\bar{v}) - \frac{w_8 r \bar{r}}{[v + (w_8 + w_9 r)\, v + w_{10}]\, [\bar{v} + (w_8 + w_9 \bar{r})\, \bar{v} + w_{10}]} \right] \\
&+ \int_\Omega (r-\bar{r})^2 \left[\frac{w_{12} v \bar{v}}{[v + (w_8 + w_9 r)\, v + w_{10}]\, [\bar{v} + (w_8 + w_9 \bar{r})\, \bar{v} + w_{10}]} \right] + \frac{1}{4\epsilon} \int_\Omega \left[\frac{w_6 w_7 (vu - \bar{v}\bar{u})(v-\bar{v})}{(u+w_7)(\bar{u}+w_7)} \right. \\
&+ \frac{v\bar{v}(r-\bar{r})(v-\bar{v}) + w_{12} v \bar{v}(r+\bar{r}) + w_{10} w_{12} (rv - \bar{r}\bar{v})(r-\bar{r}) - w_{10}(\bar{r}\bar{v} - rv)(v-\bar{v})}{[v + (w_8 + w_9 r)\, v + w_{10}]\, [\bar{v} + (w_8 + w_9 \bar{r})\, \bar{v} + w_{10}]} \\
&\left. - (u-\bar{u})(v-\bar{v})(\bar{u} w_4 + u\bar{u}) \right],
\end{aligned}
$$

where ϵ is any arbitrary positive constant.

By Poincare inequality we have,

$d_1 \int_\Omega |\nabla u|^2 + d_2 \int_\Omega |\nabla v|^2 + d_3 \int_\Omega |\nabla r|^2$

$\geq \int_\Omega d_1 \mu_2 (u-\bar{u})^2 + \int_\Omega d_2 \mu_2 (v-\bar{v})^2 + \int_\Omega d_3 \mu_2 (r-\bar{r})^2$. By Theorem 5.1, we can choose a sufficiently small ϵ_0 such that,

$$
\begin{aligned}
d_1 \mu_2 &> 1 - u - \bar{u} - \frac{w_4 v}{(u+w_4)(\bar{u}+w_4)} + \epsilon_0 \\
d_3 \mu_2 &> \frac{w_{12} v \bar{v}}{[v + (w_8 + w_9 r)\, v + w_{10}]\, [\bar{v} + (w_8 + w_9 \bar{r})\, \bar{v} + w_{10}]} - w_{11} + \epsilon
\end{aligned}
$$

Lastly, by taking

$$
\begin{aligned}
\bar{D}_2 &> \frac{1}{\mu_2} \left[\frac{w_6 u \bar{u}}{(u+w_7)(\bar{u}+w_7)} - w_5 - e(v+\bar{v}) - \frac{w_8 r \bar{r}}{[v + (w_8 + w_9 r)\, v + w_{10}]\, [\bar{v} + (w_8 + w_9 \bar{r})\, \bar{v} + w_{10}]} \right] \\
&+ \frac{1}{4\epsilon} \left[\frac{w_6 w_7 (uv - \bar{u}\bar{v})(v-\bar{v})}{(u+w_7)(\bar{u}+w_7)} + \frac{v\bar{v}(r-\bar{r})(v-\bar{v})}{[v + (w_8 + w_9 r)\, v + w_{10}]\, [\bar{v} + (w_8 + w_9 \bar{r})\, \bar{v} + w_{10}]} \right. \\
&\left. + \frac{w_{12} v \bar{v}(r+\bar{r}) + w_{10}(rv - \bar{r}\bar{v})(w_{12} + (v-\bar{v}))}{[v + (w_8 + w_9 r)\, v + + w_{10}]\, [\bar{v} + (w_8 + w_9 \bar{r})\, \bar{v} + w_{10}]} + (v-\bar{v})(\bar{u}-u)(\bar{u} w_4 + u\bar{u}) \right]
\end{aligned}
$$

□

We conclude that $(u, v, r) = (\bar{u}, \bar{v}, \bar{r})$, which completes the proof. Therefore, we conclude that the model system has no non-constant positive solution. Now, we prove that in the presence of cross diffusion inhomogeneous solution is generated. The following theorem discusses the global existence of of inhomogeneous solution of Equation (8) with respect to d_3 as the other parameters d_1, d_2 and d_4 are fixed.

Theorem 7. *Let d_1, d_2, d_4 be fixed and satisfy conditions of Theorems 2 and 3. Let $\tilde{\mu}$ be defined in Equations (23)–(25). If $\tilde{\mu} \in (\mu_n, \mu_{n+1})$ for some $n \geq 1$, and the sum $\sigma_n = \sum_{i=2}^{n} dim E_{\mu_i}$ is odd. Then, there exist a positive number d_3^* such that, if $d_3 > d_3^*$, Equation (8) has at least one inhomogeneous positive steady state solution.*

Proof. Let $\delta_1, \delta_2, \delta_3$ and δ_4 be positive constants and satisfy $\delta_1 < d_1$, $\delta_2 < d_2$, $\delta_3 < d_3$ and $\delta_4 > d_4$. By Equations (23)–(25) and Proposition 4, there exists a positive constant d_3^* such that, when $\delta_3 > d_3^*$, Equation (32) holds and

$$0 = \mu_1 < \tilde{\mu}_2 < \mu_2, \qquad \tilde{\mu}_3 \in (\mu_n, \mu_{n+1}) \tag{33}$$

□

Now, we prove that for, $\delta_3 > d_3^*$, Equation (26) has at least one inhomogeneous positive solution. The statement has been proved by contradiction and is based on homotopy invariance of topological degree. On the contrary, we assume that the assertion is not true. For $t \in [0,1]$, define

$$\Phi(t;\mathbf{u}) = (\hat{d}u + t(d - 1 - \hat{d}_1)u, \hat{d}_2 v + t(d_2 - \hat{d}_2)v \times (\hat{d}_3 + t(\hat{d}_3 - d_3))(r + td_4 vr))^T$$

where, $\hat{d}_2 = \bar{D}_2, \hat{d}_1 = \dfrac{1}{\mu_2}, \hat{d}_3 = \dfrac{c}{\mu_2}$, and consider

$$\begin{cases} -\Delta\Phi(t,\mathbf{u}) = \mathbf{G}(\mathbf{u}) & \text{in}\Omega, \quad 0 \leq t \leq 1 \\ \partial_n \mathbf{u} = 0 & \text{on} \quad \partial\Omega \end{cases} \tag{34}$$

Then, $\mathbf{u}$ is a positive inhomogeneous solution of Equation (26) if and only if it is a solution of Equation (34) for $t = 1$. $\mathbf{u}^*$ is the unique constant solution of Equation (34) for any $0 \leq t \leq 1$. Now, we have for any $0 \leq t \leq 1$, $\mathbf{u}$ is a unique positive solution of Equation (34) if and only if,

$$F(t;\mathbf{u}) \cong \mathbf{u} - (I - \Delta)^{-1}\left\{\Phi_u^{-1}[\mathbf{G}(\mathbf{u}) + \nabla\mathbf{u}.\Phi_{uu}(t;\mathbf{u}).\nabla\mathbf{u}] + \mathbf{u}\right\} = 0 \quad \text{in} \quad \mathbf{X}^+$$

As $F(1;\mathbf{u}) = F(\mathbf{u})$, by Theorem 4.2, $\mathbf{F}(\mathbf{0};\mathbf{u}) = 0$ has only positive solution $\mathbf{u}^*$ in X^+, therefore we have,

$$\mathcal{D}_u F(0,\mathbf{u}^*) = I - (I - \Delta)^{-1}\left\{\mathcal{D}^{-1}\mathbf{G}_{\mathbf{u}}(\mathbf{u}^*) + 1\right\}$$

$$\mathcal{D}_u F(1,\mathbf{u}^*) = I - (I - \Delta)^{-1}\left\{\Phi_u^{-1}\mathbf{G}_{\mathbf{u}}(\mathbf{u}^*) + I\right\} = \mathcal{D}_u F(\mathbf{u}^*).$$

where, $\mathcal{D} = \text{diag}(\hat{d}_1, \hat{d}_2, \hat{d}_3)$. Using Proposition 4 and Equation (33), it follows,

$$\begin{cases} H(\mu_1) = H(0) > 0, & \\ H(\mu_i) < 0, & 2 \leq i \leq n, \\ H(\mu_i) > 0. & i \geq n+1. \end{cases}$$

Therefore, 0 is not an eigenvalue of the matrix $\mu_i I - \Phi_u^{-1}(\mathbf{u}^*)\mathbf{G}_{\mathbf{u}}(\mathbf{u}^*)$ for all $i \geq 1$, and

$$\sum_{i \geq 1, H(\mu_i) < 0} \text{dim}E(\mu_i) = \sum_{i=2}^{n} \text{dim}E(\mu_i) = \sigma_n, \qquad \text{which is odd.}$$

By Proposition 3,

$$\text{index}(F(1;.),\mathbf{u}^*) = (-1)^{\tau} = (-1)^{\sigma_n} = -1 \tag{35}$$

Similarly,

$$\text{index}(F(0;.),\mathbf{u}^*) = (-1)^0 = 1 \tag{36}$$

By Theorems 3 and 4, there exists a positive constants C such that, for all $0 \le t \le 1$, the positive solutions of Equation (34) satisfying $\frac{1}{C} \le u, v, r \le C$. Therefore, $F(t;\mathbf{u}) \ne 0$ on ∂B for all $0 \le t \le 1$. By the homotopy invariance of the topological degree,

$$\deg(F(1;.),0,B(C)) = \deg(F(0,.),0,B(C)). \tag{37}$$

However, by our deduction, both equations $F(1;\mathbf{u}) = 0$ and $F(0;\mathbf{u}) = 0$ have only the positive solution $\mathbf{u}^*$ in $B(C)$, and hence by Equations (35) and (36)

$$\deg(F(0;.),0,B(C)) = \text{index}(F(0;.),\mathbf{u}^*) = 1$$

$$\deg(F(1;.),0,B(C)) = \text{index}(F(1;.),\mathbf{u}^*) = -1$$

which contradicts Equation (37) and we complete the proof.

6. Numerical Simulation

The numerical simulation was carried out to understand the spatiotemporal dynamics of top predator r under the influence of cross diffusion. For this purpose, in this section, we present a detailed investigation of the patterns in the top level predator for different diffusivity coefficients. The system of partial differential equations given in Equation (8) was numerically solved using semi implicit finite difference technique. Forward difference scheme was used for the reaction terms and standard five point explicit finite difference scheme was used for diffusion term. Turing patterns were obtained from the effect of nonlinear diffusion term for the top level predator r as in [18–20].

To discretize the system, we considered Taylor's series expansion about the non-trivial equilibrium point (u^*, v^*, r^*) of the cross diffusive term of the third equation describing the dynamics of r for the model system in Equation (8). We obtained a system of the following form:

$$\begin{aligned}
&\frac{\partial u}{\partial t} - d_1 \Delta u = u\left(1 - u - \frac{v}{u + w_4}\right) \\
&\frac{\partial v}{\partial t} - d_2 \Delta v = v\left(-w_5 + \frac{w_6 u}{u + w_7} - ev - \frac{r}{v + (w_8 + w_9 v)\, r + w_{10}}\right) \\
&\frac{\partial r}{\partial t} - d_3 d_4 r^* \Delta v - d_3(1 + d_4 v^*)\Delta r = r\left(-w_{11} + \frac{w_{12} v}{v + (w_8 + w_9 v)\, r + w_{10}}\right) \\
&\frac{\partial u}{\partial n} = \frac{\partial v}{\partial n} = \frac{\partial r}{\partial n} = 0 \qquad (x,y) \in \partial\Omega.
\end{aligned} \tag{38}$$

The space step size and time step size were chosen appropriately to ensure the convergence of the scheme.

We used the standard five-point approximation for the two-dimensional Laplacian with the zero-flux boundary conditions. Initially, the entire system was at the stationary state (u^*, v^*, r^*). The perturbation introduced in the initial condition was of the order 5×10^{-4}, as given in Equation (39):

$$\begin{aligned}
u(x,y) &= u^* + \epsilon_1 \sin\left(\frac{2\pi(x - x_0)}{0.2}\right) + \epsilon_2 \sin\left(\frac{2\pi(y - y_0)}{0.2}\right) \\
v(x,y) &= v^* + \epsilon_1 \sin\left(\frac{2\pi(x - x_0)}{0.2}\right) + \epsilon_2 \sin\left(\frac{2\pi(y - y_0)}{0.2}\right) \\
r(x,y) &= r^* + \epsilon_1 \sin\left(\frac{2\pi(x - x_0)}{0.2}\right) + \epsilon_2 \sin\left(\frac{2\pi(y - y_0)}{0.2}\right)
\end{aligned} \tag{39}$$

where $\epsilon_1 = \epsilon_2 = 5 \times 10^{-4}$, $x_0 = y_0 = 0.1$.

The set of parameter values at which the system would yield a locally asymptotically stable solution is:

$$w_4 = 0.25, \quad w_5 = 0.25, \quad w_6 = 0.8, \quad w_7 = 0.25, \quad w_8 = 0.01,$$
$$w_9 = 0.1, \quad w_{10} = 0.28, \quad w_{11} = 0.25, \quad w_{12} = 0.78, \quad e = 2.$$

To analyze the role of cross diffusion on r, we considered the following set of parameters:

$$w4 = 17.25, \quad w5 = 17.25, \quad w6 = 17.25, \quad w7 = 17.25, \quad w8 = 0.25,$$
$$w9 = 0.25, \quad w10 = 0.28, \quad w11 = 18.26, \quad w12 = 3.05, \quad e = 22.$$

The parameter values of diffusivity coefficients are presented in Table 1. We performed the simulation on a 50×50 grid with spatial step size 0.5 and time step size 0.1. To investigate the role of cross diffusion and self diffusion in the pattern formation of the top predator r, we performed simulations for a wide range of self diffusive coefficient d_3 and cross diffusivity coefficient d_4. The different values of self and cross diffusive coefficients used in numerical experiments of top level predator r are presented in Table 1. We carried out all the numerical simulation at time level t = 10,000 for the model system given in Equation (38).

Table 1. Values of diffusivity coefficients d_1, d_2, d_3 and d_4 used in the simulations.

Figure	d_1	d_2	d_3	d_4
Figure 1	85	75	10	0
Figure 2a	85	75	1.7	2.3
Figure 2b	85	75	1	5
Figure 3a	85	75	0.8	6
Figure 3b	85	75	0.3	8
Figure 4a	85	75	0.5	10
Figure 4b	85	75	0.1	15
Figure 5a	85	75	0.05	85
Figure 5b	85	75	0.06	125
Figure 6a	85	75	0.5	0.5
Figure 6b	85	75	1	1
Figure 7a	85	75	1.5	1.5
Figure 7b	85	75	2	2

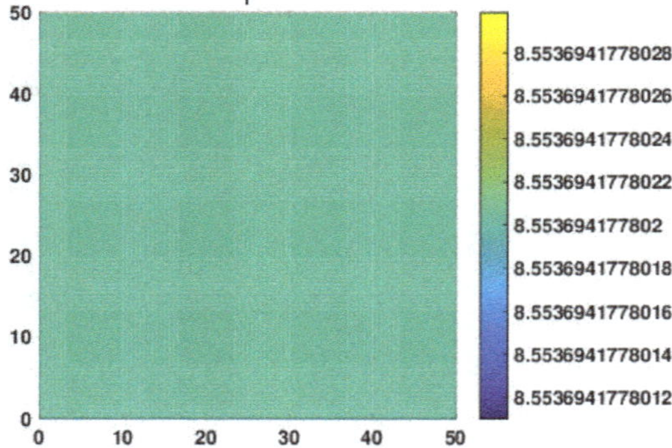

Figure 1. No Patterns for top predator of the model system in Equation (38) were obtained at time level t = 10,000 in the absence of cross diffusion, i.e., self diffusion and cross diffusion coefficients being $d_3 = 10$, $d_4 = 0$, respectively.

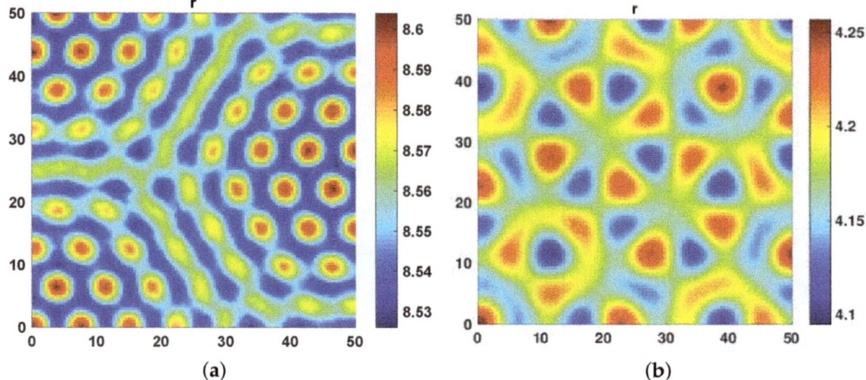

(a) (b)

Figure 2. (**a**) A mix of hot spot and labyrinth turing patterns obtained at $d_3 = 1.7$ and $d_4 = 2.3$. (**b**) Floral Turing patterns obtained at $d_3 = 1$ and $d_4 = 5$.

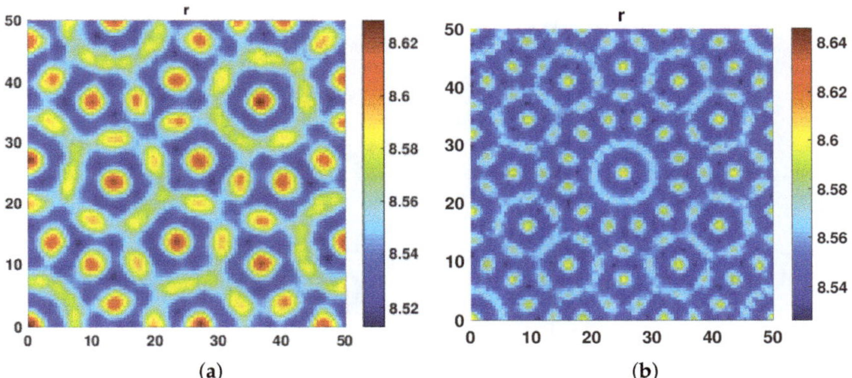

(a) (b)

Figure 3. (**a**) Pentagonal Turing patterns obtained at $d_3 = 0.8$ and $d_4 = 6$. (**b**) Floral Turing patterns obtained at $d_3 = 0.3$ and $d_4 = 8$.

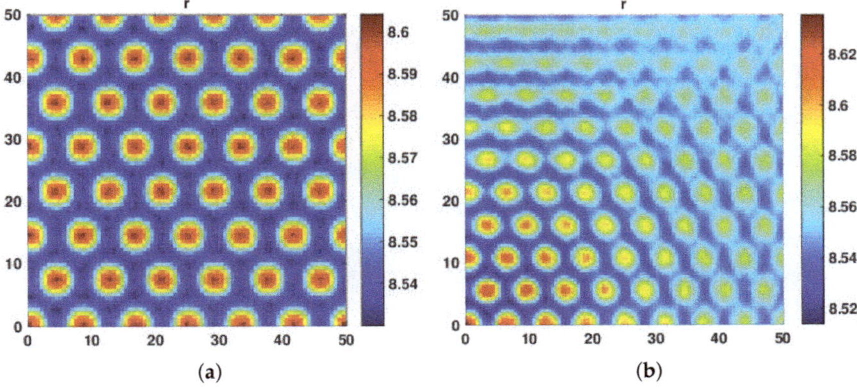

(a) (b)

Figure 4. (**a**) Another hot spot turing patterns obtained at $d_3 = 0.5$ and $d_4 = 10$. (**b**) A mix of Hot Spot and Labyrinth Turing patterns obtained at $d_3 = 0.1$ and $d_4 = 15$.

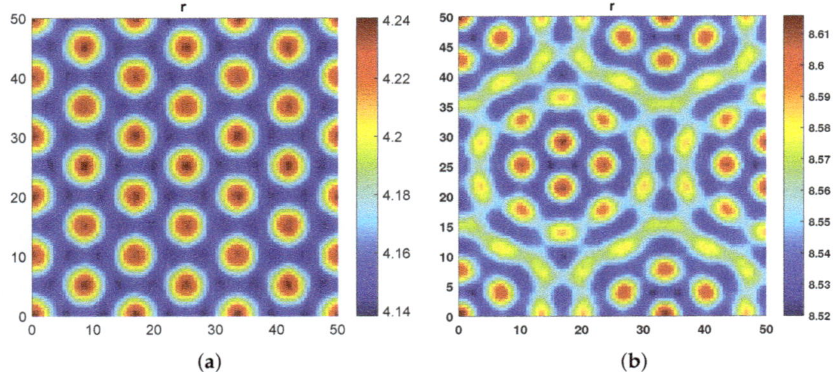

Figure 5. (**a**) Hot Spot patterns obtained at $d_3 = 0.05$ and $d_4 = 85$. (**b**) Hexagonal Spot Turing patterns obtained at $d_3 = 0.06$ and $d_4 = 125$.

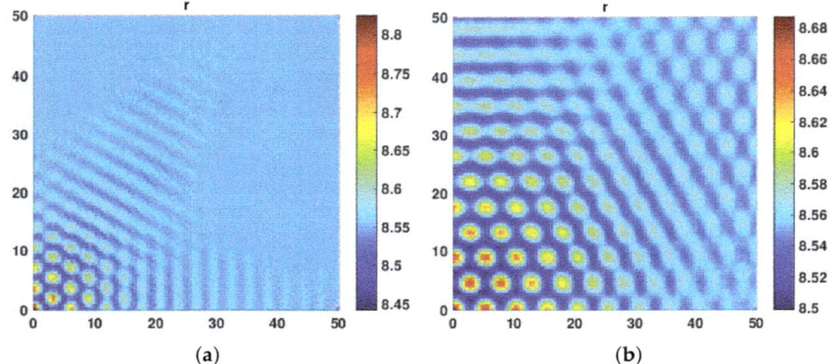

Figure 6. (**a**) A mixture of Hot Spot and Labyrinth Turing patterns obtained at $d_3 = d_4 = 0.5$. (**b**) A mix of Hot Spot and Labyrinth Turing patterns obtained at $d_3 = d_4 = 1$.

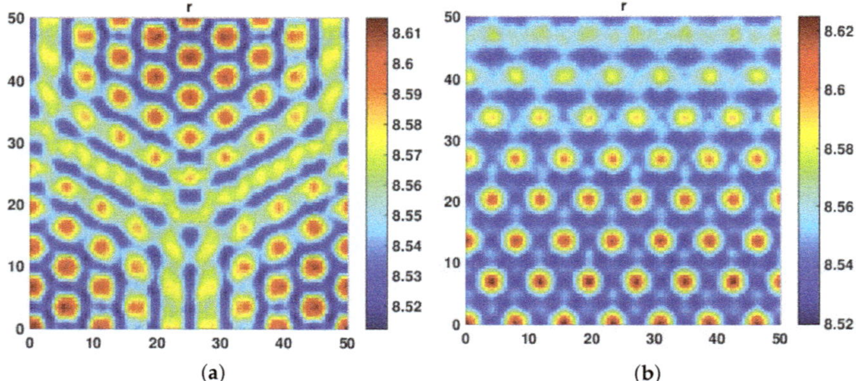

Figure 7. (**a**) A mixture of hot spot and labyrinth turing patterns obtained at $d_3 = d_4 = 1.5$. (**b**) A mixture of hot spot and labyrinth turing patterns obtained at $d_3 = d_4 = 2$.

Here, we performed numerical simulation to investigate the role of self and cross diffusion on pattern formation. In Figure 1, we notice that, in the absence of cross diffusion, i.e., $d_4 = 0$, no patterns were formed for the top level predator r, of the system in Equation (38). The behavior persists even for the higher values of self diffusivity coefficients d_3. This numerical experiment shows that in the absence of cross diffusion there is no destabilizing effect on the system even when there is self diffusion in the system.

In our second numerical experiment, we introduced the cross diffusion by gradually increasing the cross diffusivity coefficient d_4 and along with that we gradually decreased the self diffusivity coefficient d_3. The goal was to see the impact of cross diffusivity on the dynamics when there is very less role of self diffusion in the system. To this end, in Figure 2a, we increase d_4 from 0 to 2.3 and decrease d_3 from 10 to 1.7. An increase in cross diffusion coefficient results in destabilization in the top predator r dynamics leading to a mixture of hot spots and labyrinth patterns, as seen in Figure 2a. We further increased d_4 to 5 and decreased d_3 to 1 to observe floral turing patterns (Figure 2b).

On further increasing the cross diffusivity coefficient at $d_4 = 6$ and taking self diffusion coefficient at $d_3 = 0.8$, we observed pentagonal Turing patterns, as shown in Figure 3a. Further, at $d_4 = 8$ and $d_3 = 0.3$, we obtained Floral Turing Pattern.

Increasing the value of d_4 to 10 and fixing $d_3 = 0.5$, we observed hot spothl turing patterns, as given in Figure 4a at $d_4 = 15$. At $d_3 = 0.1$, a mix of hot spot and Labyrinth Turing patterns were observed, as shown in Figure 4b.

As shown in Figure 5a,b, we observed that, for a very high value of cross diffusion coefficients $d_4 = 85$ and $d_4 = 125$, and very low values of $d_3 = 0.05$ and $d_3 = 0.06$, hot spot and Hexagonal Spot Turing patterns are obtained, respectively.

From these experiments, we can conclude that the role of cross diffusion in destabilizing the dynamics is much more significant than self diffusion.

In our third numerical experiment, we started with equal values of both d_3 and d_4 fixed at 0.5. The hot spot patterns began to appear. Further, when we increased diffusivity coefficient d_3 and d_4 to 1, a mix of hot spot and Labyrinth Turing Patterns were seen. We further increased both d_3 and d_4 with the restriction that both self and cross diffusion coefficient must remain the same. We observed that the patterns became more conspicuous, leading to a mix of hot spot and Labyrinth Turing Patterns (see Figures 6a,b and 7a,b).

Through extensive simulations over a wide range of diffusivity coefficients, we observed that cross diffusion coefficient d_4 plays an important role to realize the phenomenon of pattern formation. In the absence of cross diffusion, no patterns were formed, even at higher values of d_3, as shown in Figure 1. Moreover we also determined the Turing patterns in the cases where cross diffusion coefficient equals to the self diffusion coefficient, i.e., $d_3 = d_4$. Table 1 describes the various self and cross diffusivity coefficients used for obtaining the spatial patterns of the system in Equation (38). Table 1 indicates a negative correlation between self diffusion coefficient and cross diffusion coefficients, which results in pattern formation. The above simulations results are coherent to our analytical results that non-constant positive solutions exist only in the presence of cross diffusion hence the pattern formation.

7. Conclusions

In this study, we concluded a three-dimensional continuous time ecological system, modeling a tritrophic food chain based on a hybrid type of Holling Type II and Crowley-Martin functional response. We also took into account inhomogeneous spatial distribution of the species involved and dependence of movements of the species on the population density due to the mutual interference between the individuals of the same species and included nonlinear self diffusion and cross diffusion in our discussion. We established that, under certain parametric conditions, the model is unstable only in the presence of cross diffusion. This phenomenon can be regarded as an extension to diffusion driven instability and can be referred to as cross diffusion driven instability. The phenomenon of pattern formation in spatially extended models including cross diffusion as a destabilizing mechanism was studied with varying the self diffusion and cross diffusion coefficients over a wide range, which serves as an extension to Turing pattern formations exhibited by reaction diffusion system. In addition, we analytically proved the existence of inhomogeneous steady states and gave a priori upper and lower bounds for the positive solutions of the system under steady state. We proved that no non-constant positive solutions exist in the presence of only self diffusion subject to certain conditions. Through simulation, we concluded that cross diffusion is necessary for pattern formation of models in which species interactions are governed by Crowley-Martin and Holling Type II functional response. This establishes the fact that Crowley-Martin functional response governed species interaction is stable in spatial temporal aspects when the species diffuse because of their own characteristic behavior but when under the influence of other species the system tends towards instability. Based on our analysis and simulation, we concluded that cross diffusion induces at least a inhomogeneous stationary solution and mathematically explains the segregation phenomenon of ecosystem under the influence of predator dependent functional response.

Author Contributions: The authors contributed equally to this work.

Funding: The research of the first author was funded by Science and Engineering Research Board (SERB), grant number EMR/2017/005203.

Conflicts of Interest: The authors declare no conflict of interest.

References

1. Kuang, Y.; Beretta, E. Global qualitative analysis of a ratio-dependent predator-prey system. *J. Math. Biol.* **1998**, *36*, 389–406. [CrossRef]
2. Segel, L.A. *Modeling Dynamic Phenomena in Molecular and Cellular Biology*; Cambridge University Press: Cambridge, UK, 1984.
3. Price, P.W.; Bouton, C.E.; Gross, P.; McPheron, B.A.; Thompson, J.N.; Weis, A.E. Interactions among three trophic levels: influence of plants on interactions between insect herbivores and natural enemies. *Annu. Rev. Ecol. Syst.* **1980**, *11*, 41–65. [CrossRef]
4. Hastings, A.; Powell, T. Chaos in a three-species food chain. *Ecology* **1991**, *72*, 896–903. [CrossRef]
5. McCann, K.; Yodzis, P. Biological Conditions for Chaos in a Three-Species Food Chain. *Ecology* **1994**, *75*, 561–564. [CrossRef]
6. Gakkhar, S.; Naji, R.K. Chaos in three species ratio dependent food chain. *Chaos Solitons Fractals* **2002**, *14*, 771–778. [CrossRef]
7. Rai, V.; Upadhyay, R.K. Chaotic population dynamics and biology of the top-predator. *Chaos Solitons Fractals* **2004**, *21*, 1195–1204. [CrossRef]
8. Crowley, P.H.; Martin, E.K. Functional responses and interference within and between year classes of a dragonfly population. *J. N. Am. Benthol. Soc.* **1989**, *8*, 211–221. [CrossRef]
9. Upadhyay, R.K.; Naji, R.K. Dynamics of a three species food chain model with Crowley Martin type functional response. *Chaos Solitons Fractals* **2009**, *42*, 1337–1346. [CrossRef]
10. Turing, A.M. The chemical basis of morphogenesis. *Philos. Trans. R. Soc. Lond. B* **1952**, *237*, 37–72.
11. Kondo, S. The reaction-diffusion system: a mechanism for autonomous pattern formation in the animal skin. *Genes Cells* **2002**, *7*, 535–541. [CrossRef] [PubMed]
12. Kondo, S.; Miura, T. Reaction-diffusion model as a framework for understanding biological pattern formation. *Science* **2010**, *329*, 1616–1620. [CrossRef] [PubMed]
13. Dong, Y.; Zhang, S.; Li, S.; Li, Y. Qualitative Analysis of a Predator-Prey Model with Crowley-Martin Functional Response. *Int. J. Bifurc. Chaos* **2015**, *25*, 1550110. [CrossRef]
14. Kuto, K. Stability of steady-state solutions to a prey-predator system with cross-diffusion. *J. Differ. Equ.* **2004**, *197*, 293–314. [CrossRef]
15. Kuto, K.; Yamada, Y. Multiple coexistence states for a prey-predator system with cross-diffusion. *J. Differ. Equ.* **2004**, *197*, 315–348. [CrossRef]

16. Pang, P.Y.H.; Wang, M. Strategy and stationary pattern in a three-species predator-prey model. *J. Differ. Equ.* **2004**, *200*, 245–273. [CrossRef]
17. Wang, M. Stationary patterns caused by cross-diffusion for a three-species prey-predator model. *Comput. Math. Appl.* **2006**, *52*, 707–720. [CrossRef]
18. Tian, C. Turing patterns created by cross-diffusion for a Holling II and Leslie-Gower type three species food chain model. *J. Math. Chem.* **2011**, *49*, 1128–1150. [CrossRef]
19. Tian, C.; Ling, Z.; Lin, Z. Spatial patterns created by cross-diffusion for a three-species food chain model. *Int. J. Biomath.* **2014**, *7*, 1450013. [CrossRef]
20. Medvinsky, A.B.; Petrovskii, S.V.; Tikhonova, I.A.; Malchow, H.; Li, B.-L. Spatiotemporal complexity of plankton and fish dynamics. *SIAM Rev.* **2002**, *44*, 311–370. [CrossRef]

Article

Dynamics of a Diffusive Two-Prey-One-Predator Model with Nonlocal Intra-Specific Competition for Both the Prey Species

Kalyan Manna [1], Vitaly Volpert [2,3,4] and Malay Banerjee [1,*]

1 Department of Mathematics and Statistics, Indian Institute of Technology Kanpur, Kanpur 208016, India; kalyan274667@gmail.com
2 Institut Camille Jordan, UMR 5208 CNRS, University Lyon 1, 69622 Villeurbanne, France; volpert@math.univ-lyon1.fr
3 Peoples Friendship University of Russia (RUDN University), 6 Miklukho-Maklaya St, Moscow 117198, Russia
4 INRIA Team Dracula, INRIA Lyon La Doua, 69603 Villeurbanne, France
* Correspondence: malayb@iitk.ac.in

Received: 13 December 2019; Accepted: 2 January 2020; Published: 7 January 2020

Abstract: Investigation of interacting populations is an active area of research, and various modeling approaches have been adopted to describe their dynamics. Mathematical models of such interactions using differential equations are capable to mimic the stationary and oscillating (regular or irregular) population distributions. Recently, some researchers have paid their attention to explain the consequences of transient dynamics of population density (especially the long transients) and able to capture such behaviors with simple models. Existence of multiple stationary patches and settlement to a stable distribution after a long quasi-stable transient dynamics can be explained by spatiotemporal models with nonlocal interaction terms. However, the studies of such interesting phenomena for three interacting species are not abundant in literature. Motivated by these facts here we have considered a three species prey–predator model where the predator is generalist in nature as it survives on two prey species. Nonlocalities are introduced in the intra-specific competition terms for the two prey species in order to model the accessibility of nearby resources. Using linear analysis, we have derived the Turing instability conditions for both the spatiotemporal models with and without nonlocal interactions. Validation of such conditions indicates the possibility of existence of stationary spatially heterogeneous distributions for all the three species. Existence of long transient dynamics has been presented under certain parametric domain. Exhaustive numerical simulations reveal various scenarios of stabilization of population distribution due to the presence of nonlocal intra-specific competition for the two prey species. Chaotic oscillation exhibited by the temporal model is significantly suppressed when the populations are allowed to move over their habitat and prey species can access the nearby resources.

Keywords: prey–predator; diffusion; nonlocal interaction; Turing instability; spatiotemporal pattern

1. Introduction

Generally, the ecological systems and interactions among different species in nature are too complex to be understood with ease. In order to comprehend and make predictions about the dynamics of a complex ecological system, mathematical models and quantitative mathematical methods act as useful tools. Since Paine's discovery on predation-mediated species diversity [1], several theoretical studies have been conducted in order to understand the role played by predation on the dynamics of temporal food web models (especially on two-prey-one-predator temporal models).

In this direction, Fujii [2] considered a two-prey-one-predator model with the competition between the two species of prey proposed in [3] and investigated for predation-mediated stabilization of the system. He showed the existence of a globally stable limit cycle around the unstable coexistence equilibrium point, whereas the corresponding two-species system without predator is unstable. On the other hand, Vance [4] argued that this coexistence of three species is not a result of predation alone but it occurs due to the combined effect of resource partitioning for the two prey species and predation. Working on the same model with Lotka–Volterra type linear functional responses, Takeuchi and Adachi [5] showed that chaotic coexistence is also possible apart from the already established periodic one. It was demonstrated that chaotic coexistence bifurcates from the periodic one when one prey's competitive ability is greater than the other. A criterion for permanent coexistence was established in [6] for the two-prey-one-predator model with intra-specific competition for predator.

Moreover, Matsuda et al. [7] investigated the effect of switching property of predator on the stability of the three-species model and concluded that heterogeneity in intrinsic growth rates for prey is an important component in order to achieve the stabilization. They further showed that the tendency of switching response toward the prey with lesser intrinsic growth rate enhances the stabilizing effect. Abrams and Matsuda [8] investigated the consequences of the adaptive anti-predator behavior of prey in a two-prey-one-predator system. In [9], the authors showed that the occurrence of chaotic dynamics for two-prey-one-predator food chain is independent of the particular choice of the model. Abrams showed that the amplitude of population cycles increases through enrichment in a model with saturating functional response and this enlargement of cycles favors the endangered species in [10]. In [11], the authors defined a functional response taking into account both the prey densities and showed that the switching effect of predator can potentially cause the extinction of predator species. Lee and Kajiwara [12] proposed a two-prey-one-predator model where one prey species consumes the remaining part of carcass of other prey species left by predator. They showed that this type of consumption can eventually lead to chaos. In case of non-interacting preys, some sufficient conditions for the stable coexistence equilibrium and periodic solution were provided in [13]. Recently, Groll et al. [14] examined a three-species microbial model and found the existence of chaotic regime between two regimes of stable limit cycles. They associated the reason for this type of chaotic regime to the competition between two distinct limit cycles.

Prey–predator patterns for ecological communities are ubiquitous in natural habitat such as the patchy spatial distribution in plankton ecosystem and the chaotic distribution for vole populations in northern Fennoscandia [15,16]. Generally, the spatial component acts as a crucial ingredient in shaping the spatial structure of an ecological community. Over the past few decades, many researchers have paid attention in order to understand the emergence of spatiotemporal patterns in ecological communities and their possible mechanisms since the seminal work of Turing [17]. Theoretical investigations on spatiotemporal pattern formations are generally accomplished by analyzing the associated reaction-diffusion equations. Till date, numerous works on the pattern formation for two-species models are present in literature, and some of these works can be found in [18–20] and the references therein. As a result, a wide variety of stationary patterns such as hot spots, cold spots, labyrinthine and mixture of spots and stripes, and dynamic patterns such as traveling waves, periodic traveling waves and spatiotemporal chaos etc. have been reported [18–20]. Also, Turing and Hopf–Turing bifurcations have been identified as two common mechanisms for the emergence of these different types of patterns [18–20].

However, the number of works on pattern formations for three-species food chain models is somewhat limited as compared to the two-species cases. For instance, White and Gilligan provided sufficient conditions to induce diffusion-driven instability for a generic three-species reaction-diffusion system in [21]. Petrovskii et al. [22] considered a three-species competitive spatial model of Lotka–Volterra type, and showed the existence of traveling waves and spatiotemporal chaos. Chen and Peng [23] proved the existence of stationary patterns for a competitor-competitor-mutualist model with cross-diffusion by establishing the existence of non-constant positive steady states. Also, they demonstrated that the corresponding model without cross-diffusion fails to produce any stationary pattern. In [24], the authors examined a spatially extended one-prey-two-predator system with defense switching mechanism for prey and collaborative exploitation of prey by predators. Their study also held cross-diffusion responsible for the emergence of stationary patterns. On the other hand, Wang [25] found stationary patterns for three-species models with both the prey-dependent and ratio-dependent functional responses as a result of self-diffusion. The author further showed the existence of stationary patterns for these models in presence of the cross-diffusion as well in [26]. Tian [27] presented both the theoretical and numerical results on cross-diffusion induced Turing patterns for Holling type II and Leslie-Gower type three-species food chain model. Similar types of results were presented for a predator-prey-mutualist system in [28]. Rao [29] investigated the complex spatiotemporal dynamics for an one-prey-two-predator system with ratio-dependent functional responses and showed the transition of different types of stationary patterns to chaotic wave patterns depending on the magnitudes of the maximum ingestion rate or the mortality rate of intermediate predator. In [30], the authors examined for the possible pattern formations in a three-species food chain model with the Holling type II and a modified Leslie-Gower type functional responses over a circular domain. Some other studies in this direction can be found in [31–36]. Very recently, Mukherjee et al. [37] employed weakly nonlinear analysis for a three-species model in order to obtain amplitude equations which determine the thresholds for emergence and stability of Turing patterns in the vicinity of the Turing bifurcation boundary.

The majority of the existing works on pattern formation in ecological communities have been accomplished by considering the local intra- and inter-specific interactions, that is, an individual located at a specific spatial point $\hat{x}$ can only interact with individuals present at that point. However, spatial mobility of a species gives rise to the scenario that the resource consumptions at the nearby locations and hence interactions should depend on a weighted average of the population density in a neighborhood of $\hat{x}$ instead of the pointwise values of the population density [38,39]. Nonlocal interaction can be modeled by incorporating a convolution integral with a specified non-negative kernel function which takes care of the weighting aspect [38,39]. The incorporation of nonlocality into a model results in a system of integro-partial differential equations. In the available literature, several different types of kernel functions have been considered to model nonlocal interactions. These different types of kernel functions include top-hat kernel, Gaussian kernel, Laplacian kernel, triangular kernel, parabolic kernel, etc. [40–42]. Among these kernel functions, the simplest one is the symmetric top-hat kernel which is basically a step function, and incorporation of it in the modeling approach can lead to very rich and complex spatiotemporal pattern formations [40,42].

Over the last decade, considerable amount of efforts have been made in order to understand the effects of nonlocal interactions on the spatiotemporal dynamics possessed by ecological communities. For instance, Gourley considered a nonlocal Fisher equation and established the existence of traveling wavefront solutions connecting two spatially homogeneous steady states for sufficiently weak nonlocality in [43]. Merchant and Nagata showed the destabilizing effect of nonlocal prey competition and reported various complex spatiotemporal patterns such as stationary periodic in space patterns, wave trains, and irregular spatiotemporal oscillations in [41]. In a subsequent study [44], they studied the correlation between the properties of selected wave trains and the standard deviation of nonlocality, and suggested that highly nonlocal systems are more likely to produce spatiotemporal chaos. Segal et al. [42] investigated pattern formations in a nonlocal model of competing populations and showed that the population distribution can eventually evolve into localized island structure. Further, they described analytically the structure of these islands by using step-function kernel. In [38], the authors studied a model for two competing species with asymmetric nonlocal coupling and found the existence of different types of spatiotemporal solutions such as propagating traveling waves of islands, colony formation, and modulated traveling waves. Deng and Wu [45] analyzed the global stability property of a nonlocal single species population model. In [46], the authors investigated the dynamics of a prey–predator model with nonlocal consumption of prey, and showed the existence of multiple stationary states, simple and periodic traveling waves. They further showed that the incorporation of nonlocal interactions in the classical Rosenzweig–MacArthur model can produce Turing patterns, while the local counterpart of it is unable to do so [47]. Some other interesting works on the impacts of nonlocal interactions on spatiotemporal dynamics for two-species models can be found in [40,48–54].

On the other hand, there exist very few studies on the effect of nonlocal interactions for three-species food chain models. Recently, Autry et al. [39] considered a three-species food chain model with ratio-dependent functional response and two types of nonlocality. They addressed the issue of biological pest control and showed that biological control cannot be achieved for highly diffusive pest species, however, robust partial control can be achieved if the super-predator is sufficiently diffusive and behaves nonlocally. In [55], the authors investigated the spatiotemporal pattern formation in a nonlocal intraguild predation model where the nonlocal interaction is incorporated in the growth of the shared resource. They identified the transition of stationary Turing patterns to spatiotemporal chaotic patterns through dynamic oscillatory patterns depending upon the magnitudes of the extent of nonlocality.

In this article, our aim is to provide rich spatiotemporal dynamics possessed by models of three interacting species (in particular, two preys and their common predator). Mainly, we are interested to examine the effects of random dispersal of all the three species and nonlocal intra-specific competition for two prey species on the population distributions. Apart from these, we are also interested to investigate some of the key issues in ecology such as long transient phenomenon and habitat size dependence of resulting patterns. In this regard, we investigate systematically a temporal model, a reaction-diffusion model and a nonlocal model. These systematic investigations can be useful in comparing the resulting dynamics and their appropriate causes as these three models constitute a system of nested models. The rest of this article is organized in the following fashion. In Section 2, we introduce a three-species (two-prey-one-predator) temporal model and briefly discuss about the dynamical behaviors. Then in Section 3, we extend the model by incorporating spatial components and present a wide variety of spatiotemporal dynamics possessed by it. The spatiotemporal model is further extended by considering the nonlocal intra-specific competitions for both the prey species in Section 4 with the manifestation of possible impacts of the extent of nonlocality on the spatiotemporal dynamics. Finally, we end this article with a brief discussion in Section 5.

2. Temporal Model

In this section, we introduce a three-species (two-prey-one-predator) temporal model for prey–predator interactions presented in [5] which will serve as a base model for our study. The concerned temporal model is given by the following system of three ordinary differential equations:

$$\frac{du_1}{dt} = u_1\left(b_1 - u_1 - \alpha u_2 - \epsilon v\right), \tag{1}$$

$$\frac{du_2}{dt} = u_2\left(b_2 - \beta u_1 - u_2 - \mu v\right), \tag{2}$$

$$\frac{dv}{dt} = v\left(-b_3 + d\epsilon u_1 + d\mu u_2\right), \tag{3}$$

where $u_1(t)$, $u_2(t)$ and $v(t)$ represent the densities of two prey species and one predator species at time t, respectively. The parameters b_i $(i = 1, 2, 3)$ involved in this system denote the intrinsic rates of increase or decrease in density for the considered three species. The rates of competitive effects between the two prey species are represented by the parameters α and β, while the parameters ϵ and μ account for the diminishing rates in density of both the prey species due to predation by the considered predator. The parameter d denotes the transformation rate of prey biomass into predator biomass.

Note that all the parameters involved in this prey–predator system (1)–(3) are positive constants. Also, the model (1)–(3) is subjected to the non-negative initial conditions.

The possible equilibrium points for the temporal model (1)–(3) can be explicitly derived from the system of algebraic equations arising by equating the reaction parts to zero. The system (1)–(3) admits the following six trivial and semi-trivial equilibrium points: $E_{000} = (0,0,0)$, $E_{+00} = (b_1, 0, 0)$, $E_{0+0} = (0, b_2, 0)$, $E_{++0} = \left(\frac{b_1-\alpha b_2}{1-\alpha\beta}, \frac{b_2-\beta b_1}{1-\alpha\beta}, 0\right)$, $E_{+0+} = \left(\frac{b_3}{d\epsilon}, 0, \frac{db_1\epsilon-b_3}{d\epsilon^2}\right)$ and $E_{0++} = \left(0, \frac{b_3}{d\mu}, \frac{db_2\mu-b_3}{d\mu^2}\right)$. Also, this system admits a unique coexistence equilibrium point $E_{+++} = (u_1^*, u_2^*, v^*)$, where

$$
\begin{aligned}
u_1^* &= \frac{b_3\epsilon - db_2\epsilon\mu - \alpha b_3\mu + db_1\mu^2}{d(\epsilon^2 + \mu^2 - (\alpha+\beta)\epsilon\mu)},\\
u_2^* &= \frac{b_3\mu - db_1\epsilon\mu - \beta b_3\epsilon + db_2\epsilon^2}{d(\epsilon^2 + \mu^2 - (\alpha+\beta)\epsilon\mu)},\\
v^* &= \frac{b_3(\alpha\beta - 1) + d\mu(b_2 - \beta b_1) + d\epsilon(b_1 - \alpha b_2)}{d(\epsilon^2 + \mu^2 - (\alpha+\beta)\epsilon\mu)}.
\end{aligned}
$$

The feasibility conditions for the coexistence equilibrium point E_{+++} are given by (i) $b_3\epsilon + db_1\mu^2 > \mu(db_2\epsilon + \alpha b_3)$, $b_3\mu + db_2\epsilon^2 > \epsilon(db_1\mu + \beta b_3)$, $b_3\alpha\beta + d(b_1\epsilon + b_2\mu) > b_3 + d(b_2\alpha\epsilon + b_1\beta\mu)$ and $\epsilon^2 + \mu^2 > (\alpha+\beta)\epsilon\mu$, or (ii) $b_3\epsilon + db_1\mu^2 < \mu(db_2\epsilon + \alpha b_3)$, $b_3\mu + db_2\epsilon^2 < \epsilon(db_1\mu + \beta b_3)$, $b_3\alpha\beta + d(b_1\epsilon + b_2\mu) < b_3 + d(b_2\alpha\epsilon + b_1\beta\mu)$ and $\epsilon^2 + \mu^2 < (\alpha+\beta)\epsilon\mu$.

In this study, we are mainly concerned with the dynamical behaviors of the system around the coexistence equilibrium point since the existence of this equilibrium point indicates the species diversity. For detailed theoretical and numerical investigations on the dynamics of this system in the vicinity of the coexistence equilibrium point, one can go through the references [5,56]. However, here we briefly describe the theoretical results on local stability and Hopf bifurcation of the coexistence equilibrium point E_{+++}. Linearizing the above model (1)–(3) around the equilibrium point E_{+++}, we obtain the following characteristic equation

$$\lambda^3 + a_2\lambda^2 + a_1\lambda + a_0 = 0, \tag{4}$$

where

$$
\begin{aligned}
a_2 &= u_1^* + u_2^*, &(5)\\
a_1 &= (1-\alpha\beta)u_1^*u_2^* + d(\epsilon^2 u_1^* + \mu^2 u_2^*)v^*, &(6)\\
a_0 &= d\{(\epsilon^2+\mu^2) - (\alpha+\beta)\epsilon\mu\}u_1^*u_2^*v^*. &(7)
\end{aligned}
$$

Therefore, the equilibrium point E_{+++} is locally asymptotically stable provided the following Routh–Hurwitz criteria are satisfied:

$$a_2 > 0, \;\; a_0 > 0, \;\; a_2a_1 - a_0 > 0. \tag{8}$$

The condition $a_2 > 0$ is automatically satisfied for feasible E_{+++}. However, we need to have the parametric restriction $\epsilon^2 + \mu^2 > (\alpha+\beta)\epsilon\mu$ in order to satisfy the condition $a_0 > 0$. Therefore, the parametric restriction $\epsilon^2 + \mu^2 > (\alpha+\beta)\epsilon\mu$ acts as a necessary condition for local stability. Furthermore, the model (1)–(3) undergoes Hopf bifurcation at $\epsilon = \epsilon_*$ when the following conditions are satisfied:

$$(a_2a_1 - a_0)|_{\epsilon=\epsilon_*} = 0, \;\; \frac{d}{d\epsilon}(a_2a_1 - a_0)|_{\epsilon=\epsilon_*} \neq 0, \tag{9}$$

with the parametric restriction $\epsilon^2 + \mu^2 > (\alpha+\beta)\epsilon\mu$.

Now, we encapsulate numerically the dynamical behaviors around the coexistence equilibrium point E_{+++} in the form of a bifurcation diagram with respect to the parameter accounting for the diminishing rate in first prey density due to predation (ϵ). For this purpose, we consider a fixed set of parameter values

$$b_1 = 1.0, \ \alpha = 1.0, \ b_2 = 1.0, \ \beta = 1.5, \ \mu = 1.0, \ b_3 = 1.0, \ d = 0.5, \tag{10}$$

and vary the parameter ϵ. The resulting dynamical behaviors are encapsulated in Figure 1. From this figure, we can observe that the stable coexistence equilibrium point E_{+++} loses its stability through the occurrence of Hopf bifurcation approximately at $\epsilon = 5.486$ and stable limit cycles appear for higher ϵ-values than this threshold value. The size of the limit cycle gradually increases as the ϵ-value moves away in positive direction from this threshold. For sufficiently large values of ϵ (for example, $\epsilon = 8.0$), the system admits two-periodic solutions. Further increments in the value of ϵ result in chaotic dynamics (for example, $\epsilon = 10.0$). This figure also clearly illustrates that the system takes the period-doubling route to reach chaos finally.

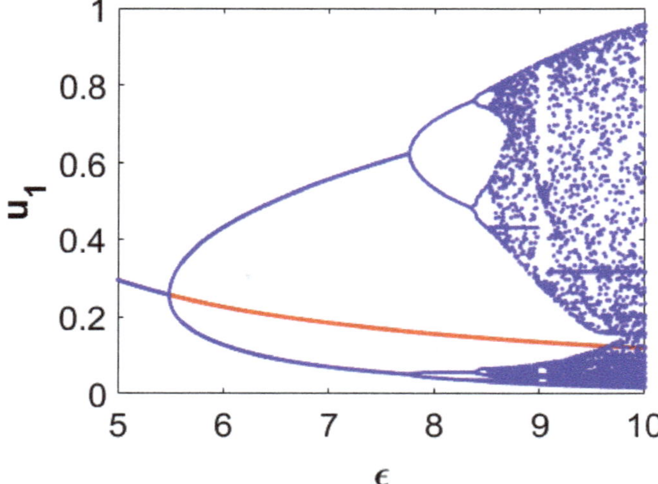

Figure 1. Single parameter bifurcation diagram for the density of first prey population (u_1) with respect to the parameter ϵ for the temporal model (1)–(3). Other parameter values are mentioned in (10).

3. Spatiotemporal Model

In this section, we extend the temporal model (1)–(3) presented in the preceding section by incorporating the random dispersal of all the three species. The corresponding spatiotemporal model is given by

$$\frac{\partial u_1}{\partial t} = d_1 \frac{\partial^2 u_1}{\partial x^2} + u_1 (b_1 - u_1 - \alpha u_2 - \epsilon v), \tag{11}$$

$$\frac{\partial u_2}{\partial t} = d_2 \frac{\partial^2 u_2}{\partial x^2} + u_2 (b_2 - \beta u_1 - u_2 - \mu v), \tag{12}$$

$$\frac{\partial v}{\partial t} = d_3 \frac{\partial^2 v}{\partial x^2} + v (-b_3 + d\epsilon u_1 + d\mu u_2), \tag{13}$$

subjected to non-negative initial conditions and periodic boundary conditions. Here, $u_1(x,t)$, $u_2(x,t)$ and $v(x,t)$ respectively denote the densities of the three species at spatial position x and time t. The bounded spatial domain is given by $[-L, L]$, where $L(> 0) \in \mathbb{R}$. The positive parameters d_i ($i = 1, 2, 3$) represent the diffusion coefficients for two prey and one predator species, respectively.

Note that the equilibrium points corresponding to the temporal model (1)–(3) also serve as the spatially homogeneous steady states for the model (11)–(13).

Turing instability occurs when the stable spatially homogeneous steady state becomes unstable due to small amplitude heterogeneous perturbations around the spatially homogeneous steady state. Introducing small amplitude spatiotemporal perturbation around the spatially homogeneous coexistence steady state $E_{+++} = (u_1^*, u_2^*, v^*)$ and then linearizing we find the following Jacobian matrix:

$$\mathcal{J}(k^2) = \begin{bmatrix} -u_1^* - d_1k^2 & -\alpha u_1^* & -\epsilon u_1^* \\ -\beta u_2^* & -u_2^* - d_2k^2 & -\mu u_2^* \\ d\epsilon v^* & d\mu v^* & -d_3k^2 \end{bmatrix}, \tag{14}$$

where k represents the wavenumber. Therefore, the characteristic equation is given by

$$\lambda^3 + A(k^2)\lambda^2 + B(k^2)\lambda + C(k^2) = 0, \tag{15}$$

where

$$\begin{aligned} A(k^2) &= (d_1 + d_2 + d_3)k^2 + u_1^* + u_2^*, \\ B(k^2) &= (d_1d_2 + d_2d_3 + d_3d_1)k^4 + \{d_1u_2^* + d_2u_1^* + d_3(u_1^* + u_2^*)\}k^2 + (1 - \alpha\beta)u_1^*u_2^* + d(\epsilon^2u_1^* + \mu^2u_2^*)v^*, \\ C(k^2) &= d_1d_2d_3k^6 + d_3(d_1u_2^* + d_2u_1^*)k^4 + \{d_1d\mu^2u_2^*v^* + d_2d\epsilon^2u_1^*v^* + d_3(1 - \alpha\beta)u_1^*u_2^*\}k^2 + \{d(\epsilon^2 + \mu^2) \\ &\quad -d(\alpha + \beta)\epsilon\mu\}u_1^*u_2^*v^*. \end{aligned}$$

Now, Re(λ) is less than zero provided the Routh–Hurwitz criteria are satisfied:

$$A(k^2) > 0,\ C(k^2) > 0,\ A(k^2)B(k^2) - C(k^2) > 0. \tag{16}$$

When all the conditions presented in (16) are satisfied, then the spatially homogeneous steady state is stable. Diffusion-driven instability occurs only when one eigenvalue is zero while other two eigenvalues still have negative real parts [21,57]. If λ_1, λ_2 and λ_3 denote the roots of the characteristic Equation (15), then from the properties of the roots of a cubic equation we have the following relations:

$$\lambda_1 + \lambda_2 + \lambda_3 = -A(k^2), \tag{17}$$

$$\lambda_1\lambda_2 + \lambda_2\lambda_3 + \lambda_3\lambda_1 = B(k^2), \tag{18}$$

$$\lambda_1\lambda_2\lambda_3 = -C(k^2), \tag{19}$$

$$-(\lambda_1 + \lambda_2)(\lambda_2 + \lambda_3)(\lambda_3 + \lambda_1) = A(k^2)B(k^2) - C(k^2). \tag{20}$$

Below the Turing bifurcation threshold, the spatially homogeneous steady state is stable and accordingly all the eigenvalues have negative real parts. At the Turing bifurcation threshold, exactly one eigenvalue becomes zero at the critical wavenumber k_T, whereas other two eigenvalues still have negative real parts [58]. Since at the critical wavenumber $k = k_T$ we have one of the roots is equal to zero, without any loss of generality we assume

$$\lambda_1 \mid_{k^2=k_T^2} = 0,\ \ \text{Re}(\lambda_2) \mid_{k^2=k_T^2} < 0 \ \text{ and } \ \text{Re}(\lambda_3) \mid_{k^2=k_T^2} < 0. \tag{21}$$

Hence, at the critical wavenumber $k = k_T$ we obtain $C(k_T^2) = 0$. Further due to the above conditions (21) for Turing instability, we get $A(k_T^2) > 0$, $B(k_T^2) > 0$ and $A(k_T^2)B(k_T^2) - C(k_T^2) = A(k_T^2)B(k_T^2) > 0$. Thus, the system remains stable if $C(k^2) > 0$ holds for all k and it becomes Turing unstable if $C(k^2) < 0$ holds for at least one k. Also, $C(k^2) < 0$ holds for a range of k-values around k_T beyond the Turing bifurcation threshold. The expression of $C(k^2)$ can be rewritten as

$$C(k^2) \equiv C_3(k^2)^3 + C_2(k^2)^2 + C_1(k^2) + C_0, \tag{22}$$

where $C_3 > 0$ since the diffusion coefficients are positive constants and $C_0 > 0$ from the local asymptotic stability condition of E_{+++}. Also, the coefficient $C_2 > 0$ as the expression of it is given by $C_2 = d_3(d_1 u_2^* + d_2 u_1^*)$. The minimum for $C(k^2)$ occurs at $k = k_T$ where

$$k_T^2 \quad = \quad \frac{-C_2 + \sqrt{C_2^2 - 3C_1C_3}}{3C_3}, \tag{23}$$

as we have $\frac{dC(k^2)}{d(k^2)} = 0$ and $\frac{d^2C(k^2)}{d(k^2)^2} > 0$ at $k = k_T$. Now from the expression (23), we can notice that k_T^2 is positive if $C_1 < 0$ or $C_2 < 0$ and $C_2^2 > 3C_1C_3$. Since $C_2 > 0$ holds for our considered system, then the condition for the positivity of k_T^2 reduces to $C_1 < 0$. Therefore, the Turing bifurcation boundary is given by

$$2C_2^3 - 9C_1C_2C_3 - 2(C_2^2 - 3C_1C_3)^{\frac{3}{2}} + 27C_0C_3^2 \quad = \quad 0. \tag{24}$$

This is an implicit expression for the Turing bifurcation curve whenever other relevant conditions are satisfied.

Numerical Results

In this subsection, we present some numerical results in order to validate the analytical predictions and to manifest the spatiotemporal dynamics which are beyond the scope of linear analysis. For this purpose, we considered a set of parameter values

$$b_1 = 1.0,\ \alpha = 1.0,\ b_2 = 1.0,\ \beta = 1.5,\ \mu = 1.0,\ d = 0.5,\ b_3 = 1.0,\ d_1 = 0.1,\ d_2 = 0.2, \tag{25}$$

and varied the parameters ϵ and d_3. For this set of parameter values, the temporal-Hopf bifurcation threshold is approximately given by $\epsilon = 5.486$. First, we present the temporal-Hopf and Turing bifurcation curves for the local spatiotemporal model (11)–(13) in (ϵ, d_3)-parameter space in Figure 2. From this figure, we can notice that these two bifurcation curves divide the two-dimensional parameter space into four different regions, such as stable region (left to the temporal-Hopf curve and below the Turing curve), Turing region (left to the temporal-Hopf curve and above the Turing curve), Hopf–Turing region (right to the temporal-Hopf curve and above the Turing curve) and pure Hopf region (right to the temporal-Hopf curve and below the Turing curve).

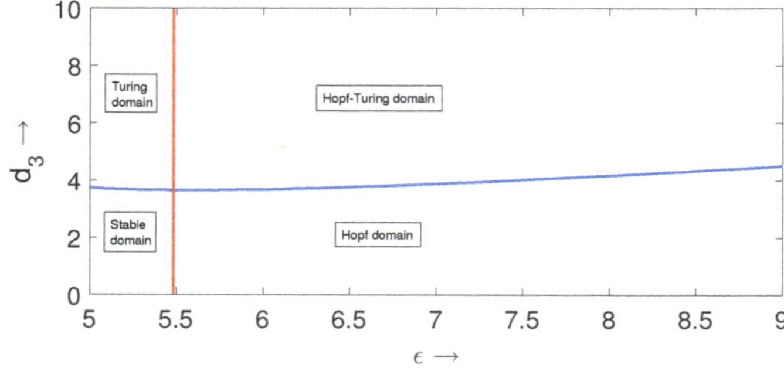

Figure 2. Plots of temporal-Hopf and Turing bifurcation curves in (ϵ, d_3)-parameter space for the local spatiotemporal model (11)–(13). Here, red vertical straight line represents the temporal-Hopf bifurcation threshold and the blue curve represents the Turing bifurcation curve. Other parameter values are mentioned in (25). The red vertical straight line has been drawn by solving the equations presented in (9) simultaneously and the blue curve has been drawn from the Equation (24).

Before proceeding further, here we briefly describe about the numerical discretization of the system (11)–(13). We have used the forward Euler scheme and central difference scheme to discretize the reaction and diffusion parts, respectively. The periodic boundary conditions have been employed at the boundary of the computational domain. For all the patterns presented in this subsection, we have used the following pulse-type initial conditions:

$$\begin{aligned} u_1(x,0) &= \begin{cases} u_1^* + \eta_1, & |x| < 10, \\ u_1^*, & 10 \leq |x| \leq L, \end{cases} \\ u_2(x,0) &= \begin{cases} u_2^* + \eta_2, & |x| < 10, \\ u_2^*, & 10 \leq |x| \leq L, \end{cases} \\ v(x,0) &= \begin{cases} v^* + \eta_3, & |x| < 10, \\ v^*, & 10 \leq |x| \leq L, \end{cases} \end{aligned} \tag{26}$$

where $|\eta_r| \ll 1$, for $r = 1,2,3$.

Now, we consider a point $(\epsilon, d_3) = (5.4, 9.0)$, which lies in the Turing region. For this choice of ϵ, we obtain the coexistence steady state $E_{+++} = (0.2641, 0.5738, 0.03)$ which is locally asymptotically stable in the absence of diffusion. Figure 3 shows the stationary Turing pattern produced by u_1-component and the corresponding temporal evolution of the average densities of all considered species. We would like to mention here that the patterns for the second prey and predator species are qualitatively similar with the pattern for the first prey species presented in Figure 3a. However, the regions with high first prey density correspond to the same for predator but to low second prey density. These scenarios for the density correlation among the three considered species are presented in Figure 4. This type of density correlation among the three considered species has also been observed for other patterns presented in this section. Also, note that we have considered the one-dimensional spatial domain $[-100, 100]$ for the clarity of understanding in Figure 4, but all the patterns presented in this section are obtained for the spatial domain $[-200, 200]$.

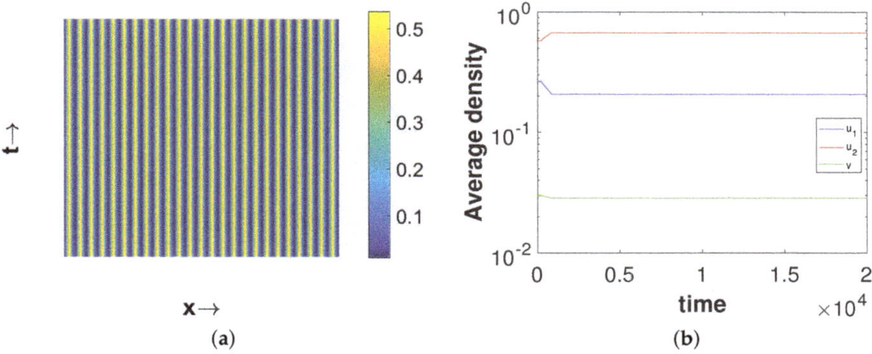

Figure 3. (**a**) Plot of stationary Turing pattern exhibited by u_1, (**b**) time evolution of the average densities of three species. $\epsilon = 5.4$ and $d_3 = 9.0$, other parameter values are mentioned in (25).

Now if we choose $\epsilon = 6.0$, then $E_{+++} = (0.2273, 0.6364, 0.0227)$ is unstable, due to temporal-Hopf bifurcation, in the absence of diffusion terms. Hence, we can observe homogeneous in space and periodic in time pattern for $d_3 = 3.0$ in Figure 5a, periodic in both space and time pattern for $d_3 = 3.75$ in Figure 5b, and periodic in space but stationary in time pattern for $d_3 = 9.0$ in Figure 5c. Phase diagrams presented in Figure 6 confirm the temporal periodicity of the patterns presented in Figure 5a,b, respectively. In Figure 7, we present the bifurcation diagram for spatially averaged density of first prey species in order to illustrate the transition of different types of patterns depending on the value of the diffusion coefficient d_3 for $\epsilon = 6.0$. In order to prepare this bifurcation diagram, we have used small pulse type perturbation about the mid-point around the spatially homogeneous steady state for each d_3-value and then plotted the maxima and minima of spatially averaged densities. From this figure, we can observe that periodic in time and homogeneous in space pattern persists up to $d_3 = 3.67$. The periodic in both space and time pattern exists for $d_3 \in [3.68, 3.91]$ and then stationary pattern emerges for $d_3 > 3.91$. Interestingly, the number of patches with high first prey density gradually decreases with the increment in the value of d_3.

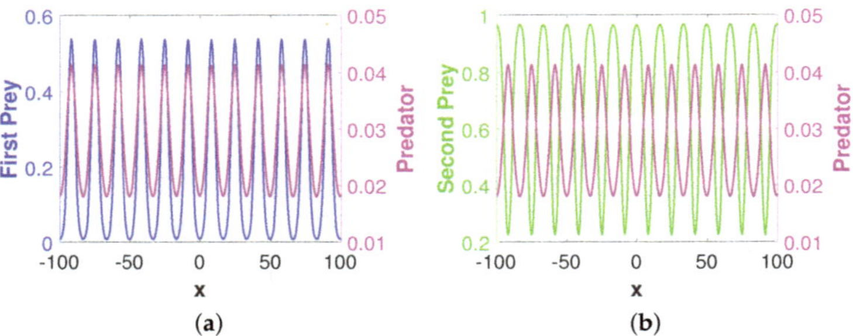

Figure 4. Correlation of densities among two prey and one predator species at a particular instant of time when the patterns are stationary in time. Left panel (**a**) shows the density correlation between first prey and predator species and right panel (**b**) shows the same between second prey and predator species. Numerical simulation has been performed for the same parameter values as in the Figure 3 over the one-dimensional spatial domain $[-100, 100]$.

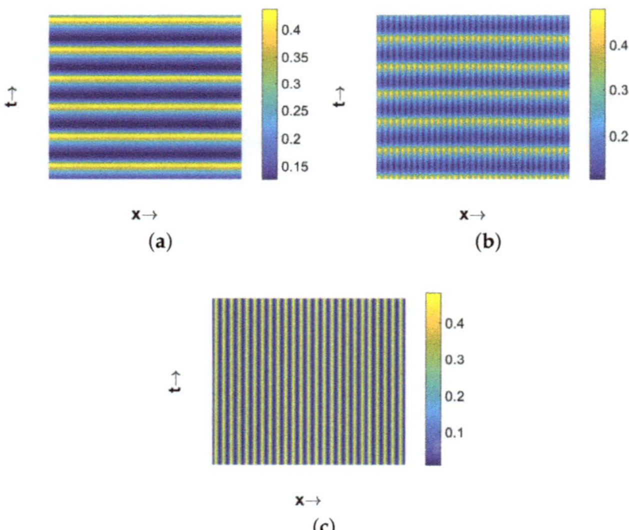

Figure 5. Plots of different types of patterns exhibited by u_1 for $\epsilon = 6.0$. Three different patterns are obtained for (**a**) $d_3 = 3.0$; (**b**) $d_3 = 3.75$; (**c**) $d_3 = 9.0$. Other parameter values are mentioned in (25).

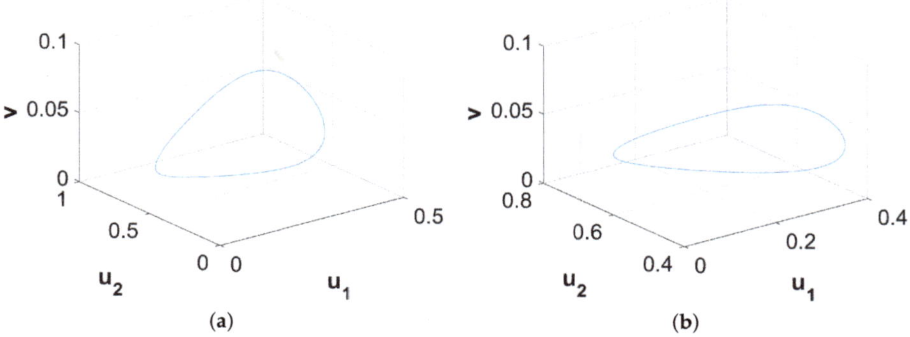

Figure 6. Phase diagrams of spatial average densities of the associated three species corresponding to (**a**) Figure 5a, (**b**) Figure 5b.

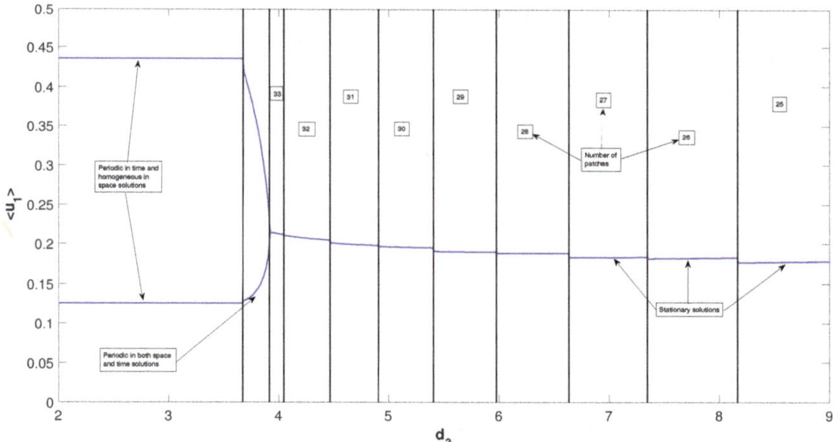

Figure 7. Single parameter bifurcation diagram for the spatially averaged density of first prey population ($< u_1 >$) with respect to the diffusion coefficient d_3. Here, $\epsilon = 6.0$ and other parameter values are mentioned in (25).

If we further increase the magnitude of ϵ to 8.0, then the system (11)–(13) admits the spatially homogeneous coexistence steady state $E_{+++} = (0.1556, 0.7556, 0.0111)$ which is temporally unstable. Then the choice $d_3 = 1.0$ leads to the spatiotemporal chaotic pattern (see Figure 8a). A close look at the chaotic pattern reveals that the distribution of the species is symmetric about the midpoint in space and this is due to our consideration of the symmetric pulse type initial conditions. The chaotic nature has been cross verified by checking the sensitivity to initial conditions as described in [59], but we do not present the results here for the sake of brevity. Interestingly, spatiotemporal chaotic pattern disappears and eventually settles to a stationary pattern for a higher value of d_3 (for example, $d_3 = 10.0$). This transition from spatiotemporal chaotic to stationary pattern takes place via a pattern where patches move periodically for intermediate values of d_3 (for example, $d_3 = 9.5$). The resulting patterns are presented in Figure 8. The phase diagrams presented in Figure 9 confirm the chaotic nature of the pattern presented in Figure 8a and the periodic nature of the pattern presented in Figure 8b.

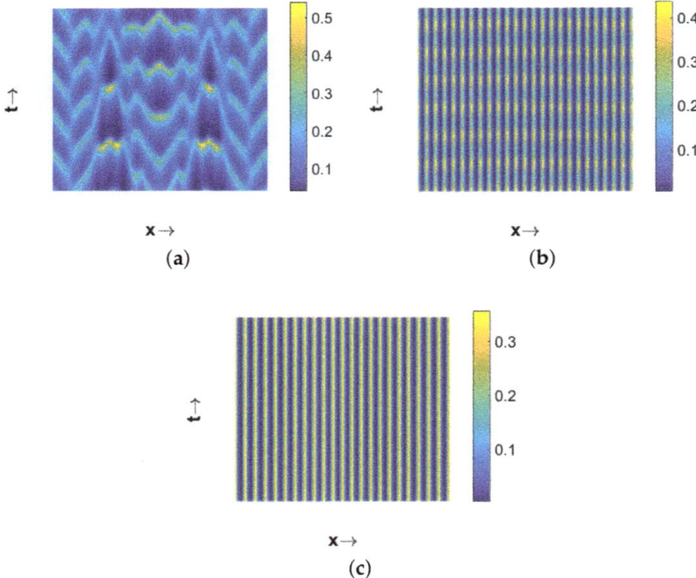

Figure 8. Plots of different types of patterns exhibited by u_1 for $\epsilon = 8.0$. Three different patterns are obtained for (**a**) $d_3 = 1.0$; (**b**) $d_3 = 9.5$; (**c**) $d_3 = 10.0$. Other parameter values are mentioned in (25).

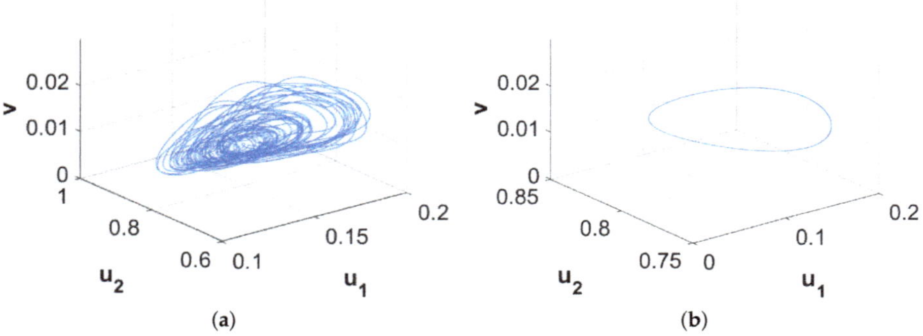

Figure 9. Phase diagrams of spatial average densities of the associated three species corresponding to (**a**) Figure 8a, (**b**) Figure 8b.

All patterns exhibited in Figures 3a, 5 and 8 are stable in the sense that their nature will not alter with further increment in time. However, the considered system (11)–(13) can produce long transient patterns in order to reach the final stable patterns. Such an instance of long transient pattern can be found for $\epsilon = 8.0$ and $d_3 = 9.0$ along with other parameter values mentioned above. Figure 10 exhibits the phase diagrams for the spatially averaged densities of the considered three species for $t \in [18,000, 20,000]$ and $t \in [58,000, 60,000]$, respectively. Since the phase diagram presented in Figure 10a is restricted to an annular region, we can term the nature of the corresponding spatiotemporal distribution as quasi-periodic which persists for approximately $t = 50,000$. On the other hand, the phase diagram presented in Figure 10b clearly shows the periodic nature of the corresponding spatiotemporal distribution and this pattern is stable for the considered parameter values. With the increment in time, the width of the annular region gradually decreases and finally the region settles into a limit cycle. However, this long time period for the existence of quasi-stable and quasi-periodic pattern can be few or even hundreds of generations for certain species. Therefore, this type of existence of long transient pattern and then finally convergence into a stable pattern can provide an alternative explanation for ecological regime shifts apart from the usual concept of exogenously sparked regime shifts.

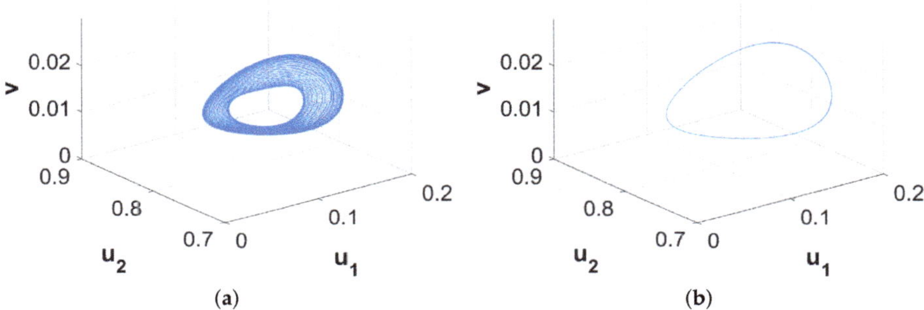

Figure 10. Phase diagrams of spatial average densities of the associated three species for $\epsilon = 8.0$ and $d_3 = 9.0$ with (**a**) $t \in [18,000, 20,000]$, (**b**) $t \in [58,000, 60,000]$. Other parameter values are mentioned in (25).

Now, we want to investigate the influence of random dispersal of the three species on the dynamics presented in Figure 1 for the temporal model. For this purpose, we considered the same parameter values as for Figure 1 with $d_1 = 0.1$, $d_2 = 0.2$ and $d_3 = 1.0$, and varied the value of the parameter ϵ. By employing spatially homogeneous initial population distributions slightly perturbed from the spatially homogeneous coexistence steady state, we found the bifurcation diagram which was more or less similar to the Figure 1 and chose not to display it here for the sake of brevity. However, consideration of pulse type initial population distributions led to different dynamical behaviors for a certain range of ϵ-values and the resulting dynamics are encapsulated in Figure 11. Comparing Figures 1 and 11, we can observe that the dynamics for both the models were same up to $\epsilon = 8.5$ approximately. For higher values of ϵ (that is, $\epsilon \in (8.5, 10.0]$), we observed the difference in chaotic dynamics. For spatiotemporal model (11)–(13), the amplitudes of chaotic oscillations were much less than that of the temporal case. We can attribute the reason for this type of phenomenon to the emergence of spatial heterogeneity in the population distributions as we have found spatially heterogeneous chaotic distributions in this case.

Further, we examined the stability of this bifurcation diagram by employing both the forward and backward continuation techniques. In order to construct a bifurcation diagram using forward continuation technique, we employed the following algorithm:

- First, consider a pulse-type initial condition for the starting value of the bifurcation parameter and run the simulation for sufficient time period such that a stable state is reached.
- Make a tiny increment in the value of the bifurcation parameter.
- Then, use the stable state obtained in the previous step as an initial condition for the present value of the bifurcation parameter and run the simulation till a stable state is reached.
- Continue the previous two steps till the last value of the bifurcation parameter is reached.

In a similar manner, one can efficiently construct a bifurcation diagram using backward continuation technique by reversing the process encapsulated in the above algorithm. Forward continuation technique results in more or less similar bifurcation diagram as the Figure 11 and accordingly we do not include it here. On the other hand, we observed significant changes in the bifurcation diagram using backward continuation technique and the resulting dynamics are encapsulated in Figure 12. From this diagram, we can observe that the period-doubling route to chaos has been destroyed and spatially heterogeneous chaotic regime survives for larger range of ϵ-values. Overall, bifurcation diagrams presented in Figures 1, 11 and 12 indicate the existence of alternative stable states for the model (11)–(13). Note that these states indicate the coexistence of all the three species and can be both the spatially homogeneous and heterogeneous population distributions.

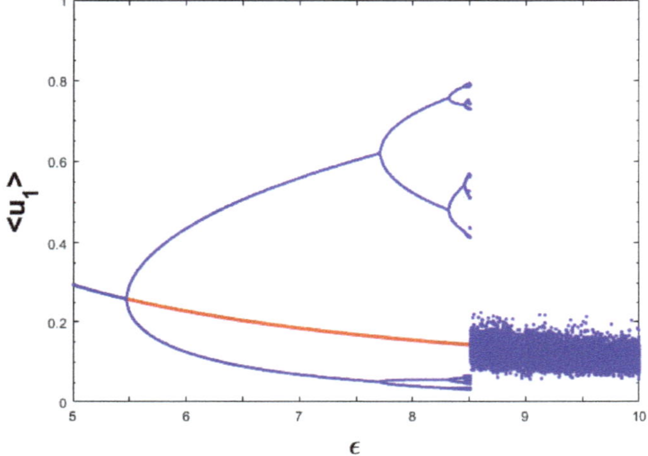

Figure 11. Single parameter bifurcation diagram for the spatially averaged density of first prey population ($< u_1 >$) with respect to the parameter ϵ for the spatiotemporal model (11)–(13). Here, we have used pulse type initial population distributions for each value of ϵ. Other parameter values are mentioned in (25) with $d_3 = 1.0$.

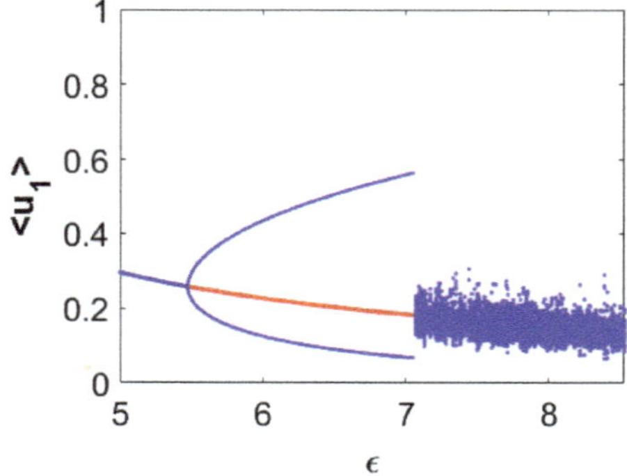

Figure 12. Single parameter bifurcation diagram for the spatially averaged density of first prey population ($< u_1 >$) with respect to the parameter ϵ for the spatiotemporal model (11)–(13). Here, backward continuation technique has been employed. Other parameter values are mentioned in (25) with $d_3 = 1.0$.

Another interesting finding is that the spatial population distributions depended on the spatial domain size when the parameter values were taken far from the temporal-Hopf bifurcation threshold. The patterns presented in Figure 13 are numerically computed over the one-dimensional spatial domain $[-50, 50]$ with the same parameter values used for Figure 8a,c. Figure 13a suggests that the smaller domain size can produce a two-periodic pattern while the larger domain size produces a chaotic one. Also, Figure 13c indicates that the smaller domain size can give rise to periodic in both the space and time population distribution while the larger domain size gives a stationary heterogeneous distribution. The two-periodic and periodic nature of the patterns presented in Figure 13a,c can be confirmed by the phase portraits of the spatially averaged densities in Figure 13b,d, respectively. However, we did not find any effect of the domain size on the resulting patterns presented in Figures 3, 5 and 8b. Therefore, our numerical findings suggest that not only certain parameters associated with a model play crucial roles in the emergence of different types of spatial population distributions but also the spatial domain size can potentially play a key role in shaping the population distributions.

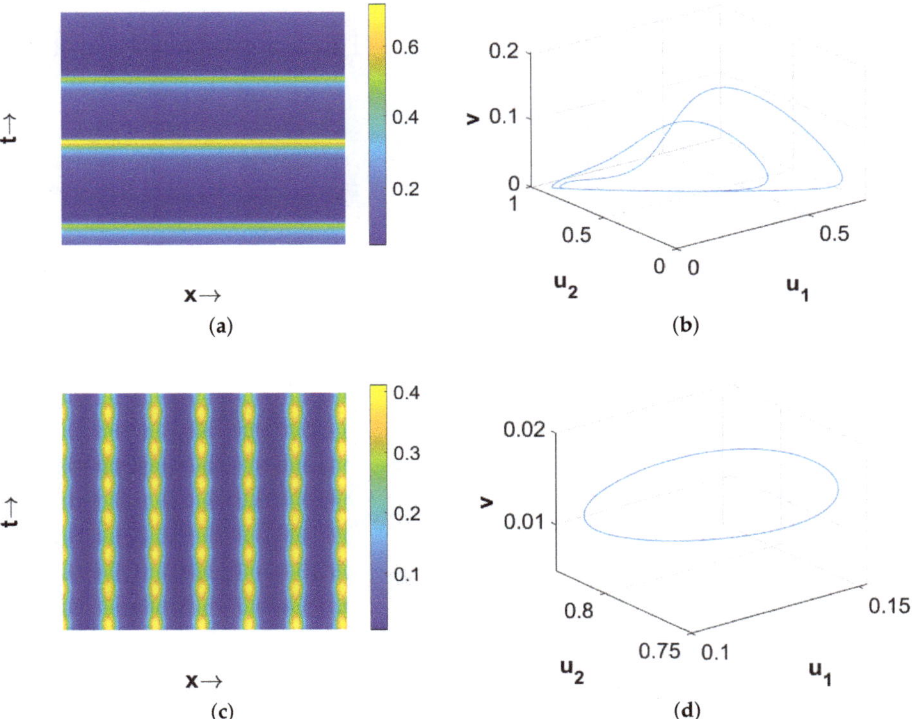

Figure 13. (**a**) Plot of spatial pattern exhibited by u_1, (**b**) corresponding phase portrait of the average densities of three species for $\epsilon = 8.0$, $d_3 = 1.0$. (**c**) Plot of spatial pattern exhibited by u_1, (**d**) corresponding phase portrait of the average densities of three species for $\epsilon = 8.0$, $d_3 = 10.0$. Spatial domain size is $[-50, 50]$ and other parameter values are mentioned in (25).

4. Nonlocal Model

In this section, we further extend the spatiotemporal model (11)–(13) by incorporating nonlocal intra-specific competition for both the prey species. Accordingly, the model (11)–(13) becomes

$$\frac{\partial u_1}{\partial t} = d_1\frac{\partial^2 u_1}{\partial x^2} + u_1\left(b_1 - J(u_1) - \alpha u_2 - \epsilon v\right), \tag{27}$$

$$\frac{\partial u_2}{\partial t} = d_2\frac{\partial^2 u_2}{\partial x^2} + u_2\left(b_2 - \beta u_1 - J(u_2) - \mu v\right), \tag{28}$$

$$\frac{\partial v}{\partial t} = d_3\frac{\partial^2 v}{\partial x^2} + v\left(-b_3 + d\epsilon u_1 + d\mu u_2\right), \tag{29}$$

where

$$J(u_r) = \int_{-\infty}^{\infty} u_r(y,t)\phi_r(x-y)dy, \; r = 1,2,$$

which take care of the nonlocality in intra-specific interactions for both the prey species arising primarily as a result of nonlocal consumption of resources. However, there exist other mechanisms such as cannibalism, space sharing, fights, etc., which can eventually lead to nonlocal intra-specific competition. The nonlocal system (27)–(29) is subjected to the same initial and boundary conditions mentioned before for the local spatiotemporal system (11)–(13). Here, the functions ϕ_r $(r = 1,2)$ represent the non-negative kernel functions. Further, in order to capture the unbiased movement of the two prey populations we assume the kernel functions ϕ_r $(r = 1,2)$ as even functions. Throughout this paper, we restrict ourselves only to the top-hat kernel functions defined as follows:

$$\phi_r(x) = \begin{cases} \frac{1}{2\delta_r}, & |x| \le \delta_r, \\ 0, & \text{otherwise}, \end{cases} \tag{30}$$

for $r = 1,2$. The newly introduced parameters δ_r $(r = 1,2)$ represent the extent of the nonlocality by controlling the width of the kernel functions ϕ_r. Also, one can easily verify that the spatially homogeneous steady states of the nonlocal system (27)–(29) are same as that of the corresponding local system (11)–(13) and accordingly we use the same notations for them.

In order to linearize the nonlocal system (27)–(29) about the spatially homogeneous coexistence steady state $E_{+++} = (u_1^*, u_2^*, v^*)$, we introduce the perturbation $u_1 = u_1^* + \varepsilon\widetilde{u}_1 e^{\lambda t + ikx}$, $u_2 = u_2^* + \varepsilon\widetilde{u}_2 e^{\lambda t + ikx}$ and $v = v^* + \varepsilon\widetilde{v}e^{\lambda t + ikx}$ with $|\varepsilon| \ll 1$. Then the Jacobian matrix corresponding to the linearized system is given by

$$\begin{aligned}\widetilde{\mathcal{J}}(k,\delta_1,\delta_2) &= \begin{bmatrix} -u_1^*\int_{-\infty}^{\infty}\phi_1(y)e^{-iky}dy - d_1k^2 & -\alpha u_1^* & -\epsilon u_1^* \\ -\beta u_2^* & -u_2^*\int_{-\infty}^{\infty}\phi_2(y)e^{-iky}dy - d_2k^2 & -\mu u_2^* \\ d\epsilon v^* & d\mu v^* & -d_3k^2 \end{bmatrix} \\ &= \begin{bmatrix} -u_1^*\frac{\sin(k\delta_1)}{k\delta_1} - d_1k^2 & -\alpha u_1^* & -\epsilon u_1^* \\ -\beta u_2^* & -u_2^*\frac{\sin(k\delta_2)}{k\delta_2} - d_2k^2 & -\mu u_2^* \\ d\epsilon v^* & d\mu v^* & -d_3k^2 \end{bmatrix}. \end{aligned} \tag{31}$$

Therefore, the characteristic equation is given by

$$\lambda^3 + \widetilde{A}(k,\delta_1,\delta_2)\lambda^2 + \widetilde{B}(k,\delta_1,\delta_2)\lambda + \widetilde{C}(k,\delta_1,\delta_2) = 0, \tag{32}$$

where

$$\widetilde{A}(k,\delta_1,\delta_2) = (d_1+d_2+d_3)k^2 + u_1^*\frac{\sin(k\delta_1)}{k\delta_1} + u_2^*\frac{\sin(k\delta_2)}{k\delta_2}, \tag{33}$$

$$\begin{aligned}\widetilde{B}(k,\delta_1,\delta_2) &= d_3k^2\left\{(d_1+d_2)k^2 + u_1^*\frac{\sin(k\delta_1)}{k\delta_1} + u_2^*\frac{\sin(k\delta_2)}{k\delta_2}\right\} + \left(d_1k^2 + u_1^*\frac{\sin(k\delta_1)}{k\delta_1}\right)\left(d_2k^2 + u_2^*\frac{\sin(k\delta_2)}{k\delta_2}\right) \\ &\quad + d(\epsilon^2u_1^* + \mu^2u_2^*)v^* - \alpha\beta u_1^*u_2^* \\ &= (d_1d_2 + d_2d_3 + d_3d_1)k^4 + \left\{d_1u_2^*\frac{\sin(k\delta_2)}{k\delta_2} + d_2u_1^*\frac{\sin(k\delta_1)}{k\delta_1} + d_3\left(u_1^*\frac{\sin(k\delta_1)}{k\delta_1} + u_2^*\frac{\sin(k\delta_2)}{k\delta_2}\right)\right\}k^2 \\ &\quad + \left(\frac{\sin(k\delta_1)\sin(k\delta_2)}{k^2\delta_1\delta_2} - \alpha\beta\right)u_1^*u_2^* + d(\epsilon^2u_1^* + \mu^2u_2^*)v^*,\end{aligned} \tag{34}$$

$$\begin{aligned}\widetilde{C}(k,\delta_1,\delta_2) &= d_3k^2\left(d_1k^2 + u_1^*\frac{\sin(k\delta_1)}{k\delta_1}\right)\left(d_2k^2 + u_2^*\frac{\sin(k\delta_2)}{k\delta_2}\right) + d\epsilon^2u_1^*v^*\left(d_2k^2 - u_2^*\frac{\sin(k\delta_2)}{k\delta_2}\right) + \\ &\quad d\mu^2u_2^*v^*\left(d_1k^2 + u_1^*\frac{\sin(k\delta_1)}{k\delta_1}\right) - d_3\alpha\beta u_1^*u_2^*k^2 - d(\alpha+\beta)\epsilon\mu u_1^*u_2^*v^* \\ &= d_1d_2d_3k^6 + d_3\left\{d_1u_2^*\frac{\sin(k\delta_2)}{k\delta_2} + d_2u_1^*\frac{\sin(k\delta_1)}{k\delta_1}\right\}k^4 + \\ &\quad \left\{d_1d\mu^2u_2^*v^* + d_2d\epsilon^2u_1^*v^* + d_3\left(\frac{\sin(k\delta_1)\sin(k\delta_2)}{k^2\delta_1\delta_2} - \alpha\beta\right)u_1^*u_2^*\right\}k^2 \\ &\quad + d\left\{\left(\epsilon^2\frac{\sin(k\delta_2)}{k\delta_2} + \mu^2\frac{\sin(k\delta_1)}{k\delta_1}\right) - (\alpha+\beta)\epsilon\mu\right\}u_1^*u_2^*v^*.\end{aligned} \tag{35}$$

Using the Routh–Hurwitz criteria, we can say that the spatially homogeneous coexistence steady state $E_{+++} = (u_1^*, u_2^*, v^*)$ will be asymptotically stable if and only if

$$\widetilde{A}(k,\delta_1,\delta_2) > 0,\ \widetilde{C}(k,\delta_1,\delta_2) > 0,\ \widetilde{A}(k,\delta_1,\delta_2)\widetilde{B}(k,\delta_1,\delta_2) - \widetilde{C}(k,\delta_1,\delta_2) > 0. \tag{36}$$

In order to attain Turing instability threshold we need to have one zero eigenvalue and other two eigenvalues with negative real parts. Following the process described in the previous section, the conditions for achieving Turing instability threshold imply the following conditions:

$$\widetilde{A}(k,\delta_1,\delta_2) > 0,\ \widetilde{B}(k,\delta_1,\delta_2) > 0,\ \widetilde{C}(k,\delta_1,\delta_2) = 0. \tag{37}$$

Note that the expressions of $\widetilde{A}$, $\widetilde{B}$ and $\widetilde{C}$ are transcendental in nature of their arguments k, δ_1 and δ_2. The associated Turing bifurcation threshold can be obtained by solving the following two equations simultaneously

$$\widetilde{C}(k,\delta_1,\delta_2) = 0,\ \frac{\partial}{\partial k}\widetilde{C}(k,\delta_1,\delta_2) = 0. \tag{38}$$

Solving the first equation of (38) for d_3, we obtain

$$d_3 = -\frac{d\left\{(d_1\mu^2u_2^* + d_2\epsilon^2u_1^*)k^2 + \left(\epsilon^2\frac{\sin(k\delta_2)}{k\delta_2} + \mu^2\frac{\sin(k\delta_1)}{k\delta_1} - (\alpha+\beta)\epsilon\mu\right)u_1^*u_2^*\right\}v^*}{d_1d_2k^6 + \left\{d_1u_2^*\frac{\sin(k\delta_2)}{\delta_2} + d_2u_1^*\frac{\sin(k\delta_1)}{\delta_1}\right\}k^3 + \left\{\frac{\sin(k\delta_1)\sin(k\delta_2)}{k^2\delta_1\delta_2} - \alpha\beta\right\}u_1^*u_2^*k^2}. \tag{39}$$

Now, differentiating $\widetilde{C}(k,\delta_1,\delta_2)$ with respect to k, we get

$$\begin{aligned}\frac{\partial\widetilde{C}}{\partial k} &= d_3\Big[6d_1d_2k^5 + \{d_1u_2^*\cos(k\delta_2) + d_2u_1^*\cos(k\delta_1)\}k^3 + 3\left\{d_1u_2^*\frac{\sin(k\delta_2)}{\delta_2} + d_2u_1^*\frac{\sin(k\delta_1)}{\delta_1}\right\}k^2 - \\ &\quad 2\alpha\beta u_1^*u_2^*k + \left\{\frac{\cos(k\delta_1)\sin(k\delta_2)}{\delta_2} + \frac{\sin(k\delta_1)\cos(k\delta_2)}{\delta_1}\right\}u_1^*u_2^*\Big] + d\,\Big[2(d_1\mu^2u_2^* + d_2\epsilon^2u_1^*)k + \\ &\quad \left\{\epsilon^2\left(\frac{\cos(k\delta_2)}{k} - \frac{\sin(k\delta_2)}{k^2\delta_2}\right) + \mu^2\left(\frac{\cos(k\delta_1)}{k} - \frac{\sin(k\delta_1)}{k^2\delta_1}\right)\right\}u_1^*u_2^*\Big]v^*.\end{aligned} \tag{40}$$

Substituting the expression for d_3 in the second equation of (38), we obtain

$$\mathcal{F}(k,\delta_1,\delta_2) = 0, \tag{41}$$

where the expression for $\mathcal{F}$, which is transcendental in nature, is given in Appendix A. Solving the Equation (41) for k, we obtain the critical wavenumber k_T. But the transcendental nature of the Equation (41) prevents us from finding an analytical solution for k_T and hence, we find the critical value numerically. Further, substituting the value of k_T into the Equation (39) we obtain the critical value of the diffusion coefficient of predator d_3^T in order to induce Turing instability.

Numerical Results

In this subsection, we first investigate the influence of the diffusion coefficients on the positioning of the Turing curve. As the expressions for Turing instability conditions for both the local and nonlocal models involve the diffusion coefficients (that is, d_1, d_2 and d_3), one can expect that variations in their magnitudes can change the corresponding positions of the Turing curves. For this purpose, we present two diagrams in $\epsilon - d_3$ parameter space for $d_2 = 0.2$ (Figure 14a) and $d_2 = 0.1$ (Figure 14b). The other parameter values used in preparation of these diagrams were $b_1 = 1.0$, $\alpha = 1.0$, $b_2 = 1.0$, $\beta = 1.5$, $\mu = 1.0$, $d = 0.5$, $b_3 = 1.0$ and $d_1 = 0.1$. In these two figures, the black vertical straight line (at $\epsilon = 5.486$) represents the temporal-Hopf bifurcation threshold. Since the expression for the temporal-Hopf bifurcation threshold did not involve any of the diffusion coefficients, we had the exact same position for this line in both the figures for the considered two different values of d_2. For the both of these figures, the blue curve represents the Turing curve for the local model (i.e., $\delta_1 = \delta_2 = 0$), magenta curve represents the Turing curve for the nonlocal model where the nonlocality is incorporated only in the first prey species (i.e., $\delta_2 = 0$) with $\delta_1 = 1.0$, green curve denotes the Turing curve for the nonlocal model where the nonlocality is incorporated only in the second prey species (i.e., $\delta_1 = 0$) with $\delta_2 = 1.0$, and red curve denotes the Turing curve for the nonlocal model (27)–(29) with $\delta_1 = \delta_2 = 1.0$. From these two diagrams (i.e., Figure 14a,b), we can observe that the lower value of the parameter d_2 enhanced the Turing instability region significantly. Also, these figures indicate that the introduction of nonlocal intra-specific competition for both the prey species led to the enlargement of the Turing instability region. For the chosen set of parameter values, we can easily notice that incorporation of nonlocality for the second prey species had greater impact on the enlargement of the Turing instability region than that of the first prey species.

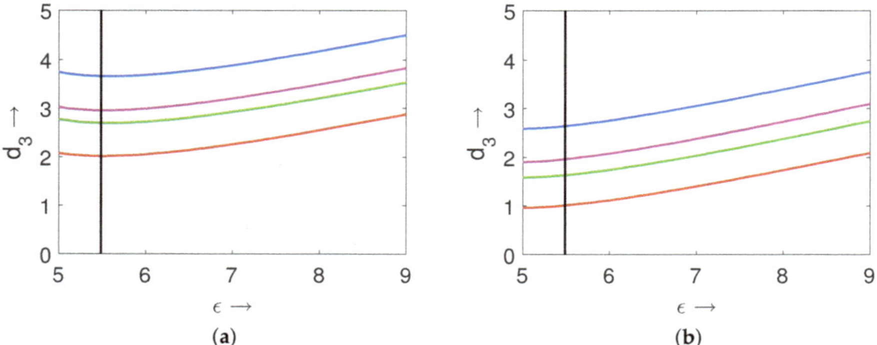

Figure 14. Plots of the temporal-Hopf and Turing bifurcation curves in (ϵ, d_3)-parameter space for (**a**) $d_2 = 0.2$, and (**b**) $d_2 = 0.1$. Here, black vertical straight line corresponds to the temporal-Hopf threshold, blue curve represents the Turing curve for $\delta_1 = \delta_2 = 0$, magenta curve represents the Turing curve for $\delta_1 = 1.0$ and $\delta_2 = 0$, green curve represents the Turing curve for $\delta_1 = 0$ and $\delta_2 = 1.0$, and red curve represents the Turing curve for $\delta_1 = \delta_2 = 1.0$. The black vertical straight line has been drawn by solving the equations presented in (9) simultaneously and the red curve has been drawn from the Equation (39). The magenta and green curves have been drawn from Equations (A4) and (A5), respectively. Brief derivations of Equations (A4) and (A5) are provided in Appendix B.

For the nonlocal system (27)–(29), we used the central difference approximation for diffusion part and forward Euler scheme for the reaction part. The nonlocal interaction terms have been numerically approximated by trapezoidal rule. For all the patterns presented in this subsection, we employed the pulse-type initial conditions (26) and the periodic boundary conditions. Now, we are interested to investigate numerically the dependence of the number of patches with high population density for the stationary patterns on the extent of nonlocality inside the Turing domain. For this purpose, we considered the parameter values $d_2 = 0.2$, $d_3 = 9.0$, $\epsilon = 5.0$ and the remaining parameter values as the same as above. Also, we took equal extents of nonlocality for both the prey species (that is, $\delta_1 = \delta_2 = \delta$). We present the corresponding numerical results for the first prey species in Figure 15 over one-dimensional spatial domain $[-50, 50]$. We employed forward continuation technique in order to prepare this figure. This figure indicates the existence of different regimes with different number of patches depending upon the value of δ. Interestingly, the number of patches initially increased with the increment in δ and then decreased.

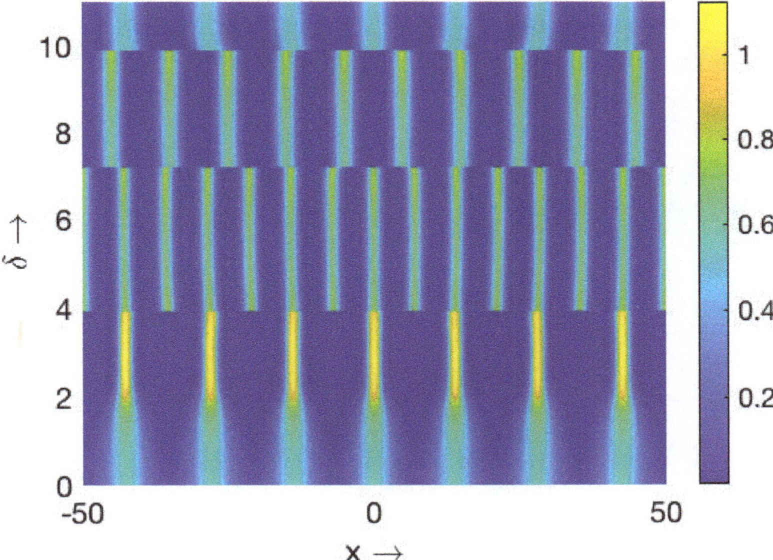

Figure 15. Color plot of density of the first prey species (u_1) using the forward continuation technique. Here, $\epsilon = 5.0$, $d_3 = 9.0$ and other parameter values are mentioned in (25).

Further, we are interested to explore the effect of the extent of nonlocality on the dynamic patterns exhibited by the corresponding local model. In order to do so, we first considered the parameter values for the homogeneous in space and periodic in time pattern presented in Figure 5a, and varied the extent of nonlocality (δ). The resulting patterns for the first prey species are presented in Figure 16. The resulting patterns suggest that small values of δ retained the spatiotemporal distribution of the populations (e.g., Figure 16a for $\delta = 0.5$), a slight increment in δ altered the spatiotemporal distribution by giving rise to periodic in both the space and time population distribution (e.g., Figure 16b for $\delta = 0.7$), and further increments in δ resulted in the emergence of the stationary periodic in space population distribution (e.g., Figure 16c for $\delta = 0.8$). The dynamic nature of the patterns presented in Figure 16a,b can be easily understood from the phase portraits illustrated in Figure 17.

Now, we considered the parameter values for the chaotic pattern presented in Figure 8a, and varied the value of δ to observe the effect of it on the chaotic dynamics of the local model.

The corresponding numerical simulations suggest that the smaller values of δ were unable to alter the chaotic nature of the population distribution, however, interestingly destroyed the symmetry observed for the local model (e.g., Figure 18a for $\delta = 1.0$). A gradual increment in δ resulted in a chaotic pattern where long patches broke into smaller patches (e.g., Figure 18b for $\delta = 2.0$), and finally settled into a stationary periodic in space pattern for higher values of δ (e.g., Figure 18c for $\delta = 2.5$). Chaotic nature of the patterns presented in Figure 18a,b can be easily observed from the phase portraits illustrated in Figure 19. The chaotic nature of these patterns have also been cross verified by checking the sensitivity to initial conditions as described in [59], but we do not present the results here for the sake of brevity. For the intermediate periodic in both the space and time population distributions presented in Figures 5b and 8b, we found that the small extents of nonlocality for both the prey species could potentially stabilize the population distributions by giving rise to the stationary periodic in space distributions and we chose not to display these results here for the sake of brevity. Overall, the above observations indicate the stabilizing effect of the extents of nonlocality for all the three considered species.

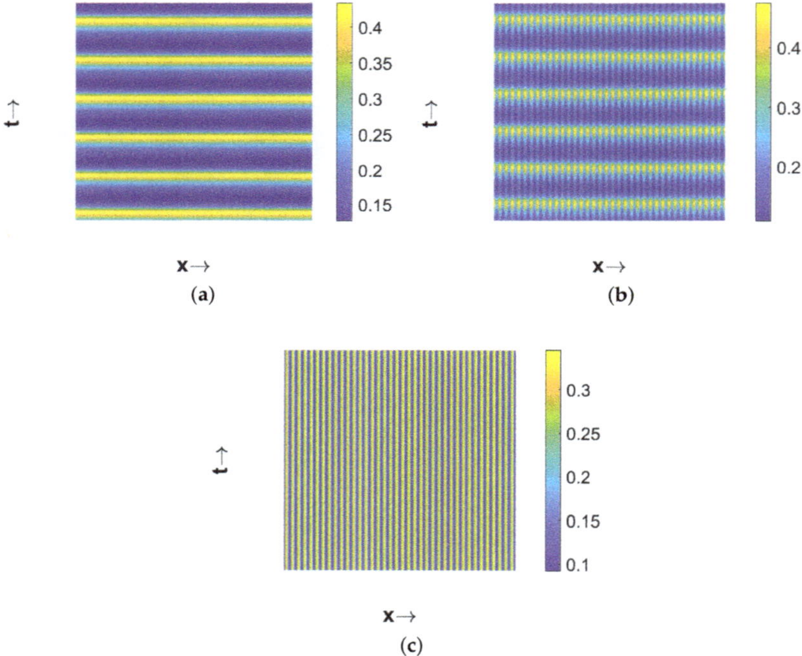

Figure 16. Plots of different types of patterns exhibited by u_1 representing the effect of the extent of nonlocality for $\epsilon = 6.0$ and $d_3 = 3.0$. Three different patterns are obtained for (**a**) $\delta = 0.5$; (**b**) $\delta = 0.7$; (**c**) $\delta = 0.8$. Other parameter values are the same as that of Figure 5a.

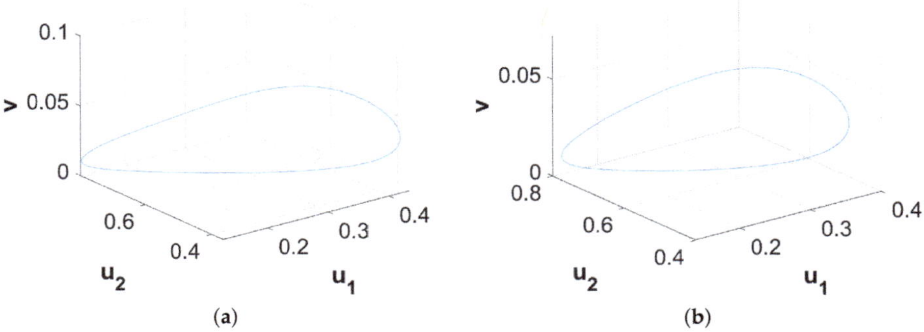

Figure 17. Phase diagrams of spatial average densities of the associated three species corresponding to (**a**) Figure 16a, (**b**) Figure 16b.

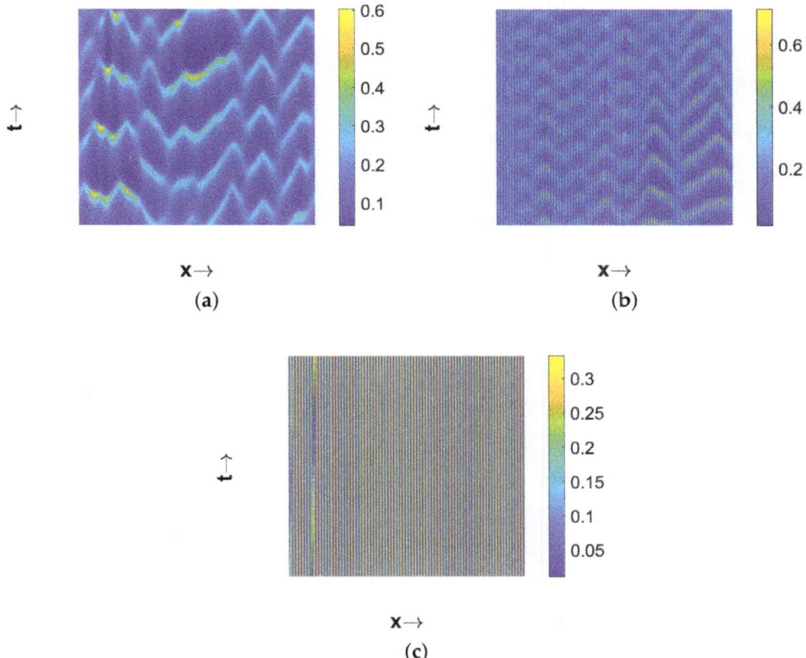

Figure 18. Plots of different types of patterns exhibited by u_1 representing the effect of the extent of nonlocality for $\epsilon = 8.0$ and $d_3 = 1.0$. Three different patterns are obtained for (**a**) $\delta = 1.0$; (**b**) $\delta = 2.0$; (**c**) $\delta = 2.5$. Other parameter values are the same as that of Figure 8a.

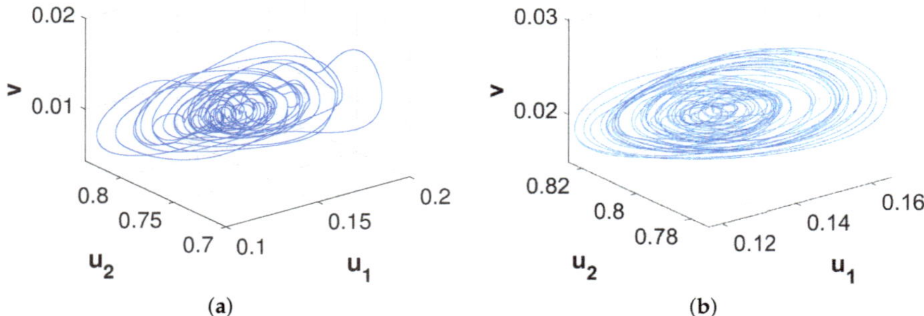

Figure 19. Phase diagrams of spatial average densities of the associated three species corresponding to (**a**) Figure 18a, (**b**) Figure 18b.

Finally, we want to explore the effects of the extent of nonlocality on the dynamics presented in Figures 1 and 11. For this purpose, we considered the same parameter values as it was used to prepare the bifurcation diagram presented in Figure 11 with $\delta_1 = \delta_2 = 2.0$. The resulting dynamics are encapsulated in Figure 20. In this figure, the red colored curve corresponds to the first prey component from the spatially homogeneous coexistence steady state and blue color stands for the simulation results of the nonlocal model (27)–(29). Here, we can observe that the average densities of the first prey species stayed almost below the steady state values throughout the region in presence of the nonlocal interactions. Also, note that spatial heterogeneity in population distribution appeared throughout the region as opposed to that for both the temporal and local spatiotemporal cases. This figure reconfirms that the spatially heterogeneous population distribution can suppress the spatially averaged density for the first prey species. The same scenario happens for the second prey species, whereas, the spatial heterogeneity enhances the spatially averaged density for the predator population. For the sake of brevity, we restrict ourselves from displaying the corresponding bifurcation diagrams for both the second prey and predator species.

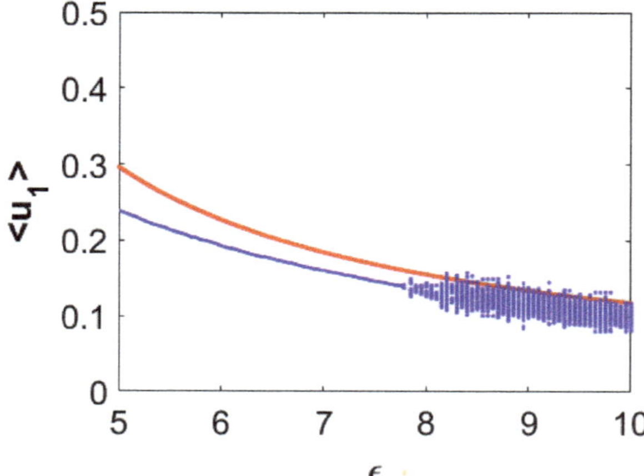

Figure 20. Single parameter bifurcation diagram for the spatially averaged density of first prey population ($< u_1 >$) with respect to the parameter ϵ for the nonlocal model (27)–(29). Here, $\delta_1 = \delta_2 = 2.0$, $d_3 = 1.0$ and other parameter values are mentioned in (25).

5. Discussion

The number of studies on spatiotemporal pattern formation for three-species models is comparatively less in existing literature and there exist even fewer works for models with nonlocal interactions. In this article, we have investigated for the possible impacts of random dispersal and nonlocal intra-specific interactions on the dynamics of a two-prey-one-predator system. For the temporal version of the model, we have mentioned the period-doubling route to chaos by means of numerical simulations. For the corresponding spatiotemporal models with both the local and nonlocal intra-specific interactions for the two considered prey species, we have derived the conditions for Turing instability through linear analysis. Further, we have carried out extensive numerical simulations in order to validate our theoretical predictions as well as to explore rich complex spatiotemporal dynamical behaviors which are beyond the scope of linear analysis. Overall, we have presented a comparative study on the dynamics possessed by the temporal, local, and nonlocal spatiotemporal models.

Local spatiotemporal model (11)–(13) along with the periodic boundary conditions possesses a wide variety of stationary and dynamic patterns. For the parameter values well inside the Turing domain, the model produces stationary periodic in space patterns for all the three considered species. However, an interesting phenomenon has been observed for these stationary patterns where the regions with high first prey density correspond to the same for predator but to low second prey density. This phenomenon arises due to the fact that the chosen values of the parameter ϵ are much large than the parameter μ and this makes the first prey species more favorable to predator than the second prey species. Apart from this, numerical simulations show the stabilizing effect of diffusion on the spatial population distributions by making the periodic in time homogeneous in space or even chaotic population distributions to stationary periodic in space one for higher values of diffusion coefficient of predator species (d_3). Both these transitions occur through the periodic in both space and time population distribution for intermediate values of d_3. We would like to mention here that the choice of other diffusion coefficients as bifurcation parameter will not reveal any additional feature as the main determining factor is the location of the parameter value with respect to the temporal-Hopf and Turing bifurcation curves.

Traditionally, the spatiotemporal dynamics of an ecological system have been investigated in the asymptotic sense. Sometimes it can take a very long time period for an ecological system to eventually settle down to a spatiotemporal state and this long time period can be as long as a few generations for certain species. Therefore, from the perspective of ecological time scales it is very important to investigate the quasi-stable transients for an ecological system and to identify the key factors for its appearance [60]. A long transient exhibits a quasi-stable dynamics for a few generations and then the ecological system experiences an abrupt transition to another regime [60]. Hence, it can provide an alternative explanation for ecological regime shifts apart from the usual concept of bifurcation [60]. However, bifurcation of dynamical behaviors for a system arises due to directional change in parameter values (for example, Figures 7 and 11) and this directional change is often considered to be mediated through external factors, such as climate change [60]. Therefore, bifurcation can provide useful explanation for exogenously triggered regime shifts whereas long transients can serve as alternative underlying mechanisms for regime shifts [60]. Hence, the bifurcation diagrams such as Figures 7 and 11 exhibit exogenously triggered regime shifts while the phase portraits presented in Figure 10 provide evidence for endogenously triggered regime shift for the local spatiotemporal model (11)–(13).

Further, we have investigated for the possible effects of random dispersal of all the considered species on the dynamics possessed by the corresponding temporal model. In this regard, we have observed that the local spatiotemporal model (11)–(13) with spatially homogeneous initial conditions preserves the temporal dynamics whereas the system with pulse type initial conditions produces smaller amplitude chaotic oscillations than the temporal chaos. This type of reduction in amplitude of chaotic oscillations arises due to the emergence of spatial heterogeneity in population distributions.

It is worthy to mention here that the spatially homogeneous perturbation over the entire domain is not ecologically realistic. Therefore, our results suggest that one should expect small amplitude chaotic oscillations for a spatially heterogeneous population as compared to a spatially homogeneous population. Also, we have shown that the backward continuation technique significantly enhances the spatially heterogeneous chaotic regime by destroying the period-doubling route to chaos. These bifurcation diagrams (that is, Figures 1, 11 and 12) vouch for the existence of alternative stable states for the spatiotemporal model without nonlocal interactions (11)–(13).

Another interesting dynamical aspect of the local spatiotemporal model (11)–(13) is that the nature of the patterns produced is not independent of the spatial domain size for certain parameter range, in particular, for parameter values in temporal-Hopf unstable region sufficiently far from the temporal-Hopf threshold. In this regard, we have shown that a chaotic pattern can become a two-periodic one in smaller spatial domain and even a stationary heterogeneous pattern can loose its stability through becoming a periodic one in both the space and time in smaller spatial domain. These spatial domain size dependence of spatiotemporal patterns have been reported in Figure 13. Therefore, our study suggests that one can get different spatiotemporal patterns depending on the spatial domain size. At this point, we would like to emphasize on the fact that this type of domain size dependence for the resulting spatiotemporal patterns is not unheard of at all. For instance, Barrio et al. [61] showed that both the domain size and shape can have significant roles in selection of different types of patterns. Also, Neville et al. [62] investigated how domain growth can influence pattern formation. Hence, our findings on domain size dependence of patterns are in accordance with literature and carry forward our understanding in this direction.

The dependence of results on the domain size can be explained as follows. If there exists a spatially periodic solution $u(x,t)$ with the wavenumber k and the length of the interval $L_1 = \frac{2\pi n_1}{k}$, where $n_1 \in \mathbb{N}$, then this solution can be naturally extended by periodicity to the interval $L_2 = \frac{2\pi n_2}{k}$, where $n_2 > n_1$. Moreover, if we increase the length of the interval, solutions with other wavenumbers can appear due to their successive bifurcations from the spatially homogeneous stationary solutions. Furthermore, solutions with a fixed wavenumber persist under the variation of the length of the interval in some range of lengths. Hence, solutions with different wavenumbers can coexist for the same length, and their number increases with the increase of L. This multiplicity of solutions is conventionally observed in bifurcation problems. Their stability and basins of attraction depend on the length L. The interaction of different solutions can result in complex dynamics including chaotic behavior. The transition from two-periodic oscillations to chaotic oscillations for a larger length (cf. Figures 8 and 13) corresponds to this understanding of the dynamics of solutions. It is also possible that the initial condition remains in the basin of attraction of the same solution, such that we observe the same dynamics for a changing interval.

Finally, we have studied the effects of nonlocal interactions for both the prey species on the spatiotemporal dynamics. Our study suggests that the introduction of nonlocal interactions for both the prey species or any one of them can lead to the enlargement of the Turing instability region where nonlocality for the second prey species possess greater impact than that for the first prey species. In order to understand the impact of nonlocality on the stationary Turing patterns, we have employed forward continuation technique which shows the existence of multiple stationary regimes with different number of patches. The number of patches initially increases and then decreases with increasing extent of nonlocality. This scenario is encapsulated in Figure 15 and it is different from the result presented in [48] where the number of patches gradually increases. Another interesting feature of the dynamics of nonlocal model (27)–(29) is that it breaks the spatial symmetry possessed by local model (11)–(13) for chaotic population distribution. Our results also indicate that the nonlocality acts as a stabilizing factor as large extent of it can suppress the oscillatory behaviors by changing them to stationary patterns. Ecologically it can be interpreted as the access to the nearby resources allow the species to be confined within a patch rather than changing its position in a regular or irregular fashion. Further, our investigation suggests that the presence of spatial heterogeneity in population distributions can diminish the spatially averaged density for both the prey species (for instance, see the Figure 20 for the first prey species) and enhances the same for the predator species.

This study leads to some other interesting and relevant problems to investigate. In this study, we have confined ourselves to only the top-hat kernel function in order to model the nonlocal interactions. However, it will be a fascinating future research prospect to investigate the spatiotemporal dynamics with other types of kernel functions such as Gaussian, Laplacian and triangular kernels etc. Another interesting problem will be to consider nonlocal consumption for predator species and examine its effect on dynamics. The present study has considered Lotka–Volterra type linear functional responses in order to capture the prey–predator interactions whereas the consideration of more suited functional responses such as Holling type-II and Beddington–DeAngelis functional responses will make the investigation ecologically more viable and challenging. These interesting as well as challenging issues will be taken care of in our future works.

Author Contributions: M.B. and V.V. proposed the formulation of the problem; K.M. performed the mathematical calculations and numerical simulations; M.B. analyzed and verified the results. All authors participated in the preparation of the manuscript. All authors have read and agreed to the published version of the manuscript.

Funding: The first author gratefully acknowledges the financial support provided by Science and Engineering Research Board, Government of India for pursuing his post-doctoral research. V.V. was supported by the "RUDN University Program 5-100" and the French-Russian project PRC2307. M.B. was supported by SERB grant MTR/2018/000527.

Conflicts of Interest: The authors declare no conflict of interest.

Appendix A

The transcendental expression for $\mathcal{F}(k,\delta_1,\delta_2)$ introduced in Section 4 is given by

$$\begin{aligned}
&\mathcal{F}(k,\delta_1,\delta_2) = \\
&4d_1d_2(d_1\mu^2u_2^* + d_2\epsilon^2u_1^*)k^9 + \{d_1^2\mu^2u_2^{*2}\cos(k\delta_2) + d_2^2\epsilon^2u_1^{*2}\cos(k\delta_1) - 6d_1d_2(\alpha+\beta)\epsilon\mu u_1^*u_2^*\}\,k^7 + \\
&\left\{8d_1d_2u_1^*u_2^*\left(\epsilon^2\frac{\sin(k\delta_2)}{\delta_2} + \mu^2\frac{\sin(k\delta_1)}{\delta_1}\right) + d_1^2\mu^2u_2^{*2}\frac{\sin(k\delta_2)}{\delta_2} + d_2^2\epsilon^2u_1^{*2}\frac{\sin(k\delta_1)}{\delta_1}\right\}k^6 - \\
&\{(\alpha+\beta)\epsilon\mu u_1^*u_2^*(d_1u_2^*\cos(k\delta_2) + d_2u_1^*\cos(k\delta_1))\}k^5 + u_1^*u_2^*\left\{2\epsilon^2\left(d_1u_2^*\frac{\sin(k\delta_2)\cos(k\delta_2)}{\delta_2} + \right.\right. \\
&\left. d_2u_1^*\frac{\sin(k\delta_1)\cos(k\delta_2)}{\delta_1} + d_2u_1^*\frac{\cos(k\delta_1)\sin(k\delta_2)}{\delta_2}\right) + 2\mu^2\left(d_2u_1^*\frac{\sin(k\delta_1)\cos(k\delta_1)}{\delta_1} + \right. \\
&\left.\left. d_1u_2^*\frac{\sin(k\delta_1)\cos(k\delta_2)}{\delta_1} + d_1u_2^*\frac{\cos(k\delta_1)\sin(k\delta_2)}{\delta_2}\right) - 3(\alpha+\beta)\epsilon\mu\left(d_1u_2^*\frac{\sin(k\delta_2)}{\delta_2} + d_2u_1^*\frac{\sin(k\delta_1)}{\delta_1}\right)\right\}k^4 \\
&+u_1^*u_2^*\left\{4\left(d_1\epsilon^2u_2^*\frac{\sin^2(k\delta_2)}{\delta_2^2} + d_2\mu^2u_1^*\frac{\sin^2(k\delta_1)}{\delta_1^2}\right) + \alpha\beta u_1^*u_2^*(\epsilon^2\cos(k\delta_2) + \mu^2\cos(k\delta_1)) + \right. \\
&\left. 2(d_1\mu^2u_2^* + d_2\epsilon^2u_1^*)\frac{\sin(k\delta_1)\sin(k\delta_2)}{\delta_1\delta_2} + 2\alpha\beta(\alpha+\beta)\epsilon\mu u_1^*u_2^*\right\}k^3 - u_1^{*2}u_2^{*2}\left\{3\alpha\beta\left(\epsilon^2\frac{\sin(k\delta_2)}{\delta_2} + \right.\right. \\
&\left.\left. \mu^2\frac{\sin(k\delta_1)}{\delta_1}\right) + (\alpha+\beta)\epsilon\mu\left(\frac{\cos(k\delta_1)\sin(k\delta_2)}{\delta_2} + \frac{\sin(k\delta_1)\cos(k\delta_2)}{\delta_1}\right)\right\}k^2 + u_1^{*2}u_2^{*2}\left\{\epsilon^2\frac{\cos(k\delta_1)\sin^2(k\delta_2)}{\delta_2^2}\right. \\
&\left. +\mu^2\frac{\sin^2(k\delta_1)\cos(k\delta_2)}{\delta_1^2}\right\}k + u_1^{*2}u_2^{*2}\left\{\left(\epsilon^2\frac{\sin(k\delta_2)}{\delta_2} + \mu^2\frac{\sin(k\delta_1)}{\delta_1}\right)\frac{\sin(k\delta_1)\sin(k\delta_2)}{\delta_1\delta_2}\right\}.
\end{aligned} \tag{A1}$$

Appendix B

Here, we briefly describe about the calculation of critical value d_3^T for inducing Turing instability with nonlocal intra-specific competition being present in only one prey species.

Case 1: $\delta_1 \neq 0, \delta_2 = 0$

Proceeding in a similar manner as described in Section 4, we have

$$\begin{aligned}
\widetilde{C}(k,\delta_1) \;=\;& d_1d_2d_3k^6 + d_3\left\{d_1u_2^* + d_2u_1^*\frac{\sin(k\delta_1)}{k\delta_1}\right\}k^4 + \left\{d(d_1\mu^2u_2^* + d_2\epsilon^2u_1^*)v^* + d_3\left(\frac{\sin(k\delta_1)}{k\delta_1} - \alpha\beta\right)u_1^*u_2^*\right\}k^2 \\
& +d\left\{\left(\epsilon^2 + \mu^2\frac{\sin(k\delta_1)}{k\delta_1}\right) - (\alpha+\beta)\epsilon\mu\right\}u_1^*u_2^*v^*.
\end{aligned} \tag{A2}$$

Then, the Turing bifurcation threshold can be obtained by solving the following two equations simultaneously

$$\widetilde{C}(k,\delta_1)=0,\ \frac{\partial}{\partial k}\widetilde{C}(k,\delta_1)=0. \quad \text{(A3)}$$

Solving the first equation of (A3) for d_3, we obtain

$$d_3 = -\frac{d\left\{(d_1\mu^2u_2^*+d_2\epsilon^2u_1^*)k^2+\left(\epsilon^2+\mu^2\frac{\sin(k\delta_1)}{k\delta_1}-(\alpha+\beta)\epsilon\mu\right)u_1^*u_2^*\right\}v^*}{d_1d_2k^6+\left\{d_1u_2^*+d_2u_1^*\frac{\sin(k\delta_1)}{k\delta_1}\right\}k^4+\left\{\frac{\sin(k\delta_1)}{k\delta_1}-\alpha\beta\right\}u_1^*u_2^*k^2}. \quad \text{(A4)}$$

Now, substituting the above expression for d_3 in the second equation of (A3) and solving for k, we obtain the critical wavenumber k_T. Further, substitution of $k=k_T$ into the Equation (A4) gives us the desired critical value d_3^T.

Case 2: $\delta_1=0,\ \delta_2\neq 0$

Proceeding in a similar manner as described above, we obtain the following critical value d_3^T to induce Turing instability in this case:

$$d_3^T = -\frac{d\left\{(d_1\mu^2u_2^*+d_2\epsilon^2u_1^*)k_T^2+\left(\epsilon^2\frac{\sin(k_T\delta_2)}{k_T\delta_2}+\mu^2-(\alpha+\beta)\epsilon\mu\right)u_1^*u_2^*\right\}v^*}{d_1d_2k_T^6+\left\{d_1u_2^*\frac{\sin(k_T\delta_2)}{k_T\delta_2}+d_2u_1^*\right\}k_T^4+\left\{\frac{\sin(k_T\delta_2)}{k_T\delta_2}-\alpha\beta\right\}u_1^*u_2^*k_T^2}. \quad \text{(A5)}$$

References

1. Paine, R.T. Food web complexity and species diversity. *Am. Nat.* **1966**, *100*, 65–75. [CrossRef]
2. Fujii, K. Complexity-stability relationship of two-prey-one-predator species system model: Local and global stability. *J. Theor. Biol.* **1977**, *69*, 613–623. [CrossRef]
3. Parrish, J.D.; Saila, S.B. Interspecific competition, predation and species diversity. *J. Theor. Biol.* **1970**, *27*, 207–220. [CrossRef]
4. Vance, R.R. Predation and resource partitioning in one predator-two prey model communities. *Am. Nat.* **1978**, *112*, 797–813. [CrossRef]
5. Takeuchi, Y.; Adachi, N. Existence and bifurcation of stable equilibrium in two-prey, one-predator communities. *Bull. Math. Biol.* **1983**, *45*, 877–900. [CrossRef]
6. Hutson, V.; Vickers, G.T. A criterion for permanent coexistence of species, with an application to a two-prey one-predator system. *Math. Biosci.* **1983**, *63*, 253–269. [CrossRef]
7. Matsuda, H.; Kawasaki, K.; Shigesada, N.; Teramoto, E.; Ricciardi, L.M. Switching effect on the stability of the prey-predator system with three trophic levels. *J. Theor. Biol.* **1986**, *122*, 251–262. [CrossRef]
8. Abrams, P.; Matsuda, H. Effects of adaptive predatory and anti-predator behaviour in a two-prey-one-predator system. *Evol. Ecol.* **1993**, *7*, 312–326. [CrossRef]
9. Klebanoff, A.; Hastings, A. Chaos in one-predator, two-prey models: General results from bifurcation theory. *Math. Biosci.* **1994**, *122*, 221–233. [CrossRef]
10. Abrams, P.A. Is predator-mediated coexistence possible in unstable systems? *Ecology* **1999**, *80*, 608–621. [CrossRef]
11. van Leeuwen, E.; Jansen, V.A.A.; Bright, P.W. How population dynamics shape the functional response in a one-predator-two-prey system. *Ecology* **2007**, *88*, 1571–1581. [CrossRef] [PubMed]
12. Lee, S.S.; Kajiwara, T. The effect of the remains of the carcass in a two-prey, one-predator model. *Discret. Contin. Dyn. Syst. Ser. B* **2008**, *9*, 353–374.
13. Zaman, G.; Saker, S.H. Dynamics and control of a system of two non-interacting preys with common predator. *Math. Methods Appl. Sci.* **2011**, *34*, 2259–2273. [CrossRef]

14. Groll, F.; Arndt, H.; Altland, A. Chaotic attractor in two-prey one-preator system originates from interplay of limit cycles. *Theor. Ecol.* **2016**, *10*, 147–154. [CrossRef]
15. Turchin, P.; Ellner, S.P. Living on the edge of chaos: Population dynamics of Fennoscandian voles. *Ecology* **2000**, *81*, 3099–3116. [CrossRef]
16. Levin, S.A.; Segel, L.A. Hypothesis for origin of planktonic patchiness. *Nature* **1976**, *259*, 659. [CrossRef]
17. Turing, A.M. The chemical basis of morphogenesis. *Philos. Trans. R. Soc. Lond. B Biol. Sci.* **1952**, *237*, 37–72.
18. Manna, K.; Banerjee, M. Stationary, non-stationary and invasive patterns for a prey-predator system with additive Allee effect in prey growth. *Ecol. Complex.* **2018**, *36*, 206–217. [CrossRef]
19. Baurmann, M.; Gross, T.; Feudel, U. Instabilities in spatially extended predator-prey systems: Spatio-temporal patterns in the neighborhood of Turing-Hopf bifurcations. *J. Theor. Biol.* **2007**, *245*, 220–229. [CrossRef]
20. Zhang, L.; Liu, J.; Banerjee, M. Hopf and steady state bifurcation analysis in a ratio-dependent predator-prey model. *Commun. Nonlinear Sci. Numer. Simul.* **2017**, *44*, 52–73. [CrossRef]
21. White, K.A.J.; Gilligan, C.A. Spatial heterogeneity in three-species, plant-parasite-hyperparasite, systems. *Phil. Trans. R. Soc. Lond. B* **1998**, *353*, 543–557. [CrossRef]
22. Petrovskii, S.; Kawasaki, K.; Takasu, F.; Shigesada, N. Diffusive waves, dynamical stabilization and spatio-temporal chaos in a community of three competitive species. *Jpn. J. Ind. Appl. Math.* **2001**, *18*, 459–481. [CrossRef]
23. Chen, W.; Peng, R. Stationary patterns created by cross-diffusion for the competitor-competitor-mutualist model. *J. Math. Anal. Appl.* **2004**, *291*, 550–564. [CrossRef]
24. Pang, P.Y.H.; Wang, M. Strategy and stationary pattern in a three-species predator-prey model. *J. Differ. Equ.* **2004**, *200*, 245–273. [CrossRef]
25. Wang, M. Stationary patterns for a prey-predator model with prey-dependent and ratio-dependent functional responses and diffusion. *Phys. D* **2004**, *196*, 172–192. [CrossRef]
26. Wang, M. Stationary patterns caused by cross-diffusion for a three-species prey-predator model. *Comput. Math. Appl.* **2006**, *52*, 707–720. [CrossRef]
27. Tian, C. Turing patterns created by cross-diffusion for a Holling II and Leslie-Gower type three species food chain model. *J. Math. Chem.* **2011**, *49*, 1128–1150. [CrossRef]
28. Tian, C.; Ling, Z.; Lin, Z. Turing pattern formation in a predator-prey-mutualist system. *Nonlinear Anal. Real World Appl.* **2011**, *12*, 3224–3237. [CrossRef]
29. Rao, F. Spatiotemporal complexity of a three-species ratio-dependent food chain model. *Nonlinear Dyn.* **2014**, *76*, 1661–1676. [CrossRef]
30. Abid, W.; Yafia, R.; Alaoui, M.A.A.; Bouhafa, H.; Abichou, A. Instability and pattern formation in three-species food chain model via Holling type II functional response on a circular domain. *Int. J. Bifurc. Chaos* **2015**, *25*, 1550092. [CrossRef]
31. Ikeda, T.; Mimura, M. An interfacial approach to regional segregation of two competing species mediated by a predator. *J. Math. Biol.* **1993**, *31*, 215–240. [CrossRef]
32. Ko, W.; Ryu, K. Analysis of diffusive two-competing-prey and one-predator systems with Beddington-DeAngelis functional response. *Nonlinear Anal. Theory Methods Appl.* **2009**, *71*, 4185–4202. [CrossRef]
33. Lv, Y.; Yuan, R.; Pei, Y. Turing pattern formation in a three species model with generalist predator and cross-difusion. *Nonlinear Anal. Theory, Methods Appl.* **2013**, *85*, 214–232. [CrossRef]
34. Xie, Z. Cross-diffusion induced Turing instability for a three species food chain model. *J. Math. Anal. Appl.* **2012**, *388*, 539–547. [CrossRef]
35. Peng, R.; Shi, J.; Wang, M. Stationary pattern of a ratio-dependent food chain model with diffusion. *SIAM J. Appl. Math.* **2007**, *67*, 1479–1503. [CrossRef]
36. Feng, W. Coexistence, stability, and limiting behavior in a one-predator-two-prey model. *J. Math. Anal. Appl.* **1993**, *179*, 592–609. [CrossRef]
37. Mukherjee, N.; Ghorai, S.; Banerjee, M. Detection of Turing patterns in a three species food chain model via amplitude equation. *Commun. Nonlinear Sci. Numer. Simul.* **2019**, *69*, 219–236. [CrossRef]
38. Tanzy, M.C.; Volpert, V.A.; Bayliss, A.; Nehrkorn, M.E. Stability and pattern formation for competing populations with asymmetric nonlocal coupling. *Math. Biosci.* **2013**, *246*, 14–26. [CrossRef]

39. Autry, E.A.; Bayliss, A.; Volpert, V.A. Biological control with nonlocal interactions. *Math. Biosci.* **2018**, *301*, 129–146. [CrossRef]
40. Pal, S.; Ghorai, S.; Banerjee, M. Effect of kernels on spatio-temporal patterns of a non-local prey-predator model. *Math. Biosci.* **2019**, *310*, 96–107. [CrossRef]
41. Merchant, S.M.; Nagata, W. Instabilities and spatiotemporal patterns behind predator invasions with nonlocal prey competition. *Theor. Popul. Biol.* **2011**, *80*, 289–297. [CrossRef] [PubMed]
42. Segal, B.L.; Volpert, V.A.; Bayliss, A. Pattern formation in a model of competing populations with nonlocal interactions. *Phys. D* **2013**, *253*, 12–22. [CrossRef]
43. Gourley, S.A. Travelling front solutions of a nonlocal Fisher equation. *J. Math. Biol.* **2000**, *41*, 272–284. [CrossRef]
44. Merchant, S.M.; Nagata, W. Selection and stability of wave trains behind predator invasions in a model with non-local prey competition. *IMA J. Appl. Math.* **2015**, *80*, 1155–1177. [CrossRef]
45. Deng, K.; Wu, Y. Global stability for a nonlocal reaction-diffusion population model. *Nonlinear Anal. Real World Appl.* **2015**, *25*, 127–136. [CrossRef]
46. Banerjee, M.; Volpert, V. Prey-predator model with a nonlocal consumption of prey. *Chaos* **2016**, *26*, 083120. [CrossRef] [PubMed]
47. Banerjee, M.; Volpert, V. Spatio-temporal pattern formation in Rosenzweig-MacArthur model: Effect of nonlocal interactions. *Ecol. Complex.* **2017**, *30*, 2–10. [CrossRef]
48. Pal, S.; Ghorai, S.; Banerjee, M. Analysis of a prey-predator model with non-local interaction in the prey population. *Bull. Math. Biol.* **2018**, *80*, 906–925. [CrossRef]
49. Bayliss, A.; Volpert, V.A. Complex predator invasion waves in a Holling-Tanner model with nonlocal prey interaction. *Phys. D* **2017**, *346*, 37–58. [CrossRef]
50. Han, B.-S.; Yang, Y.-H. On a predator-prey reaction-diffusion model with nonlocal effects. *Commun. Nonlinear Sci. Numer. Simul.* **2017**, *46*, 49–61. [CrossRef]
51. Tian, C.; Ling, Z.; Zhang, L. Nonlocal interaction driven pattern formation in a prey-predator model. *Appl. Math. Comput.* **2017**, *308*, 73–83. [CrossRef]
52. Banerjee, M.; Mukherjee, N.; Volpert, V. Prey-predator model with a nonlocal bistable dynamics of prey. *Mathematics* **2018**, *6*, 41. [CrossRef]
53. Pal, S.; Banerjee, M.; Ghorai, S. Spatio-temporal pattern formation in Holling-Tanner type model with nonlocal consumption of resources. *Int. J. Bifurc. Chaos* **2019**, *29*, 1930002. [CrossRef]
54. Ni, W.; Shi, J.; Wang, M. Global stability and pattern formation in a nonlocal diffusive Lotka-Volterra competition model. *J. Differ. Equ.* **2018**, *264*, 6891–6932. [CrossRef]
55. Han, R.; Dai, B.; Chen, Y. Pattern formation in a diffusive intraguild predation model with nonlocal interaction effects. *AIP Adv.* **2019**, *9*, 035046. [CrossRef]
56. Takeuchi, Y. *Global Dynamical Properties of Lotka-Volterra Systems*; World Scientific: Singapore, 1996.
57. Othmer, H.G.; Scriven, L.E. Interactions of reaction and diffusion in open systems. *Ind. Eng. Chem. Fundamen.* **1969**, *8*, 302–313. [CrossRef]
58. Murray, J.D. *Mathematical Biology*; Springer: Heidelberg/Berlin, Germany, 1989.
59. Banerjee, M.; Abbas, S. Existence and non-existence of spatial patterns in a ratio-dependent predator-prey model. *Ecol. Complex.* **2015**, *21*, 199–214. [CrossRef]
60. Hastings, A.; Abbott, K.C.; Cuddington, K.; Francis, T.; Gellner, G.; Lai, Y.-C.; Morozov, A.; Petrovskii, S.; Scranton, K.; Zeeman, M.L. Transient phenomena in ecology. *Science* **2018**, *361*, eaat6412. [CrossRef]
61. Barrio, R.A.; Varea, C.; Aragón, J.L.; Maini, P.K. A two-dimensional numerical study of spatial patern formation in interacting Turing systems. *Bull. Math. Biol.* **1999**, *61*, 483–505. [CrossRef]
62. Neville, A.A.; Matthews, P.C.; Byrne, H.M. Interactions between pattern formation and domain growth. *Bull. Math. Biol.* **2006**, *68*, 1975–2003. [CrossRef]

Article

Persistence for a Two-Stage Reaction-Diffusion System

Robert Stephen Cantrell [1,*], Chris Cosner [1] and Salomé Martínez [2]

1 Department of Mathematics, University of Miami, Coral Gables, FL 33124, USA; gcc@math.miami.edu
2 Departamento de Ingeniería Matemática and Centro de Modelamiento Matemático, UMI 2807 CNRS-UChile, Universidad de Chile, 8370456 Santiago, CHILE; samartin@dim.uchile.cl
* Correspondence: rsc@math.miami.edu

Received: 3 February 2020; Accepted: 5 March 2020; Published: 11 March 2020

Abstract: In this article, we study how the rates of diffusion in a reaction-diffusion model for a stage structured population in a heterogeneous environment affect the model's predictions of persistence or extinction for the population. In the case of a population without stage structure, faster diffusion is typically detrimental. In contrast to that, we find that, in a stage structured population, it can be either detrimental or helpful. If the regions where adults can reproduce are the same as those where juveniles can mature, typically slower diffusion will be favored, but if those regions are separated, then faster diffusion may be favored. Our analysis consists primarily of estimates of principal eigenvalues of the linearized system around $(0,0)$ and results on their asymptotic behavior for large or small diffusion rates. The model we study is not in general a cooperative system, but if adults only compete with other adults and juveniles with other juveniles, then it is. In that case, the general theory of cooperative systems implies that, when the model predicts persistence, it has a unique positive equilibrium. We derive some results on the asymptotic behavior of the positive equilibrium for small diffusion and for large adult reproductive rates in that case.

Keywords: reaction-diffusion; spatial ecology; population dynamics; stage structure; dispersal

AMS Classification: 92D40; 92D50; 35P15; 35K57

---

1. Introduction

The question of how dispersal interacts with spatial heterogeneity to influence population dynamics and species interactions has been studied extensively in recent years, specifically from the viewpoint of reaction-diffusion systems and related models—see, for example, [1–3] and the references cited therein. Most work on that topic assumes that each population is structured only by space and has only one mode of dispersal. However, populations are often structured by age, stage, or other attributes, and there may be variation among individuals in their dispersal rates or patterns. Here we will examine how the presence of a stage structure influences how diffusion rates influence population dynamics in a class of reaction-diffusion models for a population with two stages. In the case of a population with logistic growth, without age or stage structure, diffusing in a closed bounded spatially heterogeneous environment that is constant in time, it is well known that reaction-diffusion models predict that slower diffusion rates are advantageous relative to faster diffusion—see [4,5]. The results in [5] also hold for patch models. More broadly, a wide class of models arising in population genetics, population dynamics, and related areas display some version of the reduction principle, which says that dispersal, which causes faster mixing, typically reduces the rate of population growth—see [6]. However, the situation seems to be quite different in the case of stage structured populations. In [7], the authors considered a discrete-time patch model for a structured population and found that, in some cases, there was no selection against faster dispersal. The goal of the present paper is to use a spatially explicit reaction-diffusion model to understand how the spatial distributions of habitats that are favorable for reproduction by adults and those that are favorable for survival and growth by juveniles affect whether faster diffusion is advantageous or harmful for a stage structured population. We will see that the answer depends on the details of the spatial distribution of favorable and unfavorable habitats.

The type of reaction-diffusion model we will study is

$$\begin{cases} \dfrac{\partial u}{\partial t} = d_1 \Delta u + r(x)v - s(x)u - a(x)u - b(x)u^2 - c(x)uv \text{ in } \Omega,\ t > 0, \\ \dfrac{\partial v}{\partial t} = d_2 \Delta v + s(x)u - e(x)v - f(x)v^2 - g(x)uv \text{ in } \Omega,\ t > 0, \\ \nabla u \cdot \nu = \nabla v \cdot \nu = 0 \text{ on } \partial\Omega,\ t > 0. \end{cases} \tag{1}$$

Ω is a bounded domain in $\mathbb{R}^N$, and ν is the outward unit normal to $\partial\Omega$, such that the system has Neumann boundary conditions, which are the no-flux boundary conditions for simple diffusion. In this system, u and v represent the population densities of juveniles and individuals that have reached reproductive age, i.e. adults, respectively, of the same species. Thus, the term $s(x)$ represents the rate at which juveniles mature into adults, which is determined by the fraction of individuals that reach reproductive age and the rate at which they mature, while $r(x)$ accounts for the local fecundity of adults such that $r(x)v(x)$ describes that rate at which new juveniles are produced by an adult population with density v at location x. The terms $a(x)$, $b(x)$, $c(x)$, $e(x)$, $f(x)$, and $g(x)$ account for per-capita death rates and saturation factors due to logistic self-limitation. The diffusion coefficients d_1 and d_2 account for the the dispersal rates of juveniles and adults, respectively. The coefficients are all assumed to be nonnegative and continuous in $\overline{\Omega}$. This is the type of model for a stage structured population introduced in [8]. Related models with a different interpretation are discussed in [9,10] and in the references in those papers. The model expressed in Equation (1) is not an explicitly age-structured model. It assumes that individuals in the juvenile stage mature at some spatially dependent rate but does not track the age of individuals within each stage. Explicitly age structured models are considered in [11–13]. A different way of modeling an age-structured population, based on delayed reaction diffusion equations, is developed in [14]. Our focus here is on how spatial heterogeneity, dispersal, and stage structure interact, so we chose to use the simplest possible formulation of stage structure. In the case where $c = g = 0$, the system is cooperative, and the methods and results of [15,16] would apply to it. The linearization of Equation (1) around $(0,0)$ is cooperative, so the results of [15] apply to it; in particular, with a few technical assumptions, they imply that it has a principal eigenvalue.

The main questions we will address in this work are related to understanding the roles of the different functions and coefficients in Equation (1) in the persistence of the species. For the remainder of the paper we will focus primarily on understanding how the principal eigenvalue of the linearization of Equation (1) around $(0,0)$ depends on the coefficients and what that dependence means biologically. We will see that whether faster diffusion is harmful or helpful for the persistence of the population depends on the details of the distribution of habitats that are favorable for adult reproduction and those that are favorable for juvenile survival and maturation. In some cases slower diffusion is still an advantage, but sometimes faster diffusion turns out to be helpful, and sufficiently fast diffusion may even be necessary for persistence. The spatial distribution of habitats favorable to adult reproduction ($r(x)$ large) relative to those favorable to juvenile development ($s(x)$ large) turns out to be important in some cases. Our analysis here is similar in spirit to the sorts of results obtained for diffusive Lotka-Volterra competition models in [3,17–19]. In particular, we will examine the behavior of the system for small, large, and general diffusion rates. Related results for some epidemiological models are derived in [20,21].

The linearization of Equation (1) around $(0,0)$ has a principal eigenvalue whose sign determines whether the model predicts persistence or extinction. Since the sign of the principal eigenvalue of the linearization of Equation (1) around $(0,0)$ determines the fate that Equation (1) predicts for the population it describes, we will study in detail the following problem:

$$\begin{cases} d_1\Delta\varphi + r(x)\psi - (s(x)+a(x))\varphi & = \lambda\varphi \ \text{ in } \Omega, \\ d_2\Delta\psi + s(x)\varphi - e(x)\psi & = \lambda\psi \ \text{ in } \Omega, \\ \nabla\varphi\cdot\nu = \nabla\psi\cdot\nu & = 0 \ \text{ on } \partial\Omega. \end{cases} \tag{2}$$

2. Basic Properties

In this section, we discuss some basic properties of Equation (1). From now on, we assume that $r, s, a, b, c, e, f, g \in C^{\alpha}(\bar{\Omega})$, $\partial\Omega$ is of class $C^{2,\alpha}$, and that the following hypotheses hold:

Hypothesis 1 (H1). *$r(x), s(x) \geq 0$ in Ω, with $r(x_r) \neq 0$, $s(x_s) \neq 0$ for some $x_r, x_s \in \Omega$.*

Hypothesis 2 (H2). *$a(x), c(x), e(x), g(x) \geq 0$ in Ω.*

Hypothesis 3 (H3). *$b(x) > 0$, $f(x) > 0$ in $\overline{\Omega}$.*

The model expressed in Equation (1) has many mathematical features in common with the models discussed in [9,10] for populations where individuals can switch between two different movement modes. A key feature is that the linear part of Equation (1) is cooperative, so it will have a principal eigenvalue which determines the stability of the equilibrium $(0,0)$ and hence the persistence or extinction of the population. Another key feature of Equation (1) is that the nonlinearity is subhomogeneous. The maximum principle and existence of principal eigenvalues for cooperative linear systems such as the linear part of the right side of Equation (1) are derived in [15]. The general theory for systems such as Equation (1) is developed in [16] for the fully cooperative case (where $c = g = 0$, such that adults only compete with other adults and juveniles with other juveniles) and in the general case in [8–10,22]. As expected, the sign of the principal eigenvalue of the linearization of Equation (1) around $(0,0)$ gives us the relevant information to study the persistence of the species. If it is positive, the population will persist. If it is nonpositive, the population will go extinct. In the case where the coefficients of c and g in Equation (1) are zero such that the system is cooperative, the results and methods of [16] imply that, if the principal eigenvalue of the linear part is positive, then the system has a unique globally attractive equilibrium. If those coefficients are small, the methods of [9,10] can be applied to show that Equation (1) is asymptotically cooperative, and still has a unique globally attractive positive equilibrium. Combining results that are given in [8–10,15,16] or that follow directly by the same arguments used in those papers, we have the following:

Lemma 1. *The eigenvalue problem expressed in Equation (2) has a unique principal eigenvalue λ_1 that is characterized by having a positive eigenvector (φ, ψ).*

Lemma 2. *If $\lambda_1 > 0$, then the system expressed in Equation (1) is persistent and has at least one positive equilibrium. If $\lambda_1 \leq 0$, then $(0,0)$ is globally asymptotically stable in Equation (1).*

Lemma 3. *If $\lambda_1 > 0$ and c and g are sufficiently small, then the system expressed in Equation (1) has a unique globally attracting positive equilibrium.*

Remark 1. *In the case that $c = g = 0$, the system expressed in Equation (1) is cooperative and hence generates a monotone semi-flow on appropriate spaces.*

3. The Case of d_1, d_2 Small

Following the approach in [23], we will establish the asymptotic behavior of the principal eigenvalue of Equation (2) when d_1, d_2 are small and, in the fully cooperative case (where $c \equiv g \equiv 0$), the profile of the nonnegative solutions of the corresponding steady state system for Equation (1).

$$\begin{cases} d_1\Delta u + r(x)v - s(x)u - a(x)u - b(x)u^2 = 0 \text{ in } \Omega, \\ d_2\Delta v + s(x)u - e(x)v - f(x)v^2 = 0 \text{ in } \Omega, \\ \nabla u \cdot \nu = \nabla v \cdot \nu = 0 \text{ on } \partial\Omega, \end{cases} \tag{3}$$

as well. Related results are derived in [16]. We observe that the associated kinetic system, which corresponds to Equation (1), is given by

$$\begin{cases} u_t = r(x)V(x) - s(x)U(x) - a(x)U(x) - b(x)U^2(x) - c(x)U(x)V(x) = 0, \\ v_t = s(x)U(x) - e(x)V(x) - f(x)V^2(x) - g(x)U(x)V(x) = 0, \end{cases} \tag{4}$$

for each $x \in \Omega$.

For each x, this system shares the same properties as Equation (1) given in Lemmas 1–3, which we state for convenience.

Lemma 4. *Set $x \in \Omega$. The linearization around (0,0) of Equation (4) has a principal eigenvalue $\lambda_1(x)$. Moreover,*

(i) If $\lambda_1(x) \leq 0$, then all the solutions with nonnegative initial condition of Equation (4) converge to (0,0) when $t \to \infty$.
(ii) If $\lambda_1(x) > 0$, then Equation (4) is persistent and has at least one positive equilibrium.
(iii) If $c \equiv g \equiv 0$ and $\lambda_1(x) > 0$, then Equation (4) is cooperative and admits a unique positive equilibrium, which is the global attractor for all nonnegative, non-trivial solutions.

Observe that, when $d_1 = d_2 = 0$, the eigenvalues of the linearization around (0,0) of Equation (4) are the roots of $\det(A(x) - \lambda I)$, with

$$A(x) = \begin{bmatrix} -(s(x) + a(x)) & r(x) \\ s(x) & -e(x) \end{bmatrix} \tag{5}$$

By a simple computation, we obtain that the maximum eigenvalue is given by

$$\Lambda(x) = \frac{1}{2}\left[-(s(x) + a(x) + e(x)) + \sqrt{(s(x) + a(x) - e(x))^2 + 4r(x)s(x)}\right], \tag{6}$$

which is positive provided that $(s(x) + a(x))e(x) - r(x)s(x) < 0$. Our first result, which is a direct application of Theorem 1.4 of [23] (Theorem A1 in the Appendix A), states that this is indeed the necessary and sufficient condition to have a positive principal eigenvalue when d_1 and d_2 are small.

Proposition 1. *The principal eigenvalue λ_1 of Equation (2) satisfies*

$$\lambda_1 \to \max_{x \in \overline{\Omega}} \Lambda(x) \text{ as } d_1, d_2 \to 0. \tag{7}$$

Thus, there exists a $\delta > 0$ such that if

$$\min_{x \in \overline{\Omega}}((s(x) + a(x))e(x) - r(x)s(x)) < 0, \tag{8}$$

the principal eigenvalue of Equation (2) is positive for all $0 < d_1, d_2 < \delta$, *while if*

$$\min_{x\in\overline{\Omega}}((s(x)+a(x))e(x)-r(x)s(x)) > 0, \tag{9}$$

the principal eigenvalue is negative.

As a consequence of this result, if Equation (9) holds, the unique nonnegative equilibrium of the system expressed in Equation (3) is $(0,0)$, and that equilibrium is globally attracting, whenever d_1, d_2 are small; if Equation (8) holds, the system expressed in Equation (3) is persistent, and has a positive equilibrium for small d_1, d_2. If Equation (8) holds and $c \equiv g \equiv 0$, then by Lemma 3 the system expressed in Equation (3) has a unique globally attracting positive equilibrium, which we denote by (u_d, v_d) with $d = (d_1, d_2)$.

Throughout the remainder of this section, we will assume that $c \equiv g \equiv 0$ in Ω, in which case, the system expressed in Equation (1) is cooperative.

The next result establishes the convergence of (u_d, v_d) to the unique nonnegative steady state $(U(x), V(x))$ of the kinetic system, which satisfies

$$\begin{aligned} &(U(x),V(x)) \text{ is positive where } (s(x)+a(x))e(x)-r(x)s(x) < 0, \\ &(U(x),V(x)) = 0 \text{ where } (s(x)+a(x))e(x)-r(x)s(x) \geq 0. \end{aligned} \tag{10}$$

Theorem 1. *Suppose that Equation (8) holds. Then* $(u_d, v_d) \to (U, V)$ *as* $d \to 0$ *locally uniformly in* Ω.

To prove this theorem, we follow the proof of Theorem 1.5 of [23], specifically their Proposition 5.2 in [23] and its hypotheses, which are listed in [23] as (A1)–(A4) and are given in the Appendix A as (L1)-(L4) to avoid confusion with the equation labels there. We should point out that Assumptions (A2) and (A3) of [23] do not hold in our case, so we cannot apply that result directly. The difference is that we allow situations where the kinetic system expressed in Equation (4) has a positive equilibrium for some values of $x \in \bar{\Omega}$ but not for others, whereas Condition (A2) requires a positive equilibrium for the kinetic system for all x. For that reason, we need to construct a version of the arguments in [23] that is local in x. Condition (A3) in [23] is used only to prove the existence of a nontrivial subsolution for a system corresponding to Equation (3), which is independent of d_1, d_2. We show the existence of the analogous local subsolutions we need in our case in the next lemma.

Lemma 5. *Suppose that* $\tilde{x} \in \Omega_0$, *where*

$$\Omega_0 = \{x \in \Omega \ / \ (s(x)+a(x))e(x)-r(x)s(x) < 0\}. \tag{11}$$

Then there exists $d_0 > 0$, $\rho_0 > 0$, *and a function* $\underline{w}^0 > 0$ *in* $B(\tilde{x}, \rho) \subset \Omega_0$, *which is a subsolution of Equation (3) for all* $0 < d_1$, $d_2 < d_0$.

Proof. Let $p = (p_1, p_2)$, a positive eigenvector of $A(\tilde{x})$ with $p_1 + p_2 = 1$, associated with its principal eigenvalue $\tilde{\sigma} > 0$. Set $\varepsilon > 0$ and small. We can choose $\rho > 0$ such that $\overline{B(\tilde{x},\rho)} \subset \Omega_0$ and

$$\begin{aligned} &|a(x)-\tilde{a}| < \varepsilon, \ |r(x)-\tilde{r}| < \varepsilon, \\ &|s(x)-\tilde{s}| < \varepsilon \text{ and } |e(x)-\tilde{e}| < \varepsilon \text{ for all } x \text{ in } \overline{B(\tilde{x},\rho)}, \end{aligned} \tag{12}$$

where $\tilde{a} = a(\tilde{x})$, $\tilde{r} = r(\tilde{x})$, $\tilde{s} = s(\tilde{x})$, and $\tilde{e} = e(\tilde{x})$. Set $\eta > 0$ as the principal eigenfunction associated with $\lambda > 0$, the principal eigenvalue of

$$\Delta\eta + \lambda\eta = 0 \text{ in } B(\tilde{x},\rho), \ \eta = 0 \text{ on } \partial B(\tilde{x},\rho), \tag{13}$$

with $\max_{B(\tilde{x},\rho)} \eta = 1$. We claim that we can choose $\delta, \varepsilon, \rho, d_0 > 0$ such that $\delta\eta p$ is a subsolution of Equation (3) for all $d_1,\ d_2 < d_0$. For simplicity and to keep the notation consistent with that in [23], we define $F(x,u,v) = (F_1(x,u,v), F_2(x,u,v))$, with

$$F_1(x,u,v) = rv - su - au - bu^2 \text{ and } F_2(x,u,v) = su - ev - fv^2 = 0, \tag{14}$$

where we have omitted the variable x in $a,\ b,\ e,\ r,\ s,\ f$ to shorten the expressions. Observe that

$$\begin{aligned} F_1(x,\delta\eta p) &= \delta\eta(\tilde{\sigma}p_1 + (r-\tilde{r})p_2 - (s-\tilde{s})p_1 - (a-\tilde{a})p_1 - b\delta p_1^2\eta) \\ F_2(x,\delta\eta p) &= \delta\eta(\tilde{\sigma}p_2 + (s-\tilde{s})p_1 - (e-\tilde{e})p_2 - f\delta p_2^2\eta), \end{aligned}$$

and using Equation (12), we obtain that, if we choose $\varepsilon > 0$ and a small δ, we have that

$$\begin{aligned} F_1(x,\delta\eta p) &\geq \delta\eta(\tilde{\sigma}p_1 - \varepsilon p_2 - 2\varepsilon p_1 - b\delta p_1^2\eta) > \delta\eta\frac{\tilde{\sigma}}{2}p_1 \\ F_2(x,\delta\eta p) &\geq \delta\eta(\tilde{\sigma}p_2 - \varepsilon p_1 - \varepsilon p_2 - f\delta p_2^2) > \delta\eta\frac{\tilde{\sigma}}{2}p_2. \end{aligned} \tag{15}$$

Therefore, replacing these inequalities in Equation (3), we obtain

$$\begin{aligned} d_1\delta p_1\Delta\eta + F_1(x,\delta\eta p) &\geq \delta\eta\left(-d_1p_1\lambda + \frac{\tilde{\sigma}}{2}p_1\right) \\ d_2\delta p_2\Delta\eta + F_2(x,\delta\eta p) &\geq \delta\eta\left(-d_2p_2\lambda + \frac{\tilde{\sigma}}{2}p_2\right); \end{aligned}$$

hence, if we set $d_0 = \frac{\tilde{\sigma}}{2\lambda}$. we obtain the desired result. □

Using this lemma we can follow the proof of Proposition 5.2 in [23]. To facilitate our exposition, we will use the same notation. Set the operators $D = \operatorname{diag}(d_1, d_2)$, $\mathcal{L} = \operatorname{diag}(\Delta, \Delta)$. To prove Theorem 1, we will state the needed lemmas, discussing their relationships with the lemmas in [23] leading to the proof of Proposition 5.2.

Suppose that Equation (8) holds, setting $\underline{w}^0 = \eta\delta p$ as in Lemma 5, and $\overline{w}^0 = M$ where $M > 0$ is given in Assumption (A4) such that $F_1(x,u,v) \leq -cu$ and $F_2(x,u,v) \leq -cv$ for all $u, v \geq M$ and $x \in \Omega$, with $c > 0$ fixed. Set $K > 0$ such that $K + \partial_u F_1(x,u,v) > 0$ and $K + \partial_v F_2(x,u,v) > 0$ for all $0 \leq u, v \leq M$, and we define $z = \overline{w}^k$ as the unique solution of

$$\begin{cases} -D\mathcal{L}z + Kz = Ku + F(x,u) \text{ in } \Omega, \\ \nabla z \cdot \nu = 0 \ \text{ on } \partial\Omega, \end{cases}$$

for $u = \overline{w}^{k-1}$.

Lemma 6. *Suppose that Equation (8) holds. For every k, we have $\underline{w}^0 < \overline{w}^{k+1} < \overline{w}^k < \overline{w}^0$, and as $k \to \infty$, $\overline{w}^k$ converges uniformly to the unique positive solution w of Equation (3), which satisfies $\underline{w}^0 < w < \overline{w}^k$ in Ω for all $k \geq 0$.*

Proof. We will prove that $\underline{w}^0 < \overline{w}^k$ by induction. Suppose this is true for k. Observe that $\underline{w}^0 < \overline{w}^0$ by construction. In the set, $B(\tilde{x},\rho) \subset \Omega_0$ as in Lemma 5 $\overline{w}^{k+1}$ satisfies

$$-D\mathcal{L}(\overline{w}^{k+1} - \underline{w}^0) + K(\overline{w}^{k+1} - \underline{w}^0) = K(\overline{w}^k - \underline{w}^0) + F(x,\overline{w}^k) - F(\underline{w}^0),$$

in $B(\tilde{x},\rho)$. By the induction hypothesis $\underline{w}^0 < \overline{w}^k$, whence $K\overline{w}^k + F(x,\overline{w}^k) > 0$; hence, by the strong maximum principle applied to each component, we have that $\overline{w}^{k+1} > 0$ in $\bar{\Omega}$. Thus, we have

$$\begin{cases} -D\mathcal{L}(\overline{w}^{k+1} - \underline{w}^0) + K(\overline{w}^{k+1} - \underline{w}^0) > 0 \text{ in } B(\tilde{x},\rho), \\ \overline{w}^{k+1} - \underline{w}^0 > 0 \text{ in } \partial B(\tilde{x},\rho), \end{cases}$$

such that we have that $\overline{w}^{k+1} - \underline{w}^0 > 0$ in $\overline{B(\tilde{x},\rho)}$. The remainder of the proof is a standard monotone iteration argument, just as in the proof of Lemma 5.3 of [23]. We observe that

$$\begin{cases} -D\mathcal{L}(\overline{w}^1 - \overline{w}^0) + K(\overline{w}^1 - \overline{w}^0) = K\overline{w}^0 + F(x,\overline{w}^0) - K\overline{w}^0 < 0 \text{ in } \Omega, \\ \nabla[\overline{w}^1 - \overline{w}^0] \cdot \nu = 0 \text{ on } \partial\Omega, \end{cases}$$

Thus, by the strong maximum principle, we have $\overline{w}^1 < \overline{w}^0$.

Similarly, if $\overline{w}^k < \overline{w}^{k-1}$, then

$$\begin{cases} -D\mathcal{L}(\overline{w}^{k+1} - \overline{w}^k) + K(\overline{w}^{k+1} - \overline{w}^k) = \\ K\overline{w}^k + F(x,\overline{w}^k) - K\overline{w}^{k-1} - F(x,\overline{w}^{k-1}) < 0 \text{ in } \Omega, \\ \nabla[\overline{w}^{k+1} - \overline{w}^k] \cdot \nu = 0 \text{ on } \partial\Omega. \end{cases}$$

By induction, the sequence $\{\overline{w}^k\}$ is decreasing, and it is bounded below by $max\{0, \underline{w}^0(x)\}$, so by standard elliptic theory, it converges to a nonnegative nontrivial solution of Equation (3) as $k \to \infty$. Since by Lemma 3 the nontrivial nonnegative solution of Equation (3) is unique, it coincides with the one constructed as the limit of the sequence $\{\overline{w}^k\}$. □

Define $\overline{W}^0 = \overline{w}^0$ and $\overline{W}^{k+1} = \overline{W}^k + F(x,\overline{W}^k)$ in Ω. Following the proof of Lemmas 5.6 and 5.7 in [23], we can prove the following result.

Lemma 7. *Suppose that Equation (8) holds. For every k, we have*

$$\underline{w}^0 < \overline{W}^{k+1} < \overline{W}^k.$$

Then, $\overline{W}^k$ converges locally uniformly to W^∞ as $k \to \infty$, with

$$W^\infty = (U(x),V(x)) \text{ in } \Omega_0, \text{ and } W^\infty(x) = 0 \text{ in } \Omega \setminus \Omega_0,$$

where Ω_0 is given by Equation (11).

Proof. Observe that, by Equation (15), the function $\underline{w}^0$ is a subsolution of the kinetic system. Repeating the proof of Lemma 5.6 in [23], or following the arguments leading to the monotonicity of the proof of Lemma 6 above, we have that the sequence $\{\overline{W}^k\}$ is monotone decreasing and bounded below by $\underline{w}^0$. Therefore, $\overline{W}^k \to W^\infty$ pointwise, which satisfies $F(x,W^\infty) = 0$, i.e. a nonnegative equilibrium of the kinetic system. Therefore, if at some $x \in \Omega$ we have that $(s(x)+a(x))e(x) - r(x)s(x) \geq 0$, then $W^\infty(x) = 0$. On the other hand, $W^\infty(x) \geq \underline{w}^0(x) > 0$ in $B(\tilde{x},\rho) \subset \Omega_0$ for $x \in \Omega_0$. Since x is arbitrary and the sequence does not depend on $\underline{w}^0$, we obtain that $W^\infty(x) = (U(x),V(x))$, the unique positive kinetic equilibrium, whenever $x \in \Omega_0$. Particularly, W^∞ is continuous. Using Theorem 5.8 of [23], we obtain that the convergence is uniform in any compact set of Ω. □

Lemma 8. *For each k, as d_1, $d_2 \to 0$, we have that $\overline{w}^k$ converges to $\overline{W}^k$ uniformly in $\bar{\Omega}$.*

The proof of this result is the same as the one of Lemma 5.5 in [23].

Proof of Theorem 7. Using a diagonal argument, and Lemma (8), the unique positive solution w of Equation (3) converges to W_∞ as $d_1, d_2 \to \infty$. $\square$

4. The Case of d_1 and d_2 Large

We will start by giving a proof of a result that is well known as a "folk theorem." It is stated in slightly more generality than is needed for the specific application. For $i = 1, \ldots, N$, let L^i denote the operator

$$L^i u = \nabla \cdot \mu_i(x)[\nabla u - u\nabla \alpha_i(x)] \quad \text{for} \quad x \in \Omega \tag{16}$$

with no-flux boundary conditions

$$[\nabla u - u\nabla \alpha_i] \cdot \nu = 0 \quad \text{for} \quad x \in \partial\Omega. \tag{17}$$

Assume that $\mu_i(x) \geq \mu_0 > 0$ on $\bar{\Omega}$ for all i. Let $A = (a_{ij}(x))$ be an $N \times N$ irreducible matrix with $a_{ij} \geq 0$ if $i \neq j$. Consider the eigenvalue problem

$$d_i L^i \varphi_i + \sum_{j=1}^{N} a_{ij}\varphi_j = \lambda \varphi_j, \quad i = 1 \ldots N \tag{18}$$

where $d_i > 0$ for all i and φ_i satisfies the boundary condition expressed in Equation (17) for each i. Note that, if we let $\Phi_i = exp(-\alpha_i(x))\varphi_i$, then Φ_i satisfies Neumann boundary conditions such that the system expressed in Equation (18) rewritten in terms of the variables Φ_i is still cooperative. Because of the classical boundary conditions, the usual results on elliptic regularity and on maximum principles for cooperative systems from [15,23] can be applied to the system for the Φ_i values, such that the system and hence Equation (18) will have a principal eigenvalue under suitable conditions on the domain Ω and the coefficients. This idea has been used in models for single populations without an age structure or competing pairs of such populations—see, for example, [2,24].

Furthermore, we have $L^i(exp(\alpha_i(x)) = 0$ such that the principal eigenvalue of L^i is zero, and the eigenfunction is a multiple of $exp(\alpha_i)$. Let $\overline{A}$ be the matrix defined by

$$\overline{A}_{ij} := \frac{\int_\Omega a_{ij} exp(\alpha_i) dx}{\int_\Omega exp(\alpha_i) dx}. \tag{19}$$

Denote the principal eigenvalue of Equation (18) as $\lambda_1(\vec{d})$ where $\vec{d} = (d_1, \ldots, d_N)$. Denote the principal eigenvalue of $\overline{A}$ as $\overline{\Lambda}$.

Lemma 9. *Suppose that, for some $\gamma \in (0,1)$, the coefficients of Equation (18) satisfy $\alpha \in C^{2,\gamma}(\overline{\Omega})$, $\mu \in C^{1,\gamma}(\overline{\Omega})$, and $a_{ij} \in C^\gamma(\overline{\Omega})$ for $i, j = 1 \ldots N$, and that $\partial\Omega$ is of class $C^{2,\gamma}$. Suppose further that $\overline{A}$ is irreducible. If $min\{d_i : i = 1, \ldots N\} \to \infty$, then $\lambda_1(\vec{d}) \to \overline{\Lambda}$.*

Proof. Choose any sequence $\vec{d_n} = (d_{1n}, \ldots d_{Nn})$ such that $min\{d_{in} : i = 1, \ldots N\} \to \infty$. Choose any subsequence, then renumber it as $\vec{d_n}$. Let λ_n be the principal eigenvalue of Equation (18) corresponding to $\vec{d_n}$ and let $\varphi_{in}(x) > 0$ be the ith component of the eigenvector, where the eigenvector is normalized by $max\{\varphi_{in}(x) : x \in \overline{\Omega},\ i = 1, \ldots, N\} = 1$. Integrating the ith equation of Equation (18) over Ω and summing over i yields

$$\lambda_n \int_\Omega \sum_{i=1}^{N} \varphi_{in}(x)dx = \int_\Omega \sum_{i,j=1}^{N} a_{ij}(x)\varphi_{jn}(x)dx \leq A_1 \int_\Omega \sum_{i=1}^{N} \varphi_{in}(x)dx$$

where A_1 is a constant, depending only on A. It follows that λ_n is uniformly bounded from above. Similarly, λ_n is uniformly bounded from below. Thus, any subsequence of λ_n itself has a convergent subsequence. It then follows from dividing the ith equation of Equation (18) by d_{in} that $L^i\varphi_{in}$ is uniformly bounded, and $L_i\varphi_{in} \to 0$ as $n \to \infty$. By elliptic regularity, the sequence φ_{in} is uniformly bounded in $W^{2,p}(\Omega)$ for any $p < \infty$, then by Sobolev embedding, it has a subsequence that is convergent in $C^1(\overline{\Omega})$ and weakly convergent in $W^{2,p}(\Omega)$. This will be true for any i. Taking a further subsequence if necessary and renumbering again, we obtain a sequence where $\lambda_n \to \lambda^*$ for some λ^* and $\varphi_{in} \to \varphi_i^*$ for all i, with $L_i\varphi_i^* = 0$. We then must have $\varphi_i^* = c_i exp(\alpha_i)$ for some nonnegative constant c_i, and with $max\{\varphi_i^*(x) : x \in \overline{\Omega},\ i = 1 \dots N\} = 1$. Integrating Equation (18) over Ω and using the no-flux boundary conditions gives

$$\sum_{j=1}^{N} \left[\int_{\Omega} a_{ij}(x)\varphi_j^*(x)dx \right] = \lambda^* \int_{\Omega} \varphi_i^*, \quad i = 1 \dots N, \tag{20}$$

such that

$$\sum_{j=1}^{N} \left[\frac{\int_{\Omega} a_{ij}(x)exp(\alpha_j(x))dx}{\int_{\Omega} exp(\alpha_i(x))dx} \right] c_j = \lambda^* c_i, \quad i = 1 \dots N. \tag{21}$$

It follows that $(c_1, \dots, c_N)$ must be a nontrivial nonnegative eigenvector of $\overline{A}$ with the normalization prescribed by $max\{c_i exp(\alpha_i(x)) : x \in \overline{\Omega},\ i = 1, \dots N\} = 1$. These last conditions uniquely determine the limits of the subsequence of the original subsequence $\{\lambda(d_n), \vec{\varphi}_n\}$. Since every subsequence of the original sequence $\{\lambda(d_n), \vec{\varphi}_n\}$ has a subsequence converging to the values determined by Equation (21), the same must be true for the original sequence. Since the original sequence of values $\{\vec{d_n}\}$ could be any increasing sequence that approaches infinity as $n \to \infty$, the conclusion of the lemma follows. □

In the specific system expressed in Equation (1) that we consider, $L^i = \Delta$, such that α_i and μ_i are constants. In that case, we have $\overline{A}_{ij} = \bar{a}_{ij}$, where $\bar{a}_{ij}$ is the average of a_{ij} over Ω. Denote the averages of the coefficients in Equation (1) by $\bar{r}$, $\bar{s}$, etc. Calculations analogous to those in Equation (5), Equation (6), and the related discussion then yield the following:

Corollary 1. *Suppose that the hypotheses of Lemma 9 are satisfied. There exists a $D > 0$ such that if*

$$\bar{e}(\bar{s} + \bar{a}) - \bar{r}\bar{s} < 0,$$

the principal eigenvalue λ_1 of Equation (2) is positive for all $d_1, d_2 > D$, while if

$$\bar{e}(\bar{s} + \bar{a}) - \bar{r}\bar{s} > 0,$$

the principal eigenvalue is nonpositive.

Remark 2. *In the ODE system corresponding to Equation (1) with coefficients averaged over Ω, one can compute R_0 as $\bar{r}\bar{s}/[\bar{e}(\bar{s} + \bar{a})]$ via the methods of [25]. The first inequality in Corollary 1 is equivalent to $R_0 > 1$, while the second is equivalent to $R_0 < 1$. By writing $R_0 = [\bar{r}/(\bar{s} + \bar{a})][\bar{s}/\bar{e}]$, we can interpret the condition for persistence as saying that the products of the ratios of the growth terms over the loss terms for adults and juveniles should be greater than 1 for persistence.*

5. General Diffusion Rates

Case1: Persistence or extinction for all diffusion rates

Proposition 2. *If*

$$\int_{\Omega} \sqrt{rs}\, dx - \frac{1}{2}\int_{\Omega}(s+a+e)dx > 0 \tag{22}$$

then $\lambda_1 > 0$ *for all positive diffusion rates.*

If

$$min_{x\in\overline{\Omega}}[4(s(x)+a(x))e(x) - (r(x)+s(x))^2] > 0 \tag{23}$$

then $\lambda_1 < 0$ *for all positive diffusion rates.*

Proof. If we divide the first equation in Equation (2) by φ and integrate over Ω, using Green's formula to integrate the term $\Delta\varphi/\varphi$, we obtain the inequality

$$|\Omega|\lambda_1 \geq \int_{\Omega} r\left(\frac{\psi}{\varphi}\right) dx - \int_{\Omega}(s+a)dx. \tag{24}$$

Similarly, if we divide the second equation by ψ and integrate we obtain

$$|\Omega|\lambda_1 \geq \int_{\Omega} s\left(\frac{\varphi}{\psi}\right) dx - \int_{\Omega} e\, dx. \tag{25}$$

If we add Equations (24) and (25) and divide by 2, we obtain

$$\lambda_1 \geq \frac{1}{2|\Omega|}\left(\int_{\Omega}\left[r\left(\frac{\psi}{\varphi}\right)+s\left(\frac{\varphi}{\psi}\right)\right]dx - \int_{\Omega}(s+a+e)dx\right). \tag{26}$$

By Cauchy's inequality, $rz+sz^{-1} \geq 2\sqrt{rs}$ for all $z>0$, so from Equation (26) we obtain

$$\lambda_1 \geq \frac{1}{|\Omega|}\left[\int_{\Omega}\sqrt{rs}\, dx - \frac{1}{2}\int_{\Omega}(s+a+e)dx\right] \tag{27}$$

Thus, $\lambda_1 > 0$ if Equation (22) holds, so the first part of Proposition 2 holds. Going in the other direction, if we multiply the first equation of Equation (2) by φ and integrate, using integration by parts on the $\varphi\Delta\varphi$ term, and similarly multiply the second equation by ψ and integrate, and then add the results, we get

$$\lambda_1 \int_{\Omega}(\varphi^2+\psi^2)dx \leq \int_{\Omega}[-(s+a)\varphi^2 + (r+s)\varphi\psi - e\psi^2)dx]. \tag{28}$$

The integrand on the right side of Equation (28) is a quadratic form in φ and ψ, which will be negative definite if

$$4(s+a)e > (r+s)^2, \tag{29}$$

so $\lambda_1 < 0$ if Equation (23) holds, which proves the second part of Proposition 2. □

Remarks: Note that the first integral in Equation (22) is what appears in the formula for the Bhattacharyya coefficient [26,27], which is used to compare how well probability distributions match each other. Specifically, if two probability distributions P and Q have probability density functions $p(x)$ and $q(x)$ for $x \in U \subset \mathbb{R}^n$, the Bhattacharyya coefficient is

$$BC(P,Q) = \int_U \sqrt{p(x)q(x)}dx.$$

For any P and Q, $0 \leq BC(P,Q) \leq 1$. If $BC(P,Q)=1$, then P and Q are the same, that is, $p=q$ a.e. If $BC(P,Q)=0$, then the supports of p and q are disjoint. If we write $r(x)=r_0\rho(x)$ and $s(x)=s_0\sigma(x)$

such that $\int_\Omega \rho(x) = \int_\Omega \sigma(x) = 1$, then we can compute $r_0 = \bar{r}|\Omega|$ and $s_0 = \bar{s}|\Omega|$. We can treat ρ and σ as if they were probability density functions for distributions R and S. We then have

$$\int_\Omega \sqrt{rs}dx = |\Omega|\sqrt{\bar{r}\bar{s}}BC(R,S). \tag{30}$$

The maximum of $BC(R,S)$ is 1, corresponding to the case where r and s are multiples of each other, and the minimum is 0, corresponding to the case where the supports of r and s are disjoint. Thus, the degree to which ρ and σ match each other has a strong impact on the estimate for λ in Equation (27).

Using Equation (30) and the fact that $BC(R,S) \leq 1$ in Equation (22) shows that Equation (22) implies $2\sqrt{\bar{r}\bar{s}} > [(\bar{s}+\bar{a})+\bar{e}]$. Squaring both sides and using Cauchy's inequality implies $\bar{e}(\bar{s}+\bar{a}) - \bar{r}\bar{s} < 0$ as in the first case of Corollary 1. Similarly, if Equation (22) holds, then $2\sqrt{r(x)s(x)} > (s(x)+a(x))+e(x)$ for some $x \in \Omega$, and it then follows in the same way that the inequality in the first case of Proposition 1 holds. If Equation (23) holds, then Equation (29) holds, and then by Cauchy's inequality the second case of Proposition 1 holds. Thus, the conditions expressed in Equation (22) and Equation (23) in Proposition 2, which imply $\lambda_1 > 0$ or $\lambda_1 < 0$ for all diffusion rates, also imply some of the corresponding conditions we have obtained for either large or small diffusion rates.

In the situation where the spatial distributions of habitat quality r for reproduction by adults and s for survival and maturation of juveniles into adults are perfectly correlated, such that $r(x) = r_1 s(x)$ for some constant r_1, the eigenvalue problem expressed in Equation (2) can be rewritten as a weighted symmetric eigenvalue problem by multiplying the second equation in Equation (2) by r_1, which yields

$$\begin{aligned} d_1\Delta\varphi(x) - (s(x)+a(x))\varphi + r(x)\psi &= \lambda\varphi \\ d_2 r_1\Delta\psi + r(x)\varphi - r_1 e(x)\psi &= \lambda r_1\psi. \end{aligned} \tag{31}$$

The principal eigenvalue for Equation (31) has a variational characterization of λ_1 as

$$\lambda_1 = \max_{\varphi,\psi\in W^{1,2}(\Omega)} \frac{\int_\Omega(-d_1|\nabla\varphi|^2 - d_2 r_1|\nabla\psi|^2 - (s+a)\varphi^2 + 2r\varphi\psi - r_1 e\psi^2)dx}{\int_\Omega(\varphi^2 + r_1\psi^2)dx}. \tag{32}$$

It follows in that case that λ_1 is decreasing in both d_1 and d_2, such that slower diffusion is advantageous.

Case 2: Asymptotic behavior for large reproductive rates

Suppose that $r(x) = nr_0(x)$ and that $s(x)r_0(x) > 0$ for $x \in \Omega_0$ with $\Omega_0 \neq \emptyset$; therefore, there is a region where both the adult reproduction rate and the juvenile maturation rate are positive. The factor n scales the reproductive rate of adults in regions where $r_0(x) > 0$. For any fixed diffusion rates, it turns out that, for sufficiently large values of the scaling coefficient n, the principal eigenvalue of Equation (2) is positive, so the system expressed in Equation (1) is persistent. We will characterize the asymptotic behavior of the principal eigenvalue as $n \to \infty$. If we make the further assumption that $g \equiv c \equiv 0$, then the system expressed in Equation (1) is cooperative, so for an n large enough that the principal eigenvalue of Equation (2) is positive, Equation (1) has a unique positive equilibrium, and we will characterize the behavior of that equilibrium as $n \to \infty$ as well in that case.

Let λ_1^n denote the principal eigenvalue for Equation (2) with $r(x) = nr_0(x)$. Observe that, since $s(x)r_0(x) > 0$ for $x \in \Omega_0$ and $\Omega_0 \neq \emptyset$, Proposition 2 implies that $\lambda_1^n > 0$ for n sufficiently large, and in fact by Equation (27), $\lambda_1^n \to \infty$ as $n \to \infty$. The following proposition states the asymptotic behavior of λ_1^n as $n \to \infty$.

Proposition 3. *If $r(x) = nr_0(x)$ and $s(x)r_0(x) > 0$ for $x \in \Omega_0$ with $\Omega_0 \neq \emptyset$, then*

$$\lim_{n\to\infty} \frac{\lambda_1^n}{\sqrt{n}} \to \max_{x\in\overline{\Omega}}(\sqrt{r_0(x)s(x)}).$$

Proof. We start by noting that, if λ_1^n, then (φ_n, ψ_n) are the principal eigenvalue and corresponding eigenfunction of Equation (2) for $r(x) = nr_0(x)$; therefore, $\lambda_1^n/\sqrt{n}$, $\widehat{\varphi}_n = \varphi_n$, and $\widehat{\psi}_n = \sqrt{n}\psi_n$ are the principal eigenvalue and eigenfunction of the following problem:

$$\begin{cases} \dfrac{d_1}{\sqrt{n}}\Delta\widehat{\varphi} - \dfrac{(s(x)+a(x))}{\sqrt{n}}\widehat{\varphi} + r_0(x)\widehat{\psi} & = \widehat{\lambda}\widehat{\varphi} \text{ in } \Omega, \\ \dfrac{d_2}{\sqrt{n}}\Delta\widehat{\psi} - \dfrac{e(x)}{\sqrt{n}}\widehat{\psi} + s(x)\widehat{\varphi} & = \widehat{\lambda}\widehat{\psi} \text{ in } \Omega, \\ \nabla\widehat{\varphi}\cdot\nu = \nabla\widehat{\psi}\cdot\nu & = 0 \text{ on } \partial\Omega. \end{cases} \tag{33}$$

Considering the elliptic operators $L_1 u = d_1\Delta u - (s(x) - a(x))u$ and $L_2 v = d_2\Delta v - e(x)v$, $D = \text{diag}\left(\frac{1}{\sqrt{n}}, \frac{1}{\sqrt{n}}\right)$ and $\mathcal{L} = \text{diag}(L_1, L_2)$, the system expressed in Equation (33) satisfies the hypothesis of Theorem 1.4 of [23]. Thus, $n \to \infty$

$$\frac{\lambda_1^n}{\sqrt{n}} \to \max_{x\in\overline{\Omega}} \lambda(A(x)),$$

where

$$A(x) = \begin{pmatrix} 0 & r_0(x) \\ s(x) & 0 \end{pmatrix},$$

which has eigenvalues $\pm\sqrt{r_0(x)s(x)}$, from whence the result follows. □

In the case where $g \equiv c \equiv 0$ such that Equation (1) is cooperative, Lemma 3 implies that Equation (1) has a unique positive equilibrium if the principal eigenvalue of Equation (2) is positive. The next result states the asymptotic behavior of the unique positive equilibrium of Equation (1) for n large in that case.

Proposition 4. *Suppose the hypotheses of Proposition 3 are satisfied and that $g \equiv c \equiv 0$. Let (u_n, v_n) be the unique positive solution of Equation (3). Then $n^{-\frac{2}{3}}(u_n, v_n) \to (U^\infty, V^\infty)$ uniformly in $\bar{\Omega}$ where*

$$\begin{aligned} &U^\infty(x) = \frac{r_0(x)^{\frac{2}{3}}}{b(x)^{\frac{2}{3}}}\frac{s(x)^{\frac{1}{3}}}{f(x)^{\frac{1}{3}}}, \quad V^\infty(x) = \frac{s(x)^{\frac{2}{3}}}{f(x)^{\frac{2}{3}}}\frac{r_0(x)^{\frac{1}{3}}}{b(x)^{\frac{1}{3}}} \textit{ when } s(x)r_0(x) > 0 \\ &U^\infty(x) = 0, V^\infty(x) = 0 \textit{ when } s(x)r_0(x) = 0. \end{aligned} \tag{34}$$

Proof. To prove this result, we use a different scaling. After some simple computations, we find that

$$(w_n, z_n) = \left(u_n n^{-\frac{2}{3}}, v_n n^{-\frac{1}{3}}\right),$$

is the unique positive solution of the scaled system

$$\begin{cases} n^{-\frac{2}{3}}[d_1\Delta w - (s(x)+a(x))w] + r_0(x)z - b(x)w^2 = 0 \text{ in } \Omega, \\ \\ n^{-\frac{1}{3}}[d_2\Delta z - e(x)z] + s(x)w - f(x)z^2 = 0 \text{ in } \Omega, \\ \nabla w\cdot\nu = \nabla z\cdot\nu = 0 \text{ on } \partial\Omega. \end{cases} \tag{35}$$

We set the operators

$$L_1 w = d_1\Delta w - (s(x)+a(x))w \text{ and } L_2 z = d_2\Delta z - e(x)z,$$

$$D = \text{diag}\left(n^{-\frac{2}{3}}, n^{-\frac{1}{3}}\right) \text{ and } \mathcal{L} = (L_1, L_2), \text{ and}$$

$$F(x, w, z) = (r_0(x)z - b(x)w^2, s(x)w - f(x)z^2).$$

To prove the proposition, we can follow the same steps as in the proof of Theorem 1. Indeed, the principal eigenvalue of the linearization around $(0,0)$ of the associated kinetic system of Equation (35) is $\sqrt{s(x)r_0(x)}$, and when it is positive, the kinetic equilibrium is given by the right hand side of Equation (34). This concludes the proof. □

6. Conclusions

The most fundamental conclusion from our analysis is that reaction-diffusion models for populations with a stage structure in spatially heterogeneous environments do not necessarily predict that slower diffusion is advantageous for persistence. This is in contrast to the case where populations are structured only by spatial location, where a version of the reduction principle [6] applies; in a competition between otherwise identical populations with different diffusion rates, the prediction is that "the slower diffuser wins" [4,5]. The mechanism underlying this observation is that, in our structured model, the regions where it is possible for adults to produce offspring may be separated from those where juveniles can survive and mature into adults. The conditions we find that imply persistence generally require that the product $r(x)s(x)$ of the reproductive rate of adults and the maturation rate of juveniles be sufficiently large relative to their death rates. For slow diffusion, the condition for persistence is that $r(x)s(x) > e(x)(s(x) + (a(x))$ at some point $x \in \Omega$. For fast diffusion, it is $\bar{r}\bar{s} > \bar{e}(\bar{s} + \bar{a})$ where $\bar{r}, \bar{s}, \bar{e}$, and $\bar{a}$ are the spatial averages of those quantities. If the spatial distributions of r and s are closely correlated and are large in a few places but small in most, such that the maximum of rs is large but the averages $\bar{r}$ and $\bar{s}$ are small, the condition for persistence with slow diffusion may be satisfied, while the condition with fast diffusion may fail. In that type of environment, slow diffusion is clearly favored. Furthermore, if r and s are perfectly correlated in the sense that they are multiples of each other, the principal eigenvalue determining the growth rate of the population at low density is decreasing with respect to the diffusion rates, as in the case of unstructured populations in heterogeneous environments. On the other hand, if both r and s are large on some regions but very small outside of them, and the regions where they are large are disjoint (that is, separated from each other), then the product rs could be small everywhere, but the averages $\bar{r}$ and $\bar{s}$ could be large. In that case, the condition for persistence with small diffusion may fail, but the condition with fast diffusion may be satisfied, such that fast diffusion is favored.

We found that a sufficient condition for persistence for all diffusion rates is

$$\int_\Omega \sqrt{rs}\, dx - \frac{1}{2}\int_\Omega (s + a + e) dx > 0.$$

The first term can be written as $\sqrt{\bar{r}\bar{s}}|\Omega| BC(r(x)/\bar{r}, s(x)/\bar{s})$, where BC denotes the Bhattacharrya coefficient (see [26,27]), which measures how closely probability densities match each other. For distributions that are equal to each other, $BC = 1$, but for distributions that are mutually exclusive in the sense that the regions where they are positive do not intersect, $BC = 0$. This observation again shows that the degree to which the spatial distributions of r and s match each other is significant in determining the predictions of the model expressed in Equation (1).

Finally, we found that, if we scale the adult reproductive rate as $r(x) = nr_0(x)$ and there is some overlap between the distributions of r and s such that $s(x)r_0(x) > 0$ on some subset of Ω with positive measure, then for any fixed diffusion rates the system expressed in Equation (1) will be persistent if n is sufficiently large. This means a population with any diffusion rates can persist if there is even a modest overlap between the regions where adults can reproduce and where juveniles can mature, provided that the reproductive rate of adults is sufficiently large. We characterized the asymptotic behavior as $n \to \infty$ of the principal eigenvalue of Equation (2). In the cooperative case where $g \equiv c \equiv 0$, the system expressed in Equation (1) will have a unique positive equilibrium if it is persistent, and in that case we also characterized the asymptotic behavior as $n \to \infty$ of the equilibrium.

There are several directions for further research on the general topic of this paper. It would be of interest to take the approach of [4] and consider the competition between two stage structured populations described by systems such as Equation (1) that differ only in their diffusion rates. That would be somewhat challenging because it would involve systems of four equations, but at least in the cooperative case where $c = g = 0$ the general theory of monotone dynamical systems as in [28] and some of the ideas and methods of [10] would apply. It would also be interesting to consider models with an explicit age structure, as introduced in [11] and studied in [12,13]. Finally, it would be interesting albeit challenging to consider the case of time-periodic environments with spatial heterogeneity. Temporal variation alone is sufficient to cause faster diffusion to be favored in such environments in some cases (see [29]), but even without a stage structure, the time-dependent case is challenging, and there are many open questions.

Author Contributions: Conceptualization, R.S.C. C.C., and S.M.;mathematical analysis, R.S.C., C.C. and S.M.; writing–original draft preparation, C.C. and S.M.; writing–review and editing, R.S.C., C.C., and S.M.; funding acquisition, R.S.C., C.C. and S.M. All authors have read and agreed to the published version of the manuscript.

Funding: This research received external funding as noted in the subsequent Acknowledgments .

Acknowledgments: R.S.C. and C.C. are supported in part by NSF Awards DMS 15-14792 and 18-53478. S.M. is supported by CONICYT + PIA/Concurso de Apoyo a Centros Científicos y Tecnológicos de Excelencia con Financiamiento Basal AFB170001.

Conflicts of Interest: The authors declare no conflict of interest.

Appendix A

In [23], the authors considered the equilibria and dynamics of the system

$$\begin{cases} \frac{\partial u}{\partial t} = D\mathcal{L}u + F(x,u) \ \text{ in } \Omega \times (0,\infty), \\ \mathcal{B}u = 0 \ \text{ on } \partial\Omega \times (0,\infty) \end{cases} \tag{A1}$$

where $u = (u_1, \ldots, u_n)^T$ is a vector of smooth functions, Ω is a bounded domain in $\mathbb{R}^N$ with smooth boundary, $u = (u_1(x), \ldots u_n)^T$ is a vector of smooth functions, $D = diag(d_1, \ldots d_n)$ is a diagonal matrix of positive constants, $\mathcal{L} = diag(L^1, \ldots L^n)$ is a diagonal matrix of second order uniformly strongly elliptic operators of the form

$$L^i = \sum_{j,k=1}^{N} \alpha^i_{jk} \frac{\partial^2}{\partial x_j \partial x_k} + \sum_{j=1}^{N} \beta^i_j \frac{\partial}{\partial x_j} + \gamma^i$$

with smooth coefficients, and $\mathcal{B} = (B_1, \ldots B_n)$, where for each i, B_i defines a Dirichlet, Neumann, or Robin boundary condition. (They include Neumann as a case of Robin.) They also considered the associated linearized problem, which they wrote as

$$\begin{cases} D\mathcal{L}\phi + Au\phi = -\lambda\phi \ \text{ in } \Omega, \\ B\phi = 0 \ \text{ on } \partial\Omega, \end{cases} \tag{A2}$$

where $A = (a_{ij})$ is an $n \times n$ matrix of smooth functions with $a_{ij} \geq 0$ for $i \neq j$, and $\phi = (\phi_1(x), \ldots \phi_n(x))^T$ is a vector of smooth functions.

Note that, in our notation, we use the opposite sign convention to the one used in [23], such that what they denote as $-\lambda$, we denote as λ.

For details on what the specific smoothness assumptions require, see [23]. Under those assumptions, by the Perron-Frobenius theorem, for each $x \in \Omega$, the matrix A has a principal eigenvalue, which in our notation we denote as $\Lambda(x)$. The first major result of [23] is their Theorem 1.4, which can be stated as follows:

Theorem A1. *(Theorem 1.4 of [23]) The principal eigenvalue λ_1 of the system expressed in Equation (A2) with Dirichlet, Neumann, or Robin boundary conditions satisfies*

$$\lim_{max\{d_1,\ldots,d_n\}\to 0} \lambda_1 = -\max_{x\in\bar{\Omega}} \Lambda(x).$$

Except for the adjustments needed for our different notation, that theorem applies directly to our system in all cases.

The second major result of [23] gives conditions under which the system expressed in Equation (A1) with small diffusion rates has the same dynamics as the kinetic system

$$\frac{dU_i}{dt} = F_i(x, U_1, \ldots, U_n) \text{ for } i = 1, \ldots, n. \tag{A3}$$

The conditions can be stated as follows, noting that we have replaced the A used in [23] with L to avoid any confusion with equation numbers in this Appendix:

(L1) $\partial F_i/\partial U_j \geq 0$ (i.e. the systems expressed in Equations (A1) and (A3) are cooperative).

(L2) For each $x_0 \in \bar{\Omega}$, the system expressed in Equation (A3) has a unique positive equilibrium $\alpha(x_0)$, which is globally asymptotically stable among positive solutions and is locally linearly stable, and $\alpha(x)$ depends continuously on x.

(L3) There is a $\delta_0 > 0$ such that, for $j = 1, \ldots, n$, $F_j(x, U)/U_j > \delta_0$ for all $x \in \bar{\Omega}$ provided $0 < U_i \leq \delta_0$ for $i = 1, \ldots, n$.

(L4) There is a $\delta_0', M > 0$ such that for $j = 1, \ldots, n$, $F_j(x, U)/U_j < -\delta_0'$ for all $x \in \bar{\Omega}$ provided $U_i \geq M$ for $i = 1, \ldots, n$.

The second result of [23] that we use is Proposition 5.2, which can be stated as follows:

Theorem A2. *(Proposition 5.2 of [23]) For any positive steady state w_d of Equation (A1), we have that $w \to \alpha$ uniformly in Ω as $max\{d_1, \ldots, d_n\} \to 0$, with the α given in Assumption (L2).*

References

1. Cantrell, R.S.; Cosner, C. *Spatial Ecology via Reaction-Diffusion Equations*; John Wiley and Sons: Chichester, UK, 2003.
2. Cosner, C. Reaction-diffusion-advection models for the effects and evolution of dispersal. *Discret. Cont. Dyn. Syst. A* **2014**, *34*, 1701–1745. [CrossRef]
3. He, X.; Ni, W.-M. Global dynamics of the Lotka-Volterra competition-diffusion system: Diffusion and Spatial Heterogeneity I. *Commun. Pure Appl. Math.* **2016**, *69*, 981–1014. [CrossRef]
4. Dockery, J.; Hutson, V.; Mischaikow, K.; Pernarowski, M. The evolution of slow dispersal rates: A reaction diffusion model. *J. Math. Biol.* **1998**, *37*, 61–83. [CrossRef]
5. Hastings, A. Can spatial variation alone lead to selection for dispersal? *Theor. Popul. Biol.* **1983**, *24*, 244–251. [CrossRef]
6. Altenberg, L. Resolvent positive linear operators exhibit the reduction phenomenon. *Proc. Natl. Acad. Sci. USA* **2012**, *109*, 3705–3710. [CrossRef] [PubMed]
7. Greenwood-Lee, J.; Taylor, P.D. The evolution of dispersal in spatially varying environments. *Evol. Ecol. Res.* **2001**, *3*, 649–665.
8. Brown, K.J.; Zhang, Y. On a system of reaction–diffusion equations describing a population with two age groups. *J. Math. Anal. Appl.* **2003**, *282*, 444–452. [CrossRef]
9. Cantrell, R.S.; Cosner, C.; Yu, X. Dynamics of populations with individual variation in dispersal on bounded domains. *J. Biol. Dyn.* **2018**, *12*, 288–317. [CrossRef]
10. Cantrell, R.S.; Cosner, C.; Yu, X. Populations with individual variation in dispersal in heterogeneous environments: Dynamics and competition with simply diffusing populations. *Sci. China Math.* **2020**, *63*, 441-464, arXiv:2001.03686.

11. Gurtin, M.E. A system of equations for age-dependent population diffusion. *J. Theor. Biol.* **1973**, *40*, 389–392. [CrossRef]
12. Kunisch, K.; Schappacher, W.; Webb, G.F. Nonlinear age-dependent population dynamics with random diffusion. *Comput. Math. Appl.* **1985**, *11*, 155–173. [CrossRef]
13. Langlais, M. A nonlinear problem in age-dependent population diffusion. *SIAM J. Math. Anal.* **1985**, *16*, 510–529. [CrossRef]
14. Yan, S.; Guo, S. Dynamics of a Lotka-Volterra competition-diffusion model with stage structure and spatial heterogeneity. *Discret. Contin. Dyn. Syst. B* **2018**, *23*, 1559–1579. [CrossRef]
15. López-Gómez, J.; Molina-Meyer, M. The maximum principle for cooperative weakly coupled elliptic systems and some applications. *Differ. Int. Equ.* **1994**, *7*, 383–398.
16. Molina-Meyer, M. Global attractivity and singular perturbation for a class of nonlinear cooperative systems. *J. Differ. Equ.* **1996**, *128*, 347–378. [CrossRef]
17. Cantrell, R.S.; Cosner, C. On the effects of spatial heterogeneity on the coexistence of competing species. *J. Math. Biol.* **1998**, *37*, 103–145. [CrossRef]
18. Fernández-Rincón, S.; López-Gómez, J. A singular perturbation result in competition theory. *J. Math. Anal. Appl.* **2017**, *445*, 280–296. [CrossRef]
19. Hutson, V.; López-Gómez, J.; Mischaikow, K.; Vickers, G. Limit Behavior for a Competing Speces Problem with Diffusion. In *Dynamical Systems and Applications*; Agarwal, R.P., Ed.; World Sci. Publishing: Hackensack, NJ, USA, 1995; pp. 343–358.
20. Chen, S.; Shi, J. Asymptotic profiles of basic reproduction number for epidemic spreading in heterogeneous environment. *arXiv* **2019**, arXiv:1909.10107.
21. Magal, P.; Webb, G.F, Wu,Y.-X. On the basic reproduction number of reaction-diffusion epidemic models. *SIAM J. Appl. Math.* **2019**, *79*, 284–304. [CrossRef]
22. Hei, L.J.; Wu, J.H. Existence and stability of positive solutions for an elliptic cooperative system. *Acta Math. Sin. Engl. Ser.* **2005**, *21*, 1113–1120. [CrossRef]
23. Lam, K.-Y.; Lou, Y. Asymptotic behavior of the principal eigenvalue for cooperative elliptic systems and applications. *J. Dyn. Differ. Equ.* **2016**, *29*, 29–48. [CrossRef]
24. Cantrell, R.S.; Cosner, C.; Lou, Y. Evolution of dispersal and the ideal free distribution. *Math. Biosci. Eng.* **2010**, *7*, 17–36. [PubMed]
25. Van den Driessche, P.; Watmough, J. Reproduction numbers and sub-threshold endemic equilibria for compartmental models of disease transmission. *Math. Biosci.* **2002**, *108*, 29–48. [CrossRef]
26. Bhattacharyya, A. On a measure of divergence between two statistical populations defined by their probability distribution. *Bull. Calcutta Math. Soc.* **1943**, *35*, 99–110.
27. Derpanis, K.G. The Bhattacharyya measure. *Mendeley Comput.* **2008**, *1*, 1990–1992.
28. Smith, H. *Monotone Dynamical Systems. An Introduction to the Theory of Competitive and Cooperative Systems*; Mathematical Surveys and Monographs, 41; American Mathematical Society: Providence, RI, USA, 1995.
29. Hutson, V.; Mischaikow, K.; Polácik, P. The evolution of dispersal rates in a heterogeneous time-periodic environment. *J. Math. Biol.* **2001**, *43*, 501–533. [CrossRef]

Article

Using *G*-Functions to Investigate the Evolutionary Stability of Bacterial Quorum Sensing

Anne Mund, Christina Kuttler and Judith Pérez-Velázquez *

Zentrum Mathematik, Technische Universität München Boltzmannstr. 3, 85748 Garching, Germany; mund@ma.tum.de (A.M.); kuttler@ma.tum.de (C.K.)

* Correspondence: cerit@ma.tum.de

Received: 30 October 2019; Accepted: 13 November 2019; Published: 15 November 2019

Abstract: In ecology, *G*-functions can be employed to define a growth function *G* for a population b, which can then be universally applied to all individuals or groups b_i within this population. We can further define a strategy v_i for every group b_i. Examples for strategies include diverse behaviour such as number of offspring, habitat choice, and time of nesting for birds. In this work, we employ *G*-functions to investigate the evolutionary stability of the bacterial cooperation process known as quorum sensing. We employ the *G*-function ansatz to model both the population dynamics and the resulting evolutionary pressure in order to find evolutionary stable states. This results in a semi-linear parabolic system of equations, where cost and benefit are taken into account separately. Depending on different biological assumptions, we analyse a variety of typical model functions. These translate into different long-term scenarios for different functional responses, ranging from single-strategy states to coexistence. As a special feature, we distinguish between the production of public goods, available for all subpopulations, and private goods, from which only the producers can benefit.

Keywords: Evolutionary dynamics; G-function; Quorum Sensing; Public Goods; semi-linear parabolic system of equations

---

1. Introduction

Since its discovery in 1977 [1], quorum sensing (QS) has received increasing attention as a mechanism of bacterial communication. Nowadays, we know that QS coordinates a range of behaviours at the population level [2]. One of those is the production of iron-scavenging molecules, so-called siderophores, in *Pseudomonas aeruginosa*. The bacteria secrete the siderophores and other virulence factors into the surrounding medium, where their benefit can be shared by all cells in the local population, leading to the term *public goods* [3,4], e.g., with the focus on kin selection in [5] or the conflict between individual and group interests in [6]. However, such a cooperative system is vulnerable to exploitation by non-cooperative cheaters. Gaining all the benefits of QS without incurring the metabolic costs, these cheater mutants have a growth advantage compared to the wild-type. Studies have shown that such mutants arise in laboratory conditions as well as in infection models [7–13] and act as social cheats. A few examples of these social cheats can be found in [3,14,15] (all considering the Gram-negative *P. aeruginosa*), while Pollitt et al. [16] provided an example from the Gram-positive *S. aereus*.

This gives rise to questions about the evolutionary stability of QS. Once cheaters arise (e.g., through loss-of-function mutation), they should theoretically outgrow the producers as they have more resources available to invest in cell division. Cheaters have been shown to outcompete producers both in vitro and in vivo, but QS seems to be evolutionary stable in natural systems nevertheless. In contrast to the aforementioned public goods, *private goods* are only accessible to the producing cell itself. They are hence innately protected from cheaters and provide their benefit exclusively for cells with a functioning QS system. Apart from extracellular molecules, QS also controls the production of proteins which act within the cell. In *P. aeruginosa*, one such protein is nucleoside hydrolase (Nuh). As Nuh is involved in metabolising adenosine, only bacteria with intact signal receptors can digest this carbon source. In this way, cooperation via QS provides a private fitness benefit to cooperating cells if adenosine is available as carbon source [17]. Schuster et al. [18] discussed the benefits attained from private goods. Specifically, on the question why public goods are ultimately stable, they suggested that, in fact, the private good is, directly or indirectly, involved in cooperative behaviour.

Extracellular molecules do not provide benefit indiscriminately, but are limited by diffusion and habitat structure. Kümmerli et al. [19] found a negative correlation between habitat structure and water solubility of siderophores, a class of secreted enzymes under QS control in a wide range of bacteria. For highly structured environments such as animal tissues, water solubility of siderophores is high, while microstructures in the environment naturally limit the resulting diffusion. Conversely, water solubility of siderophores is low in unstructured environments such as water habitats. This leads to siderophores clinging to each other as well as to lipid membranes. In this way, a fraction of the siderophores stay with their producer and provide some private benefit.

Several mechanisms, such as kin selection [20] and policing [21], have been described that could explain the evolutionary stability of cooperation and QS despite the advantages cheaters have in such a system.

In addition to experimental approaches, there have been many different mathematical modelling approaches to study QS, from stochastic cellular automata [22] to classic individual-based models [23], including differential equations [24]. In this paper, we develop a mathematical model to investigate the evolutionary stability of QS, which results in a semi-linear parabolic system of equation, following Frank [25] and Mund et al. [26], who used an age-dependent model as basis to check evolutionary stability. We explore the population dynamics between secretors and cheaters and are also interested in the evolutionary pressure that arises through those population dynamics. As such, we employ the *G*-function ansatz that was introduced by Cohen et al. [27] and further developed since then [28–31]. With the *G*-function ansatz, one finds a function G which describes the growth rate of a subpopulation. This function is dependent on the *strategy* of this subpopulation and both population numbers and strategies of other existing subpopulations (see Figure 1). A *strategy* can be taken to be any behaviour of interest which can be assigned a real number.

In our case, we take the QS-intensity of the subpopulation as the strategy. A strategy value of zero therefore denotes a QS cheater, while we assume a value of one to be the normal output of QS active secretors. If we take v_i as the strategy of the subpopulation, v is the vector of strategy values for all subpopulations. Accordingly, b_i and b represent the population numbers of subpopulation i and the complete vector, respectively. The population dynamics for subpopulation i thus follows the equation

$$\dot{b}_i(t) = G(v_i(t), v(t), b(t)) \cdot b_i(t). \tag{1}$$

At the same time, the per-capita growth function G also governs the evolutionary dynamics of the population. If we calculate the derivative of G with respect to the subpopulation strategy, we find the direction of higher per-capita growth. This is the direction evolutionary forces will drive the population, leading to

$$\dot{v}_i(t) = \partial_u G(u, v(t), b(t))|_{u=v_i(t)}. \tag{2}$$

To simplify notation from here on, we define

$$\partial_1 G(v_i, v, b) := \partial_u G(u, v, b)|_{u=v_i}. \tag{3}$$

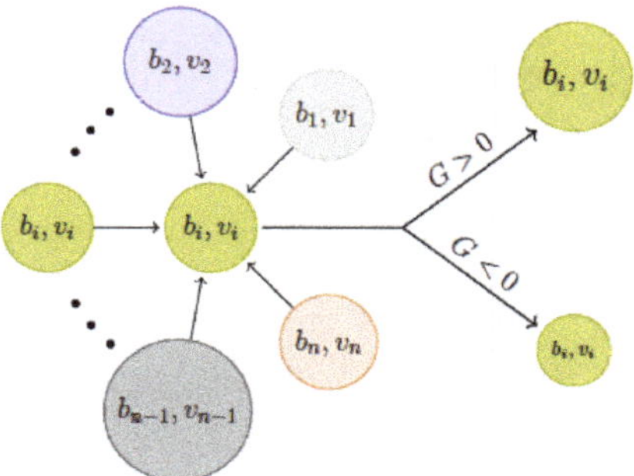

Figure 1. Schematic representation of influences on the growth rate of a population: the growth rate G of a subpopulation will be mainly influenced by three factors: its own strategy v_i, the environment defined by the complete strategy set v and the whole population b.

This framework allows us to find strategy values which are resistant against others in such a way that, under the influence of natural selection, no other strategy can invade the population. These strategies are called evolutionary stable strategies (ESS) [32].

Many experiments have shown the importance of spatial arrangements when considering bacterial interaction [33,34]. As such, we aim to include spatial variables in our model. This can be done in various ways, but we concentrate on the one that was introduced by Kronik and Cohen [31]. This leads to a system of partial differential equations.

In [35], we showed the existence and uniqueness of weak and classical solutions to the associated semi-linear parabolic system in Equations (1) and (2) with Robin boundary conditions using the coupled upper-lower solution approach. In this paper, we analyse a variety of typical model functions. These translate into different long-term scenarios for different functional responses, ranging from single-strategy states to coexistence. In Section 2, we introduce the basic model and some assumptions we work with, before considering private goods models in Section 3. These models are then analysed numerically in Section 4. In Section 5, we summarise and interpret our findings.

2. The Basic Model: Using the G-Function Approach

In this case G takes the form:

$$G(v_i, v, b) = B(v, b) \cdot C(v_i) - \mu \|b\|_1 . \tag{4}$$

For a formulation with carrying capacity instead of population-dependent death rate, we can transform this equation into

$$G(v_i, v, b) = B(v, b) C(v_i) \left(1 - \frac{\mu \|b\|_1}{B(v, b) C(v_i)}\right), \tag{5}$$

which gives us $B(v,b)C(v_i)/\mu$ as the capacity for the population. We can now look at the derivative of G with respect to v_i.

$$\partial_1 G(v_i, v, b) = B(v,b) \cdot C'(v_i). \tag{6}$$

We make some regularity assumptions on G and its derivative to simplify the mathematical analysis. These do not restrict our biological application, as our model growth functions do not include discontinuous behaviour and are smooth.

(I) $G(v_i, v, b)$ is Lipschitz continuous in all variables.
(II) $\partial_1 G(v_i, v, b)$ is Lipschitz continuous in $(v, b)^T$, where ^{T} denotes transpose. Additionally, we make some assumptions on G from the biological background.
(III) $G(v_i, v, b)$ has a (direct or indirect) negative feedback loop in b_i. The population number cannot go towards infinity.
(IV) $\exists\, \overline{u}\, \forall (v,b)\, \partial_1 G(v_i, v, b) \leq 0 \quad \forall v_i > \overline{u}$. From a certain threshold onwards, higher strategy values no longer not a higher net growth.
(V) $\exists\, \underline{u} < \overline{u}\, \forall\, (v,b) : \partial_1 G(\underline{u}, v, b) = 0$. Reducing the strategy further than the value $\underline{u}$ provides no higher growth rate.

These assumptions ensure that Equation (21) does not exhibit divergent behaviour ($v \to \infty$ and/or $b \to \infty$), which is not biologically plausible. If, e.g., $\partial_1 G(v_i, v, b) \geq \gamma > 0\, \forall v_i$, then $c + \lambda t$ with $\lambda = \gamma \cdot \varepsilon$ is a lower solution of Equation (21b), as

$$\varepsilon \partial_1 G(\lambda t + c, v, b) - \lambda \geq 0.$$

It follows that $v_i \geq c + \lambda t$ and thus $v_i \to \infty$. This would mean ever-increasing strategy values without some kind of trade-off, which we do not find in nature.

In the special case of QS, we take $\underline{u}$ to be 0, requiring $\partial_1 G(0, v, b) = 0$. This keeps v from leaving the biologically meaningful parameter range $\mathbb{R}_0^+$ (production cannot be lower than 0).

2.1. Cost and Benefit

To model the G-function for QS, one of the avenues we can take is to divide the impact of v_i, v, and b into two parts: a growth term influenced by v_i and a benefit provided by v, b, and possibly also v_i. This reflects the fact that production of QS molecules is costly to the individual, while the resulting factors are public goods and therefore provide benefit to all bacteria. The additional dependence on v_i can be seen as a form of private benefit and is discussed in detail when it occurs.

One important thing to note is that the growth term is actually *reduced* with rising v_i, as increased public good production incurs increased metabolic costs. In this way, less energy is retained for reproduction. We denote the cost term by $C : \mathbb{R} \to \mathbb{R}_+$, and make the following assumptions:

(VI) As public good production is costly, C is strictly monotonically decreasing for $v_i \in [0, \infty)$.
(VII) When producing public goods, the growth rate is reduced by a certain factor:

$$0 < C(v_i) < 1 \quad \text{for } v_i \neq 0, \qquad C(0) = 1. \tag{7}$$

While Item (VI) is clear from our assumptions on QS, Item (VII) is not as immediately clear. We introduce Item (VII) because we use this factor multiplicatively for G. Thus, a value of 1 would signify unimpeded growth, while a value between 0 and 1 reduces growth. In this way, we assume that QS costs alone do not lead to negative growth rates.

The benefit of QS is provided by secreting extracellular proteins. We denote it by $B : \mathbb{R}^m \times \mathbb{R}^m \to \mathbb{R}_+$ and make two main assumptions here:

(VIII) There is a limit to how much benefit can be obtained,

$$\lim_{v \to \infty} B(v, b) = B_{\max}. \tag{8}$$

(IX) There is no benefit if no public goods are produced,

$$B(v, b) = 0, \quad \text{if } b = 0, \sum_i b_i v_i = 0. \tag{9}$$

Equation (8) models a saturation behaviour for the benefit—even if the cells were producing an infinite amount of extracellular protein, the benefit that can be derived is still capped through saturation of enzymes or similar phenomena. Equation (9) ensures that there is no benefit from QS when there are no living bacteria, or all of them have stopped participating in QS.

2.2. Analysis

We know from Assumption (VI) that $C(v_i)$ is strictly monotonically decreasing for positive v_i. Combined with Assumption (V), we have $C'(v_i) < 0 \, \forall v_i > 0$, $C'(0) = 0$ and there exists only $v_i = 0$ as an ESS. As the only stable stationary point, $v_i = 0$ will attract all other solutions, leading to the decline of QS.

3. Models with Private Benefit

Following up on the *private goods*-hypothesis explained in Section 1, we now assume that there is a private benefit associated with producing the public goods, e.g. a small percentage of the produced enzymes may cling to the producing bacteria. We model this by adding a term $B(v_i)$, which is a benefit term that is solely dependent on the strategy of the subpopulation itself. This $B(v_i)$ should fulfil similar assumptions to those we made of $B(v, b)$, which is why we choose to denote it by the same letter. Overall, we can write for G:

$$G(v_i, v, b) = (B(v, b) + B(v_i)) \cdot C(v_i) - \mu \|b\|_1 . \tag{10}$$

It follows that

$$\partial_1 G(v_i, v, b) = \big(B(v, b) + B(v_i)\big) \cdot C'(v_i) + B'(v_i) \cdot C(v_i), \tag{11}$$

hence for an ESS it must hold that

$$-C'(v_i) \, \big(B(v, b) + B(v_i)\big) = B'(v_i) C(v_i). \tag{12}$$

To proceed with a more in-depth analysis, we choose a specific cost function which satisfies all our assumptions,

$$C(v_i) = \exp\left(-K v_i^2\right), \tag{13}$$

and look at two specific types of model terms for $B(v_i)$.

3.1. First Type of Model Terms with Monotonicity Property

The first type of model terms consists of those functions $B(v_i)$, for which $B'(v_i)/v_i$ is monotonically decreasing in v_i. This is the case for all possible concave functions (e.g., those with $B(0) = 0, B(\infty) = \bar{B} > 0$), as well as for

$$B(v_i) = \frac{v_i^2}{v_i^2 + a^2}, \qquad \text{and } B(v_i) = -e^{-\omega v_i^2} + 1.$$

Equation (13) allows us to write Equation (12) as

$$\left(B(v,b) + B(\bar{v}_i)\right)(2K) = \frac{B'(\bar{v}_i)}{\bar{v}_i} =: R(v_i) \tag{14}$$

for $\bar{v}_i \neq 0$. Since we know the left-hand side of this equation to be monotonically increasing, a monotonically decreasing right-hand side means that there can only be one solution to Equation (14). Whether or not a solution exists at all can be determined by looking at the limit behaviour.

We know that $B(v_i)$ should exhibit a saturation for $v_i \to \infty$, forcing

$$\lim_{v_i \to \infty} \frac{B'(\bar{v}_i)}{\bar{v}_i} = 0. \tag{15}$$

We also take a look at the limiting values at 0:

$$\lim_{v_i \to 0} \left(B(v,b) + B(v_i)\right)(2K) = 2KB(v,b) =: B_0, \tag{16a}$$

$$\lim_{v_i \to 0} \frac{B'(v_i)}{v_i} = \begin{cases} B''(0) & \text{if } B'(0) = 0 \\ +\infty & \text{if } B'(0) > 0 \end{cases}. \tag{16b}$$

It follows that for $B'(0) = 0$ there exists a positive stationary $\bar{v}_i$ if and only if $B''(0) > 2KB(v,b)$, while it always exists if $B'(0) > 0$. If such a positive $\bar{v}_i$ exists, it is stable (see Appendix B.2.1). If, on the other hand, $B'(0) = 0$ but $B''(0) < 2KB(v,b)$, then $\bar{v}_i = 0$ is an ESS.

3.2. Second Type of Model Terms: Hill Equations

For a second type of model terms, we consider Hill equations. These relatively simple functions are often used in mathematical modelling for cooperative binding of molecules. In the case of QS, Dockery and Keener [36] employed them in the modelling of QS signal regulation. We look at terms of the form

$$B(v_i) = \frac{v_i^h}{v_i^h + a^h} \tag{17}$$

with $h \geq 3$, as the cases $h = 1$ and $h = 2$ fall into the first type of model terms. Then, we have

$$R(v_i) = \frac{B'(v_i)}{v_i} = \frac{h a^h v_i^{h-2}}{(v_i^h + a^h)^2}. \tag{18}$$

Instead of being monotonically decreasing as before, $R(v_i)$ has the following properties:

1. $R(0) = 0$.
2. $\lim_{v_i \to \infty} R(v_i) = 0$.
3. $R(v_i) \geq 0 \quad \forall v_i \geq 0$, with $R(v_i) > 0$ if $v_i > 0$.
4. $R(v_i)$ has exactly one maximum, $R(v_i^*)$.
5. $R^{(n)}(v_i) = \frac{v_i^{h-(n+2)} f(v_i)}{(v_i^h + a^h)^{n+2}}$ for $n \leq h-2$, with a function $f(v_i)$ for which $f(0) > 0$.

For a better understanding of the ideas behind, we assume for a moment that $B(v,b) > 0$. If we can now show that $R(v_i^*) > 2K(B(v,b) + B(v_i^*))$, two positive stationary solutions v_i to Equation (12) exist.

$$R(v_i^*) = \frac{h \cdot \left(\frac{h-2}{h+2}\right)^{\frac{h-2}{h}}}{a^2 \cdot \left(\frac{2h}{h+2}\right)^2} \qquad\qquad B(v_i^*) = \frac{h-2}{2h}$$

In that way, $R(v_i^*) > 2K(B(v,b) + B(v_i^*))$ is equivalent to the condition

$$\begin{aligned} &(B(v,b)4h + 2(h-2))\, 2Ka^2 < (h-2)^{\frac{h-2}{h}} (h+2)^{\frac{2}{h}} \\ \Rightarrow \quad & Ka^2 < \frac{(h-2)^{\frac{h-2}{h}} (h+2)^{\frac{2}{h}}}{4(B(v,b)2h + h - 2)}, \end{aligned} \tag{19}$$

which unfortunately cannot be simplified much further, without assumptions on the parameters K and a. We remind ourselves that K is the cost of cooperation, while a denotes the amount of signalling necessary to gain half of the maximal (private) benefits. Equation (19) thus gives upper limits to these terms, dependent on the steepness of the activation curve h as well as the public benefit $B(v,b)$. A larger public benefit is actually detrimental to the existence of positive stationary strategies v_i.

At the same time, there cannot be more than two intersections in $[0, +\infty)$ by Rolle's theorem (see, e.g., [37], p. 134), since

$$\begin{aligned} (2KB(v,b) + B(v_i))' - R'(v_i) &= 2KB'(v_i) - R'(v_i) \\ &= \frac{a^h h v_i^{h-3}}{(v_i^h + a^h)^3} \left(2Kv_i^2(v_i^h + a^h) + (h+2)v_i^h - (h-2)a^h\right) \\ &= \frac{a^h h v_i^{h-3}}{(v_i^h + a^h)^3} \left(2Kv_i^{h+2} + (h+2)v_i^h + 2Ka^h v_i^2 - (h-2)a^h\right), \end{aligned} \tag{20}$$

where the term in parenthesis is strictly monotonically increasing and the equation therefore only has one $\xi \in (0, +\infty)$ for which $(2KB(v,b) + B(\xi))' - R'(\xi) = 0$. Both $\bar{v}_i = 0$ and the larger of both positive stationary solutions are stable regardless of parameter values (see Appendix B.2.1).

If $B(v,b) = 0$, which occurs if for example the existing subpopulations do not engage in QS, the situation is different. As both $R(0) = 0$ and $2KB(v,b) + B(0) = 0$, there can only be one positive intersection. By using Property 5, we know that $R(v_i)$ is increasing more quickly than $2KB(v,b) + B(v_i)$ for small v_i. It follows that a positive intersection exists. As we know that for $B(v,b) = 0$ there is an intersection at $v_i = 0$ and from Equation (20) that there can only be two intersections in $[0, +\infty)$, there can be no other positive intersection. Additionally, one can prove this positive intersection to be stable, while the zero solution is unstable (see Appendix B.2.1).

In biological terms, our results indicate that the surrounding subpopulations have a profound impact on the evolution of cooperativity. If one of the subpopulations is cooperating ($B(v,b) \neq 0$), the others will experience bistable behaviour—they might cooperate themselves or not participate in QS, depending on starting strategy. If, on the other hand, all of the subpopulations do not cooperate at first ($B(v,b) = 0$), then the evolutionary pressure will drive them towards the positive ESS, meaning they pick up QS with time.

4. Numerical Simulations

The full set of equations we are working with therefore reads

$$\dot{b_i}(x,t) = G(v_i(x,t), v(x,t), b(x,t)) \cdot b_i(x,t) + D \triangle b_i(x,t) \tag{21a}$$

$$\dot{v_i}(x,t) = \varepsilon \partial_1 G(v_i(x,t), v(x,t), b(x,t)) + D\left(\frac{2\nabla b_i(x,t) \cdot \nabla v_i(x,t)}{b_i(x,t)} + \triangle v_i(x,t)\right), \tag{21b}$$

noting that $\dot{b_i}(x,t) = \frac{\partial b_i}{\partial t}$.

We used our own experimental data from [34] to derive values for the parameters in our model, which can be found in Table A1. Our aim here was to obtain rough estimates which replicate the general qualitative behaviour, as opposed to fit the experimental data perfectly. To solve the system numerically, we employed the method of lines [38] and used an explicit Runge–Kutta solver on the discretised system. The simulations were run in Matlab.

We assumed a closed system in which the existing subpopulations are mixed but not completely homogeneous in the beginning. This leads to Neumann boundary conditions and the initial distribution displayed in Figure 2.

Figure 2. Visual representation of initial conditions for numerical simulations. For initial conditions, we used patches. The two subpopulations (dark and light patches) were mixed but not homogeneously distributed.

We also added in a feature of QS that we have neglected before: the costs of QS depend on the signal level, as QS products are only produced when the signal level is high enough. We can achieve this by modifying the cost term from Equation (13) to

$$C(v_i, v, b) = \exp\left(-K \cdot (B(v,b) \cdot v_i)^2\right), \tag{22}$$

as $B(v,b)$ is a measure for the level of QS. We have not done so before, since it does not change the analysis qualitatively, but leads to unnecessarily long and obfuscating terms.

4.1. Basic Model

We have seen in our calculations that $v_i = 0$ is the only stable strategy in this situation. We should therefore have both $v_i \to 0$ and $b_i \to 0$. However, when looking at the simulation results for 3000 h as displayed in Figure 3, we notice that this convergence is very slow indeed. A quick calculation of $\partial_1 G$ and G for our parameter values confirms that both are close to zero. Thus, even though QS is theoretically unstable, both cheaters and producers persist alongside each other for a very long time, albeit at different densities.

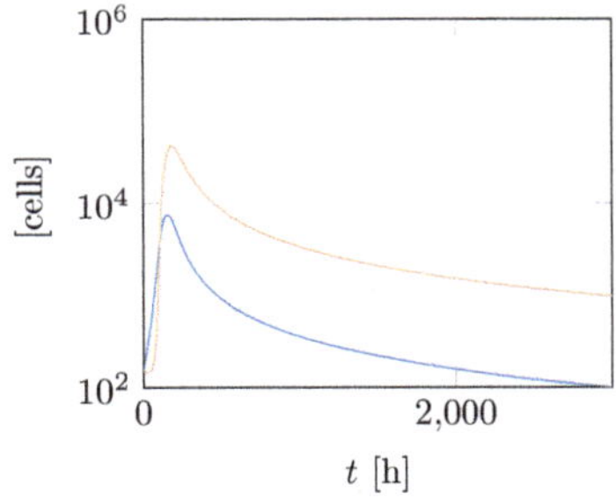

(**a**) Population dynamics

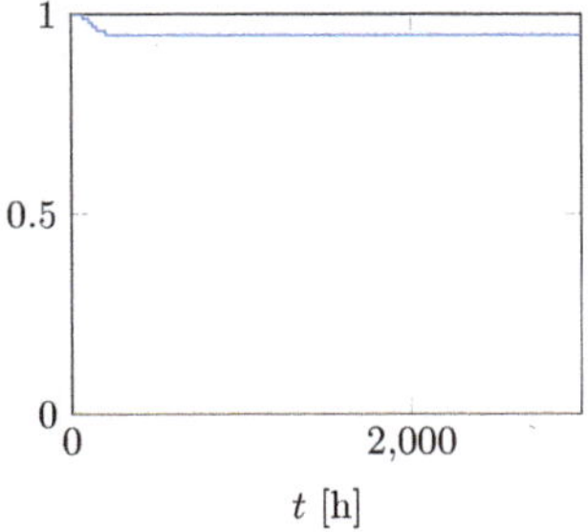

(**b**) Strategy dynamics

Figure 3. Long-term behaviour of two populations with different start strategies, using a G-function without additions. Both populations numbers are slowly converging in time course towards zero, with producers (dark line) having the lower numbers.

4.2. Models with Private Benefit

In Section 3 we distinguish between two types of model functions. Depending on the Hill coefficient, Hill-terms can fall into either one. We investigated the long-term behaviour for $h = 1, 2$ or 3.

As predicted, the behaviour in these cases is quite different. For $h \leq 2$, $B(v_i)$ falls into the first type of model functions. For this type, we postulate that there can only be one positive stationary point for v and if it is stable, and then the zero solution is unstable. This is the case for the parameters we calculated. Additionally, we found that for $h = 1$ the zero solution is not a stationary point. Thus, both strategies converge towards the stable positive equilibrium, one from above and the other from below. Population 2, which started out as cheaters, gains QS functionality, while there is reduced production from Population 1. Since reducing production is slower to reach the stable point in this parameter constellation, Population 1 succumbs to the population pressure from Population 2 and dies out (see Figure 4). This happens on a rather short time frame of less than 200 h.

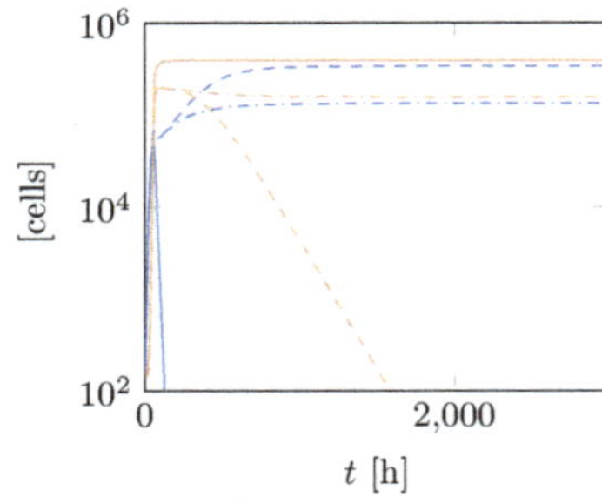

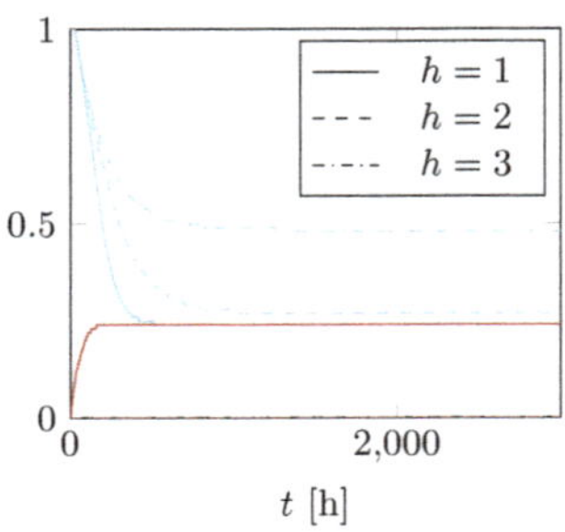

(**a**) Population dynamics (**b**) Strategy dynamics

Figure 4. Evolution of two populations with different start strategies, using a G-function with private benefit and hill factors of $h = 1$ (line), 2 (dashed) and 3 (dash-dotted). For $h = 1$, Population 1 (dark colour) dies out rather quickly, while Population 2 (light colour) gains the QS functionality, albeit at a low level, and remains at a stable population level. If $h = 2$, Population 1 reduces its QS strategy to a lower, but stable value. Population 2 is unable to compete and dies out. For $h = 3$, Populations 1 and 2 coexistt in a stable way with similar numbers. One population is QS active while the other is not.

For $h = 2$, we can see in Figure 4 that once again the strategies converge to a positive value. However, this time the zero solution is a stationary point, although an unstable one. Hence, Population 2 cannot gain QS functionality by starting out with $v_2 = 0$. The end result is the extinction of Population 2, while Population 1 remains stable.

Overall, for $h \leq 2$, we found that one population is driven to extinction, while the other stays at a stable level with QS intact at lower levels.

When $h > 2$, $B(v_i)$ falls into the second type of model functions. That means there might be a stable positive strategy in addition to a stable zero solution. We found this to be the case for $h = 3$ for our parameters, as one can see in Figure 4. This means that Populations 1 and 2 remain at stable population and strategy levels, with Population 1 being QS active while Population 2 consists of cheaters. In this scenario, producers and non-producers can live side-by-side indefinitely.

5. Discussion

We have applied the G-function ansatz to the QS dynamics of *P. aeruginosa* and analysed the resulting model both analytically and numerically. We have found that the type of G-function considered changes the expected outcome drastically.

If we assume that there is no private benefit to QS at all, then the long-term outcome is preassigned and independent of parameters. Whether QS is inactive in all bacteria after some time, or the subpopulation of secretors dies out, the effect remains the same: QS is evolutionary unstable.

As soon as we assume that there is some kind of private benefit, the situation changes. Depending on parameter values, QS might still be unstable, but there can also be positive evolutionary stable secretion values. For the first type of model functions, there is only one stable secretion value. It follows that in the long term the population will unitise (Figure 4). The situation remains much the same if, instead of a direct benefit, we consider a reduction in death rate (results not shown). In contrast, private benefit functions of the second type admit bistable behaviour. This allows secretors and cheaters to coexist indefinitely in addition to single subpopulation states. We have also found that bacteria are driven towards QS if there is no pre-existing cooperation in their environment.

Since the left-hand-side of Equation (12) has a value of zero for $v_i = 0$, whereas the right-hand-side has a value greater than zero for $v_i = 0$, there could be any number of stable strategies (or none) if one chooses different functional responses.

Our model functions have simplified the actual process of QS a great deal, but one could also include direct dependencies on the signal molecules and enzymes produced. Such models with abiotic components were considered by Cohen et al. [30]. While their short-term dynamics might differ from the scenarios we have considered, their long-term outcome does not. They are also difficult to analyse without resorting to numerical methods, which is why we have concentrated on the simplified functions.

The G-function has allowed us to determine the long-term evolution of QS by considering the short-term population dynamics. By analysing different functional responses, we have explored some possible outcomes. One could also further refine the population dynamics for different real world scenarios. The G-function serves to always model the resulting evolutionary pressure, which tells us something about the way mutation and selection are going to drive the population. As both are happening on a fast time-scale in bacteria, they have a big impact on the interaction between bacterial populations as well as on their interaction with humans (see, e.g., [3]). The G-function can thus be a valuable tool in modelling bacterial interactions.

For completeness, in Appendix C, we show the spatial evolution of two populations with different start strategies, using a G-function without additions (Figure A1) and with private benefit and a hill factor of $h = 1$ (Figure A2).

Lastly, we would like to refer the interested reader to the following associated publications which complement this study: Mund et al. [34] and Mund et al. [35].

Author Contributions: Conceptualization, C.K. and J.P.-V.; Data curation, A.M.; Formal analysis, A.M.; Funding acquisition, C.K.; Investigation, A.M., C.K. and J.P.-V.; Methodology, A.M.; Software, A.M.; Supervision, C.K. and J.P.-V.; Validation, A.M.; Visualization, A.M.; Writing—original draft, A.M., C.K. and J.P.-V.; Writing—review and editing, A.M., C.K. and J.P.-V.

Funding: This work was supported by the German Research Foundation (DFG) and the Technical University of Munich within the funding programme Open Access Publishing.

Conflicts of Interest: The authors declare no conflict of interest.

Appendix A. Variables and Parameters

Table A1. Table of all occurring parameters.

Name	Value		Stands for
a	7.186×10^{-1}	1	threshold value for private benefit
$B_{\max}$	0.30	1/h	maximal bacterial growth rate
K	2.459	1	cost of participating in QS
μ	7.24×10^{-7}	1/(cells h)	bacterial death rate
$\mu_{\min}$	7.24×10^{-8}	1/(cells h)	minimal bacterial death rate for strategy-dependent death rate
τ	6.06×10^{3}	cells	threshold value for public benefit

Appendix B. Proofs

Appendix B.1. Properties of $R(V_i)$ for the Second Type of Model Terms

We first prove that $R(v_i)$ has exactly one maximum as well as calculate its maximal value. To that end, we take the derivative of R

$$R'(v_i) = \frac{a^h h v_i^{h-3}\left(-(h+2)v_i^h + (h-2)a^h\right)}{\left(v_i^h + a^h\right)^3} \tag{A1}$$

and search for critical points

$$0 = a^h h v_i^{h-3}\left(-(h+2)v_i^h + (h-2)a^h\right). \tag{A2}$$

For $h > 3$, there exists a critical point at $v_i = 0$. However, since $R(0) = 0$ and $R(v_i) > 0$ for $v_i > 0$, this critical point is clearly a minimum on the bounded space $v_i \in \mathbb{R}_0^+$. The other critical point satisfies

$$v_i^* = \sqrt[h]{\frac{h-2}{h+2}} \cdot a$$

and is thus unique in $\mathbb{R}_0^+$. It remains to show that v_i^* is indeed a maximum. However, knowing that $R(0) = 0$ and $\lim_{v_i \to \infty} R(v_i) = 0$, we can conclude that it can only be a maximum.

We prove Property 5 by induction.

base case We can derive from the definition of $R(v_i)$ that

$$\begin{aligned} R'(v_i) &= \frac{a^h h v_i^{h-3}\left(-(h+2)v_i^h + (h-2)a^h\right)}{\left(v_i^h + a^h\right)^3} \\ &= \frac{v_i^{h-(1+2)} f(v_i)}{\left(v_i^h + a^h\right)^{1+2}}, \end{aligned}$$

where $f(v_i) = a^h h\left(-(h+2)v_i^h + (h-2)a^h\right)$ and thus $f(0) = ha^{2h}(h-2) > 0$.

inductive step Assuming that the statement holds for $R^{(n)}(v_i)$ and that $n+1 \le h-2$, we have

$$\begin{aligned} R^{(n+1)}(v_i) = &\Big((v_i^h + a^h)^{n+2}(h-(2+n))v_i^{h-(n+3)} f(v_i) \\ &+ (v_i^h + a^h)^{n+2} f'(v_i) v_i^{h-(n+2)} \\ &- (n+2)h(v_i^h + a^h)^{n+1} f(v_i) v_i^{2h-(n+3)}\Big) \Big/ \left(v_i^h + a^h\right)^{2n+4} \\ = &\frac{v_i^{h-(n+1+2)} g(v_i)}{(v_i^h + a^h)^{n+1+2}}, \end{aligned}$$

where

$$g(v_i) = (v_i^h + a^h)\Big((h-(2+n))f(v_i) + f'(v_i)v_i\Big) - (n+2)hf(v_i)v_i^h.$$

and thus $g(0) = a^h(h-(2+n))f(0) > 0$ since $n < h-2$ and $f(0) > 0$.

It follows that $R^{(n)}(0) = 0$ for $n < h-2$ and $R^{(h-2)}(0) > 0$. At the same time, a similar argument shows that

$$\left(2K(B(v,b)+B(v_i))\right)^{(n)} = 2KB^{(n)}(v_i) = \frac{v_i^{h-n}2Kf(v_i)}{(v_i^h+a^h)^{1+n}}.$$

As such, $\left(2K(B(v,b)+B(v_i))\right)^{(n)} = 0$ for $n < h$.

Appendix B.2. Stability of $\bar{v}_i$

Appendix B.2.1. Models with Private Benefit

First Type of Model Terms

For $B(v_i)$ in the first type of model terms, we assume that $B'(v_i)/v_i$ is monotonically decreasing. It follows that

$$\begin{aligned} 0 > \left(\frac{B'(v_i)}{v_i}\right)' &= \frac{v_iB''(v_i)-B'(v_i)}{v_i^2} \\ \Rightarrow \quad 0 > B''(v_i) - \frac{B'(v_i)}{v_i} &\quad \forall v_i > 0. \end{aligned} \tag{A3}$$

Looking at the second derivative of G, we have

$$\begin{aligned} \partial_1^2 G(v_i,v,b) &= (B(v,b)+B(v_i))\,C''(v_i) + 2B'(v_i)C'(v_i) + B''(v_i)C(v_i) \\ &= \left((B(v,b)+B(v_i))\left(-2K+4K^2v_i^2\right) + 2B'(v_i)(-2K)v_i + B''(v_i)\right)\mathrm{e}^{-Kv_i^2} \\ \partial_1^2 G(0,v,b) &= B(v,b)(-2K) + B''(0), \end{aligned}$$

from which we can gather that $\bar{v}_i = 0$ is stable if $B''(0) < 2KB(v,b)$. We use Equation (14) and substitute for $B(v,b)+B(\bar{v}_i)$

$$\begin{aligned} \partial_1^2 G(\bar{v}_i,v,b) &= \left(\frac{B'(\bar{v}_i)}{2K\bar{v}_i}\left(-2K-4K^2\bar{v}_i^2\right) + 2B'(\bar{v}_i)(-2K)\bar{v}_i + B''(\bar{v}_i)\right)\mathrm{e}^{-K\bar{v}_i^2} \\ &= \left(-\frac{B'(\bar{v}_i)}{\bar{v}_i} + B''(\bar{v}_i) - 2K\bar{v}_iB'(\bar{v}_i)\right)\mathrm{e}^{-K\bar{v}_i^2}. \end{aligned} \tag{A4}$$

As $B'(v_i) > 0$, stability of positive $\bar{v}_i$ immediately follows from Equation (A3).

Second Type of Model Terms

We start out by assuming $B(v,b) > 0$. We know that $v_i = 0$ is stable if $B''(0) < 2KB(v,b)$ and that for $h \geq 3$ it holds that $B''(0) = 0$. Hence, $v_i = 0$ is an ESS regardless of parameter values.

A positive stationary solution $\bar{v}_i$ is stable, if

$$
\begin{aligned}
& 0 > -\left(\frac{1}{\bar{v}_i} + 2K\bar{v}_i\right) B'(\bar{v}_i) + B''(\bar{v}_i) \\
\Rightarrow \quad & 0 > -\left(\frac{1}{\bar{v}_i} + 2K\bar{v}_i\right) \frac{ha^h\bar{v}_i^{h-1}}{(\bar{v}_i^h + a^h)^2} + \frac{ha^h\bar{v}_i^{h-2}\left(-(h+1)\bar{v}_i^h + (h-1)a^h\right)}{(\bar{v}_i^h + a^h)^3} \\
\Rightarrow \quad & 0 > -\left(\frac{1}{\bar{v}_i} + 2K\bar{v}_i\right) \bar{v}_i(\bar{v}_i^h + a^h) - (h+1)\bar{v}_i^h + (h-1)a^h.
\end{aligned}
$$

We can rearrange this inequality to

$$
\left(\frac{1}{\bar{v}_i} + 2K\bar{v}_i\right) \bar{v}_i(\bar{v}_i^h + a^h) + (h+1)\bar{v}_i^h > (h-1)a^h.
$$

The larger of both positive stationary solutions fulfils $\bar{v}_i > v_i^* = \sqrt[h]{\frac{h-2}{h+2}} \cdot a$, if it exists. It follows that

$$
\begin{aligned}
&\left(\frac{1}{\bar{v}_i} + 2K\bar{v}_i\right) \bar{v}_i(\bar{v}_i^h + a^h) + (h+1)\bar{v}_i^h \\
&\qquad > \left(1 + 2K\bar{v}_i^2\right)\left(\frac{h-2}{h+2} \cdot a^h + a^h\right) + (h+1) \cdot \frac{h-2}{h+2} \cdot a^h \\
&\qquad > \frac{2h}{h+2} \cdot a^h + \frac{(h+1)(h-2)}{h+2} \cdot a^h \\
&\qquad = \frac{2h + h^2 - h - 2}{h+2} \cdot a^h \\
&\qquad = (h-1) \cdot a^h.
\end{aligned}
$$

Hence, the larger of both positive solutions is always stable, if it exists. It follows immediately from the stability of the zero solution that the smaller of both positive solutions must be unstable.

In the other case, we first note that both $B''(0) = 0$ and $B(v, b) = 0$, which means we cannot gain insight into the stability of the zero solution directly through Equation (A4). However, we know that, for small enough $\varepsilon > 0$, $R(\varepsilon) > 2KB(\varepsilon)$ and that there exists only one positive ξ for which $2KB'(\xi) = R'(\xi)$. Additionally, we know that ξ lies in the open interval between zero and the positive intersection $\bar{v}_i$ and as such is unequal to $\bar{v}_i$. It can thus follow that

$$
\begin{aligned}
R'(\bar{v}_i) &< 2KB'(\bar{v}_i) \\
R'(\bar{v}_i) = \left(\frac{B'(\bar{v}_i)}{\bar{v}_i}\right)' &= \frac{B''(\bar{v}_i)}{\bar{v}_i} - \frac{B'(\bar{v}_i)}{\bar{v}_i^2} \\
\Rightarrow \quad \frac{B''(\bar{v}_i)}{\bar{v}_i} - \frac{B'(\bar{v}_i)}{\bar{v}_i^2} &< 2KB'(\bar{v}_i) \\
-2KB'(\bar{v}_i)\bar{v}_i - \frac{B'(\bar{v}_i)}{\bar{v}_i} + B''(\bar{v}_i) &< 0,
\end{aligned}
$$

which is exactly the condition for stability from Equation (A4). For $B(v, b) = 0$, we thus have an unstable zero solution and a stable positive strategy.

Appendix C. Spatial Evolution of Two Populations with Different Start Strategies

In this section, we show the spatial evolution of two populations with different start strategies, using a G-function without additions (Figure A1) and with private benefit and a hill factor of $h = 1$ (Figure A2).

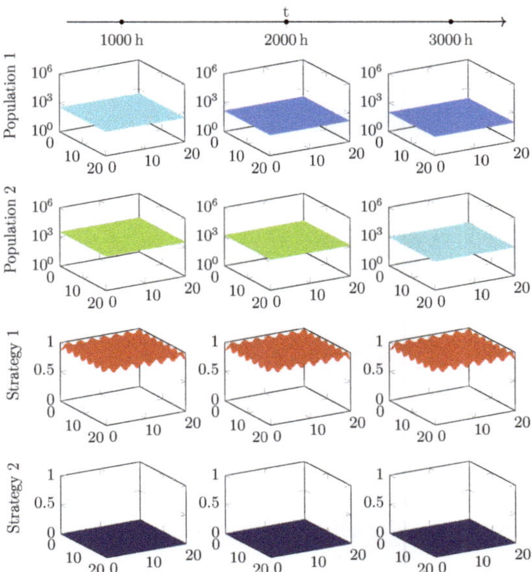

Figure A1. Long-term behaviour of two populations with different start strategies, using a G-function without additions.

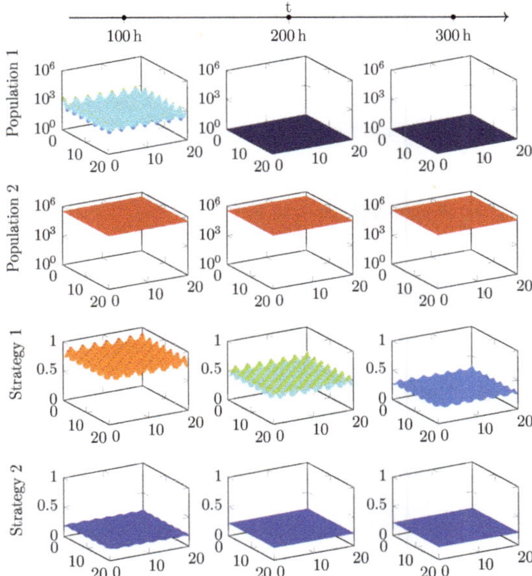

Figure A2. Evolution of two populations with different start strategies, using a G-function with private benefit and a hill factor of $h = 1$. Note the shorter timescale in this plot.

References

1. Hastings, J.; Nealson, K. Bacterial bioluminescence. *Annu. Rev. Microbiol.* **1977**, *31*, 549–595. [CrossRef] [PubMed]
2. Rutherford, S.T.; Bassler, B.L. Bacterial quorum sensing: Its role in virulence and possibilities for its control. *Cold Spring Harbor Perspect. Med.* **2012**, *2*, a012427. [CrossRef] [PubMed]
3. Rumbaugh, K.P.; Diggle, S.P.; Watters, C.M.; Ross-Gillespie, A.; Griffin, A.S.; West, S.A. Quorum sensing and the social evolution of bacterial virulence. *Curr. Biol.* **2009**, *19*, 341–345. [CrossRef] [PubMed]
4. West, S.A.; Griffin, A.S.; Gardner, A.; Diggle, S.P. Social evolution theory for microorganisms. *Nat. Rev. Microbiol.* **2006**, *4*, 597–609. [CrossRef] [PubMed]
5. Diggle, S.P.; Griffin, A.S.; Campbell, G.S.; West, S.A. Cooperation and conflict in quorum-sensing bacterial populations. *Nature* **2007**, *450*, 411–414. [CrossRef] [PubMed]
6. Brown, S.P.; Johnstone, R.A. Cooperation in the dark: Signalling and collective action in quorum-sensing bacteria. *Proc. R. Soc. B* **2001**, *268*, 961–965. [CrossRef] [PubMed]
7. Bjarnsholt, T.; Jensen, P.; Jakobsen, T.H.; Phipps, R.; Nielsen, A.K.; Rybtke, M.T.; Tolker-Nielsen, T.; Givskov, M.; Høiby, N.; Ciofu, O.; et al. Quorum sensing and virulence of *Pseudomonas aeruginosa* during lung infection of cystic fibrosis patients. *PLoS ONE* **2010**, *5*, e10115. [CrossRef]
8. Dénervaud, V.; TuQuoc, P.; Blanc, D.; Favre-Bonté, S.; Krishnapillai, V.; Reimmann, C.; Haas, D.; Delden, C.V. Characterization of cell-to-cell signaling-deficient *Pseudomonas Aeruginosa* Strains Colon. Intubated Patients. *J. Clin. Microbiol.* **2004**, *42*, 554–562. [CrossRef]
9. Köhler, T.; Buckling, A.; van Delden, C. Cooperation and virulence of clinical *Pseudomonas Aeruginosa* Popul. *Proc. Natl. Acad. Sci. USA* **2009**, *106*, 6339–6344. [CrossRef]
10. Schaber, J.A.; Carty, N.L.; McDonald, N.A.; Graham, E.D.; Cheluvappa, R.; Griswold, J.A.; Hamood, A.N. Analysis of quorum sensing-deficient clinical isolates of *Pseudomonas Aeruginosa. J. Med. Microbiol.* **2004**, *53*, 841–853. [CrossRef]
11. Smith, E.E.; Buckley, D.G.; Wu, Z.; Saenphimmachak, C.; Hoffman, L.R.; D'Argenio, D.A.; Miller, S.I.; Ramsey, B.W.; Speert, D.P.; Moskowitz, S.M.; et al. Genetic adaptation by *Pseudomonas Aeruginosa* Airways Cyst. Fibros. Patients. *Proc. Natl. Acad. Sci. USA* **2006**, *103*, 8487–8492. [CrossRef] [PubMed]
12. West, S.A.; Winzer, K.; Gardner, A.; Diggle, S.P. Quorum sensing and the confusion about diffusion. *Trends Microbiol.* **2012**, *20*, 586–594. [CrossRef] [PubMed]
13. Wilder, C.N.; Allada, G.; Schuster, M. Instantaneous within-patient diversity of *Pseudomonas Aeruginosa* quorum-sensing populations from cystic fibrosis lung infections Lung Infect. *Infec. Immun.* **2009**, *77*, 5631–5639. [CrossRef] [PubMed]
14. Popat, R.; Crusz, S.A.; Messina, M.; Williams, P.; West, S.A.; Diggle, S.P. Quorum-sensing and cheating in bacterial biofilms. *Proc. R. Soc. Lond. B Biol. Sci.* **2012**, *479*. [CrossRef]
15. Wilder, C.N.; Diggle, S.P.; Schuster, M. Cooperation and cheating in *Pseudomonas Aeruginosa*: The roles of the *las, rhl* and *pqs* quorum-sensing systems. *ISME J.* **2011**, *5*, 1332–1343. [CrossRef]
16. Pollitt, E.J.; West, S.A.; Crusz, S.A.; Burton-Chellew, M.N.; Diggle, S.P. Cooperation, quorum sensing, and evolution of virulence in *Staphylococcus Aureus.Infection Immun.* **2014**, *82*, 1045–1051. [CrossRef]
17. Dandekar, A.A.; Chugani, S.; Greenberg, E.P. Bacterial quorum sensing and metabolic incentives to cooperate. *Science* **2012**, *338*, 264–266. [CrossRef]
18. Schuster, M.; Sexton, D.J.; Hense, B.A. Why quorum sensing controls private goods. *Front. Microbiol.* **2017**, *8*, 885. [CrossRef]
19. Kümmerli, R.; Schiessl, K.T.; Waldvogel, T.; McNeill, K.; Ackermann, M. Habitat structure and the evolution of diffusible siderophores in bacteria. *Ecol. Lett.* **2014**, *17*, 1536–1544. [CrossRef]
20. Buckling, A.; Brockhurst, M.A. Kin selection and the evolution of virulence. *Heredity* **2008**, *100*, 484–488. [CrossRef]
21. Wang, M.; Schaefer, A.L.; Dandekar, A.A.; Greenberg, E.P. Quorum sensing and policing of *Pseudomonas Aeruginosa* Soc. Cheaters. *Proc. Natl. Acad. Sci. USA* **2015**, *112*, 2187–2191. [CrossRef] [PubMed]

22. Czárán, T.; Hoekstra, R.F. Microbial communication, cooperation and cheating: Quorum sensing drives the evolution of cooperation in bacteria. *PLoS ONE* **2009**, *4*, e6655. [CrossRef] [PubMed]
23. Cremer, J.; Melbinger, A.; Frey, E. Growth dynamics and the evolution of cooperation in microbial populations. *Sci. Rep.* **2012**, *2*, 281. [CrossRef] [PubMed]
24. Pérez-Velázquez, J.; Hense, B.A. *Differential Equations Models to Study Quorum Sensing*; Springer: New York, NY, USA, 2018; pp. 253–271. ISBN 978-1-4939-7309-5. [CrossRef]
25. Frank, S.A. Microbial secretor–cheater dynamics. *Philos. Trans. R. Soc. Lond. B Biol. Sci.* **2010**, *365*, 2515–2522. [CrossRef] [PubMed]
26. Mund, A.; Kuttler, C.; Pérez-Velázquez, J.; Hense, B. An age-dependent model to analyse the evolutionary stability of bacterial quorum sensing. *J. Theor. Biol.* **2016**, *405*. [CrossRef] [PubMed]
27. Cohen, Y.; Vincent, T.; Brown, J. A G-function approach to fitness minima, fitness maxima, evolutionary stable strategies and adaptive landscapes. *Evol. Ecol. Res.* **1999**, *1*, 923–942.
28. Apaloo, J.; Brown, J.S.; Vincent, T.L. Evolutionary game theory: ESS, convergence stability, and NIS. *Evol. Ecol. Res.* **2009**, *11*, 489–515.
29. Brown, J.S.; Vincent, T.L. G-functions for the hermeneutic circle of evolution. *Comput. Oper. Res.* **2006**, *33*, 479–499. [CrossRef]
30. Cohen, Y.; Pastor, J.; Vincent, T.L. Evolutionary strategies and nutrient cycling in ecosystems. *Evol. Ecol. Res.* **2000**, *2*, 719–743.
31. Kronik, N.; Cohen, Y. Evolutionary games in space. *Math. Model. Nat. Phenom.* **2009**, *4*, 54–90. [CrossRef]
32. Müller, J.; Kuttler, C. *Methods and Models in Mathematical Biology*; Springer: Berlin, Germany, 2015; ISBN 978-3-642-27251-6.
33. Kümmerli, R.; Griffin, A.S.; West, S.A.; Buckling, A.; Harrison, F. Viscous medium promotes cooperation in the pathogenic bacterium *Pseudomonas Aeruginosa*. *Proc. R. Soc. Lond. B Biol. Sci.* **2009**, *276*, 3531–3538. [CrossRef] [PubMed]
34. Mund, A.; Diggle, S.P.; Harrison, F. The fitness of *Pseudomonas aeruginosa* quorum sensing signal cheats is influenced by the diffusivity of the environment. *mBio* **2017**, *8*. [CrossRef] [PubMed]
35. Mund, A.; Kuttler, C.; Perez-Velazquez, J. Existence and uniqueness of solutions to a family of semi-linear parabolic systems using coupled upper-lower solutions. *Discret. Contin. Dyn. Syst. B* **2019**, *24*, 5695. [CrossRef]
36. Dockery, J.D.; Keener, J.P. A mathematical model for quorum sensing in *Pseudomonas Aeruginosa*. *Bull. Math. Biol.* **2001**, *63*, 95–116. [CrossRef] [PubMed]
37. Königsberger, K. *Analysis 1*; Springer: Berlin/Heidelberg, Germany, 1992; ISBN 978-3-540-55116-4.
38. Schiesser, W. *The Numerical Method of Lines: Integration of Partial Differential Equations*; Elsevier Science: Amsterdam, The Netherlands, 2012; ISBN 9780128015513.

Article

Modelling Population Dynamics of Social Protests in Time and Space: The Reaction-Diffusion Approach

Sergei Petrovskii [1,2,*], Weam Alharbi [3], Abdulqader Alhomairi [4] and Andrew Morozov [1,5]

1. School of Mathematics and Actuarial Science, University of Leicester, Leicester LE1 7RH, UK; am379@le.ac.uk
2. Peoples Friendship University of Russia (RUDN), 6 Miklukho-Maklaya St, Moscow 117198, Russia
3. Mathematics Department, Faculty of Science, Tabuk University, Tabuk 71491, Saudi Arabia; wgalharbi@ut.edu.sa
4. Department of Curriculum and Teaching Methods, Faculty of Education and Art, Tabuk University, Tabuk 71491, Saudi Arabia; ahomari@ut.edu.sa
5. Institute of Ecology and Evolution, Russian Academy of Sciences, 33 Leninskii pr., Moscow 119071, Russia

* Correspondence: sp237@le.ac.uk

Received: 1 November 2019; Accepted: 20 December 2019; Published: 3 January 2020

Abstract: Understanding of the dynamics of riots, protests, and social unrest more generally is important in order to ensure a stable, sustainable development of various social groups, as well as the society as a whole. Mathematical models of social dynamics have been increasingly recognized as a powerful research tool to facilitate the progress in this field. However, the question as to what should be an adequate mathematical framework to describe the corresponding social processes is largely open. In particular, a great majority of the previous studies dealt with non-spatial or spatially implicit systems, but the literature dealing with spatial systems remains meagre. Meanwhile, in many cases, the dynamics of social protests has a clear spatial aspect. In this paper, we attempt to close this gap partially by considering a spatial extension of a few recently developed models of social protests. We show that even a straightforward spatial extension immediately bring new dynamical behaviours, in particular predicting a new scenario of the protests' termination.

Keywords: social dynamics; wave of protests; long transients; ghost attractor

---

1. Introduction

Understanding of the dynamics of riots, protests, and social unrest more generally is important in order to ensure a stable, sustainable development of various social groups, as well as the society as a whole. To achieve this goal, mathematical modelling has been gaining growing recognition recently as a powerful research approach [1–14]. Indeed, since replicated sociological experiments with large groups of people are rarely possible (and it is hardly possible at all to simulate social unrest under controlled conditions), capturing the complexity of the social dynamics through tractable experiments is not feasible. Mathematical modelling and computer simulations create a "virtual society" where hypotheses can be tested and different scenarios investigated in detail, thus providing a feasible alternative to the experiment.

A variety of modelling approaches and techniques has been developed and used such as the costs-benefits analysis [9], social network models [5,10], agent based models [3,7], and behavioural and epidemiological models [2,11,13,14], both in spatial and non-spatial systems. The models range from relatively simple, allowing only for some basic feature of the phenomenon [3], to more complicated ones that take into account more details such as, for instance, heterogeneity of social norms and behavioural responses [9,15]. The case studies include a few well known events such as the Russian revolution of 1905–1907 [1], the French riots of 2005 [13], the London riots of 2011 [9], the Arab

Spring [16], and the Yellow Vest Movement in 2018–2019 [17], in particular aiming to describe the intensity and timing of the unrest and to analyse the processes and mechanisms involved.

Social protests are known to exhibit a variety of different scenarios. For instance, the total number of people involved in protests (to which we refer to as "crowd") can change through time in different ways, e.g., staying approximately constant for some time and then decaying fast, or decreasing steadily but slowly. Furthermore, the social structure of the crowd can change significantly, e.g., due to the individual variability of people's perception and expectations [18]. Whilst a comprehensive description of protests requires several state variables and a multivariate analysis (e.g., to account for the individual variability, cf. [15]), one quantity of immediate importance is the total number of people participating in the given event at the given time. Arguably, understanding how the total number may evolve with time provides crucial information needed to ensure security measures and to minimize possible damage and violence. An inspection of the data on the number of people participating in protests and riots available for several events of social unrest [9,13,19–21] immediately reveals that there are some generic properties of the dynamics. There usually exists the main peak, i.e., the maximum in the participants number (sometimes followed or preceded by secondary peak(s) of smaller magnitude) followed, in the course of time, by a tail where the number eventually decays to either zero or to a small background value. However, the rate of decay at the tail can be significantly different for different events. Whilst in many cases, the number of protesters decays fast, so that the unrest is effectively over in a matter of weeks [9,13], there have been several cases where the rate of decay is very slow. One such case is given by the Russian revolution of 1905–1907 [19,20], and another case is given by the Yellow Vest Movement [21]. As a result of the slow decay in the number of protesters, the duration of unrest is considerable (about two years for the Russian revolution, about one year for the Yellow Vest Movement, at the time when this paper was written). A question arises here as to whether the events with significantly different decay rates are different altogether (e.g., controlled by different factors) or they are the manifestation of the same social dynamical phenomenon, but occurring in a different parameter range. A related question is whether it might be possible to develop a "universal" mathematical model that would embrace the protests with a significantly different decay rate. Addressing this question is one of the goals of this paper.

Along with the changes in time, in many cases, riots and protests also have a clear spatial aspect. For instance, the timing and the intensity of protests can be significantly different at different locations. Experience gained from studies on dynamical systems of a different origin, e.g., in ecology [22], indicates that coupling between local protests (e.g., in the context of social dynamics, by the information exchange and/or spread of ideas) can organize them into a global spatiotemporal pattern. The simplest example of a spatiotemporal pattern is the travelling wave (sometimes also referred to as the "spreading wave"). Interestingly, the spreading wave of protests has indeed been observed in some of the events of social unrest [13,14].

There is a growing appreciation that the spatial aspect can shape the social protests in a number of ways; however, most of the previous modelling studies on protests' dynamics focused on non-spatial or spatially implicit systems. Apparently, the consensus as to whether an adequate modelling framework should necessarily be spatial or can be non-spatial has not been reached yet. Whilst it is well known in other sciences that the dynamical properties of the spatially explicit system often are significantly different from those of its non-spatial counterpart [22], it remains unclear how the modelling predictions can differ in the context of the social dynamics. A more specific question is: How different can the dependence of the protester number on time be in spatial and non-spatial models for equivalent parameter values? Our paper intends to bridge this gap partially by considering a spatial extension of two non-spatial models recently developed in [17]. We will show that the dynamics of the spatial models can be qualitatively different, e.g., predicting the growth in the protester number for the parameter values where the non-spatial models predict decay. We will also show that the spatial model brings new dynamical features, in particular predicting a new scenario of the protests' termination.

2. Mathematical Models in the Non-Spatial Case

We describe the dynamics of protests in terms of average values, hence disregarding any explicit effects of stochasticity: arguably, this approach works well when applied to sufficiently large systems [23], i.e., in the context of our study, to sufficiently large groups of people. We assume that the dynamics can be described by continuous, sufficiently smooth functions of time, so that in the non-spatial case, the mathematical model consists of ordinary differential equation(s).

We mention here that any group of people, e.g., the "population" of protesters or the larger community, is not uniform, but is structured according to various traits [15]. For instance, in the context of social unrest, the population of protesters can be structured according to the probability for a given individual to express a violent behaviour. Consider the Yellow Vest Movement as an example: whilst most participants protested peacefully, there was often a certain fraction that tended to behave aggressively. The wider community is structured with regard to the probability for a given individual to join the protests; this, in its turn, correlated with age, income, education, social background, etc. Whilst some people can be ignited easily (e.g., students [24,25]), people from other social groups might not so likely join the street actions in any circumstances. Aiming to build tractable models and to avoid unnecessary complexity (cf. "Occam's razor"), in this paper, we disregard the effect of individual differences; in particular, we treat the protesters as a uniform group. We also assume that the total number of people in the society that could, under certain conditions, join the protests is sufficiently large and hence is not a factor that limits the growth of the number of protesters.

2.1. Single Species Model

We describe the civil unrest by a single variable, say N, which is the number of people actively participating in the street actions [26]. In order to build the model that describes the dynamics of N with time, we consider the following assumptions:

- The rate of change in the number of people attending the event is a result of the interplay between two processes, recruitment (people joining) and withdrawal (people quitting);
- Following earlier studies [18], we consider recruitment to be a collective phenomenon so that the recruitment rate is a nonlinear function of the number of people currently involved in the event;
- Decision of withdrawal is made individually, so that the withdrawal rate is a linear function of the number of people participating in the event, where the per capita withdrawal rate (say m) depends on time.

Under the above assumptions, we arrive at the following equation for the "population size" of the demonstrators:

$$\frac{dN(t)}{dt} = F(N) - m(t)N, \tag{1}$$

where the first and second terms in the right hand side are the recruitment rate and withdrawal rate, respectively, and $t \geq 0$ is time. We regard the per capita withdrawal rate m as a measure of the dissatisfaction with the effect of the protests, i.e., when the demands of the demonstrators are not being met by the authorities. In our model, higher dissatisfaction therefore means a higher rate at which people quit the movement. In case of an unchanging policy, the dissatisfaction can be expected to increase with time; hence, m is an increasing function of time.

We consider the recruitment rate in the following form:

$$F(N) = \epsilon_0 + \epsilon N + \frac{aN^2}{h^2 + N^2}, \tag{2}$$

where ϵ_0, ϵ, a, and h are nonnegative parameters.

Parametrization (2) requires further discussion. We first notice that $F(0) = \epsilon_0 \geq 0$. That reflects the observation that, in almost any society, there is often a fraction of people (albeit normally small) who are always ready and willing to rebel, e.g., to join a strike or a civil protest. Such people can be thought of as zealots that unconditionally support some ideas of ultimate social justice.

The per capita recruitment rate is given by the following expression:

$$\frac{1}{N}F(N) = \frac{\epsilon_0}{N} + \epsilon + \frac{aN}{h^2 + N^2}. \tag{3}$$

For an intermediate range of N defined by conditions $\epsilon_0/\epsilon \ll N \ll h$ (if parameter values allow for the existence of such range), the per capita recruitment rate is therefore a monotonously increasing (approximately linear) function of N. This means that the rate at which new members are joining the movement tends to increase along with the number of people already in the movement. This is a typical behaviour response observed not only in humans [27], but also in many animals; in particular, it is believed to be one of the mechanisms resulting in the Allee effect [28]. However, when the number of people in the riots becomes large, $N \gg h$, the per capita recruitment rate is approximately constant, $\frac{1}{N}F \approx \epsilon$. That reflects the assumption that there is a certain scale-independent probability of an average citizen to join protests, which agrees with observations [18,29].

Before considering the properties of Model (1) and (2) where the withdrawal rate m is a function of time, it is instructive to start with a simpler case where m is a constant parameter. In this case, it is readily seen that there exists two critical values, say m_* and m^*, so that in an intermediate range $m_* < m < m^*$, the system possesses three steady states, two of them stable and the third one unstable; see Figure 1. The system undergoes a saddle-node bifurcation when m passes any of the two critical values; for $m < m_*$ and for $m > m^*$, there is only one steady state, large population state and small population state, respectively. In the general case $\epsilon_0 \geq 0$, an analytical expression for m_* and m^* is not available. In a special case $\epsilon_0 = 0$, they are given as:

$$m_* = \epsilon \quad \text{and} \quad m^* = \epsilon + \frac{\sqrt{2a-1}}{2h}.$$

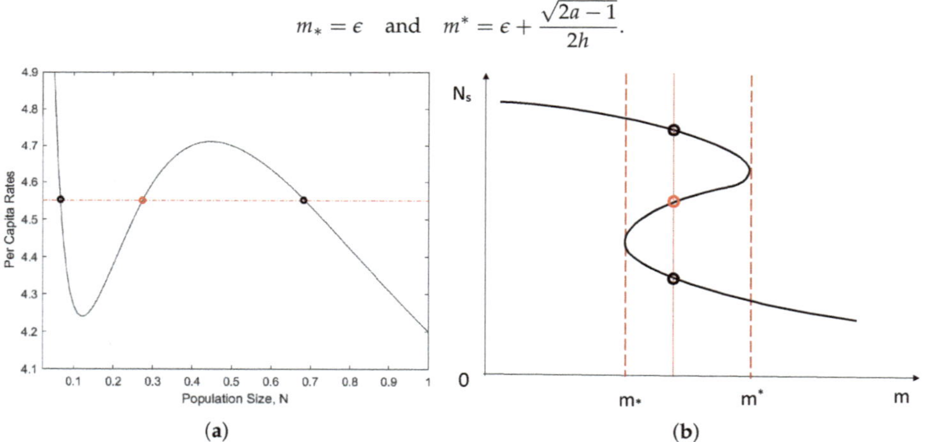

Figure 1. Properties of the model given by Equations (1) and (2) in the baseline case of constant m. (**a**) An example of the relative position of the per capita recruitment rate (black curve) as a function of N (in abstract units) and the per capita withdrawal rate (red line). Any intersection between the two lines is a steady state. The two stable steady states are shown by the small black circles, and the unstable steady state is shown by the red circle. (**b**) A sketch of the bifurcation diagram: The thick black curve shows the steady state value as a function of m. The two long-dashed vertical lines correspond to the bifurcation values of m; for any value of m such as $m_* < m < m^*$ (see the vertical dotted line), the system has three steady states, two stable and one unstable.

In the case of a variable m, in order to derive the specific expression for $m(t)$, we take into account that, in the modern society, even in the absence of major events, a number of small scale protests are always happening. We assume that there is a certain background rate, say m_0, at which the participants of those protests quit the actions. We then further assume that the society is in a globally stable state, so that any perturbation from the background withdrawal rate should decay with time. Trying to keep the mathematical framework as simple as possible, the simplest equation that describes the corresponding evolution of m is given as follows:

$$\frac{dm(t)}{dt} = b(m_0 - m), \tag{4}$$

where b is a coefficient. Equation (4) is readily solved, resulting in:

$$m(t) = m_1 + (m_0 - m_1)\left(1 - e^{-bt}\right), \tag{5}$$

where $m_1 = m(0)$ is the initial value of the withdrawal rate. In case the initial value is small, which can occur as a reflection of high social tensions, $m(t)$ is an increasing function. This seems to be in agreement with heuristic arguments. Note that, since m has the dimension of inverse time, it can be written as $m = 1/\tau$ where τ is a characteristic duration of the individual participation. In a general case, τ can change in the course of the protests, hence being a function of time. At the beginning of the protests, the people's expectations are high, the morale is high, and people are ready to spend a long time participating in the event. The characteristic duration of individual involvement is long, and the initial value of the withdrawal rate $m(0) = m_1$ is small. In the course of the protests, in case there is no sign of changes, the protesters' morale may start falling, and the characteristic duration of individual participation could start decreasing.

Equations (1) with (2) and (5) are solved numerically. We consider m_0 and m_1 as the controlling parameters. Figure 2 shows the solutions obtained for different values of m_0 and m_1 and other parameters chosen as $\epsilon_0 = 0.1$, $\epsilon = 3$, $a = 146$, $h = 35$, and $b = 0.01$. (We mention here that our choice of parameter values was not entirely random or hypothetical: it was shown in [17] that, with this parameter set, Model (1)–(5) provided a good description of the Yellow Vests protests.) We therefore observe that there are solution properties that are rather robust to parameter variation. In particular, in all cases shown in Figure 2, the population size creates a peak (which can be rather flat for m_1 large enough; see Curves 5 and 6 in Figure 2b) followed by a gradual decay. This behaviour is in qualitative agreement with the dynamics of protests often observed in real life. The larger is m_0, the faster the rate of decay is. For a subcritical value of m_0, i.e., for $m_0 < m^*$, the population size never falls to the low background value (which is zero for $\epsilon_0 = 0$), but instead stabilizes at a higher positive value (cf. the magenta curve in Figure 2a), which corresponds to the upper stable steady state of the system; see the upper black circle in Figure 1b. For a supercritical value $m_0 > m^*$, the upper steady state does not exist, so that in the large time limit N eventually converges to its lower steady state value (lower black circle in Figure 1b). However, if $0 < m_0 - m^* \ll 1$, the rate of convergence at intermediate time can be slow because of the "ghost attractor" [30,31]; e.g., see the middle part of the curves in Figure 2b.

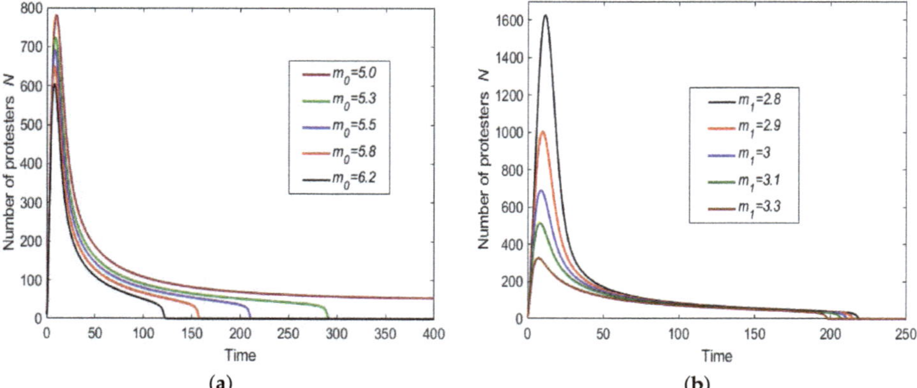

Figure 2. Solution of Equations (1) with (2) and (5) obtained for different parameter values of m_0 and m_1. (**a**) $m_1 = 3$, the values of m_0 are given in the figure legend. (**b**) $m_0 = 5.5$, the values of m_1 are given in the figure legend. Other parameters are $\epsilon_0 = 0.1$, $\epsilon = 3$, $a = 150$, $h = 35$, and $b = 0.01$, which corresponds to $m_* = 3.22$ and $m^* = 5.15$.

2.2. Two Component Model

In the previous section, the withdrawal rate was described by the self-contained Equation (4) where the right hand side depends on m only. One weakness of this approach is that it completely disregards the possible dependence of the right hand side on N. Even in a more general case where the right hand side in Equation (4) could be a nonlinear function, m is effectively just a given function of time. As an immediate yet nontrivial extension of the above model, we now consider the case where the population dynamics of riots is described by the following equations:

$$\frac{dN(t)}{dt} = F(N) - mN, \qquad \frac{dm(t)}{dt} = G(m, N), \tag{6}$$

where G is a certain function to be specified.

Apparently, Model (6) is more general than Model (1) and (2) and hence can describe a broader variety of situations. In order to find an appropriate parametrization of G, we have to consider more closely the reasons why people may quit the protests. In doing this, we first observe that street actions usually bring some economic damage. That can include direct damage (such as broken shop windows, burnt cars, etc.) and indirect damage to the economy (e.g., through the disruption of public transport). We assume that the amount of damage caused is proportionate to the number of people involved in the street actions. We then further assume that an ordinary protest participant eventually develops the feeling of guilt for causing damage to the community and that this feeling is likely to cut short the duration of his/her participation in the street actions. Correspondingly, this means that the rate of change in m should be an increasing function of the damage, which we consider to be linear.

Correspondingly, we now arrive at the following model of the protests' dynamics:

$$\frac{dN}{dt} = \left(\epsilon_0 + \epsilon N + \frac{aN^2}{h^2 + N^2}\right) - mN, \qquad \frac{dm}{dt} = \beta N, \tag{7}$$

where β is a coefficient. Unless the protests are extremely violent, it seems reasonable to assume that $\beta \ll 1$.

Figure 3 shows the phase plane of Model (7). It is readily seen that the properties of Model (7) are significantly different from those of Model (1) and (2). In particular, there is no steady state in the interim of the first quarter of the plane, i.e., for $N > 0$ and $m > 0$.

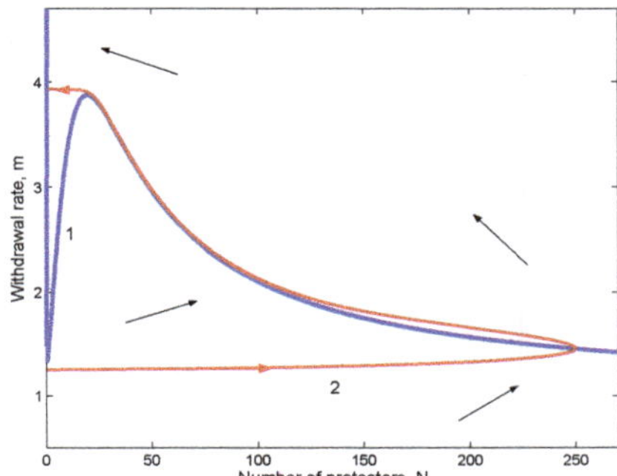

Figure 3. Phase plane of the system (7). The blue Curve 1 shows the N-isocline; note its part that lies close to the vertical axis. The red Curve 2 shows the solution of the system (obtained for parameters $\epsilon_0 = 0.1$, $\epsilon = 1$, $a = 115$, $h = 20$, and $\beta = 0.00025$) where red arrows indicate the direction of the system's evolution along the trajectory. The black arrows show the generic direction of the phase flow in different parts of the plane.

In order to reveal the solution properties, the Equations (7) are solved numerically. We consider β as the controlling parameter. Typical results are shown in Figure 4. At the beginning, the number of protesters promptly becomes very large, thus creating a high peak (see Figure 4a). Past the peak, the number of protesters starts falling, first fast, then at a decreasing rate. The number of protesters keeps decreasing with time at a lower and lower rate until, at a certain moment, it promptly decays to zero. The corresponding dependence $m(t)$ is shown in Figure 4b.

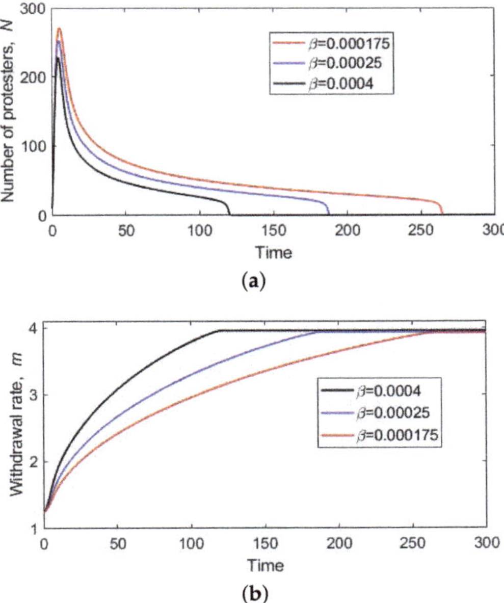

Figure 4. Solutions of Model (7) obtained for different values of β: (**a**) the number of protesters N vs. time and (**b**) the per capita withdrawal rate m vs. time. Other parameters are $\epsilon_0 = 0.1$, $\epsilon = 3$, $a = 115$, and $h = 20$.

The behaviour of the solution of Model (7) is thus apparently similar to that of the solution of the single component Model (1) and (2); cf. Figure 2. There are, however, some important differences. In Model (1) and (2), there exists a finite critical value m^*, so that for $m_0 < m^*$, the number of protesters never falls to the small equilibrium value, but stabilizes at the higher steady state (see Curve 1 in Figure 2a). On the contrary, in Model (7), there is no finite critical value: for any $\beta > 0$, the solution $N(t)$ will eventually converge to the lower steady state. We also note here that the shape of the graph of $N(t)$ shown in Figure 4a, albeit similar to Figure 2, is not the effect of the ghost attractor, but a property of the so-called slow fast dynamics (when $\beta \ll 1$) [31].

3. Spatially Explicit Model

In the previous section, we completely disregarded the existence of the spatial dimension in the dynamics of the social protests, hence implicitly assuming that the population density of the protesters is distributed uniformly over space. In reality, it is rarely so, especially when the dynamics is considered over a sufficiently large spatial scale such as a region, a county, a province, or the whole state. There are several reasons for the spatial heterogeneity to emerge. Firstly, the overall population density itself is distributed highly non-uniformly, i.e., reaching very high values in urban areas, but falling to low values in the countryside. Secondly, people's behavioural response and their social activity can be significantly different at different locations in space, e.g., as a result of different living conditions and different accessibility to social media. Therefore, although the consideration of non-spatial models makes the first necessary step in the study, a more adequate mathematical framework for modelling social protests must take space into account explicitly, as has indeed been recognized in a number of studies, e.g., see [8,12–14]. In this section, we consider the models of the social protests dynamics in space and time that consist of reaction-diffusion equations, where the "reaction" part is given by the corresponding non-spatial model (as in Section 2) and the spread of the protests in space (described by the diffusion terms) is regarded as a certain "contagion" [13,32].

We mention here that the spatial spread of information and emotions and hence the spread of social tensions are known to have two different spatial modes [12,32]. In one mode, people communicate in direct encounters due to their geographical proximity: arguably, people living and/or working in the same area would normally meet in person more frequently than people living far apart. The importance of this mode of contagion is indirectly confirmed by the well established fact that most of everyday human travel occurs within a short range from peoples' homes [33]. In the second mode, the spread of information and emotions is not related to travel and occurs due to the connection by social media. For the first mode, the spread of protests can be described by the standard diffusion [12,14]; for the second mode, kernel based integro-differential equations are thought to be more adequate [13] as they better describe the effect of long distance connections [34]; see also [35–37].

Note that, whilst the effect of the first mode is obviously limited to short distances, the second mode is not restricted to any particular spatial scale: indeed, technically, social media can be accessed from anywhere in the world. Hence, one can think, respectively, about the local coupling vs. the global coupling. In the case of the local coupling due to direct encounters, the effect of space is likely to be more pronounced; in particular, the level of social tensions should be a function of space. On the contrary, in the case of the global coupling due to social media, the social tension is likely to be distributed over space more uniformly.

3.1. Single Species Model

We begin with the single species system. In terms of the above discussion, it corresponds to the global coupling where the social tension is distributed over space uniformly, and hence, the withdrawal rate does not depend on space (but depends on time). The distribution of the "population" of protesters over space is then described by a single dynamical variable, i.e., by the population density, say $u(x,t)$. The spatially explicit counterpart of Model (1) and (2) is described, in the 1D case, by the following equation:

$$\frac{\partial u(x,t)}{\partial t} = \left(\epsilon_0 + \epsilon u + \frac{au^2}{h^2+u^2}\right) - m(t)u + D\frac{\partial^2 u}{\partial x^2}, \tag{8}$$

where x is the position in space, $-L \leq x \leq L$, D is the diffusion coefficient, and the dependence of the withdrawal rate m on time is given by Equation (5).

We assume that protests start at a certain location or a small area. Correspondingly, for the initial conditions, we consider a piece-wise constant distribution of finite support described as follows:

$$u(x,0) = u_0 \quad \text{for } -\Delta \leq x \leq \Delta, \qquad u(x,0) = 0 \quad \text{otherwise.} \tag{9}$$

At the ends of the domain, we use the Dirichlet boundary condition, $u(-L,t) = u(L,t) = 0$.

Equation (8) with (9) was solved numerically by finite differences. Typical results are shown in Figure 5. We readily observe that the system dynamics are characteristic for reaction-diffusion systems with compact initial conditions, albeit being modified by the dependence of the withdrawal rate on time. At an early stage of the dynamics, the evolution of the initial distribution results in the formation of two travelling fronts propagating away from the location of the initial condition (see the yellow, violet, green, and light-blue curves in Figure 5a): the protests spread fast over space. In the course of time, the speed of the fronts decreases as a result of the increase in m. Eventually, the front stops and reverses (cf. the dark-blue and brown curves in Figure 5a). The population then retreats to the place of its original introduction, which finally results in its collapse and extinction, i.e., the termination of the protests.

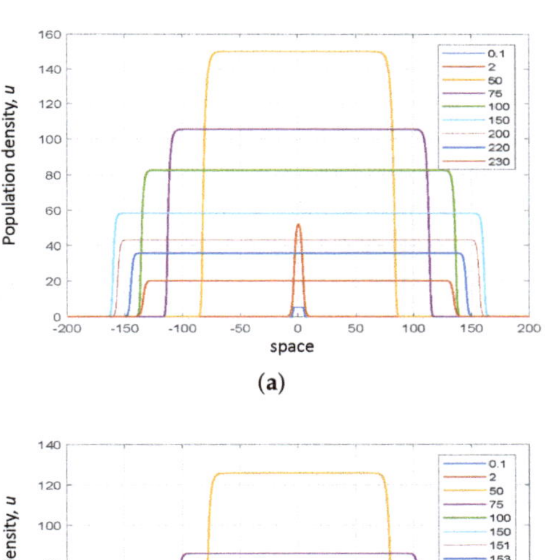

(a)

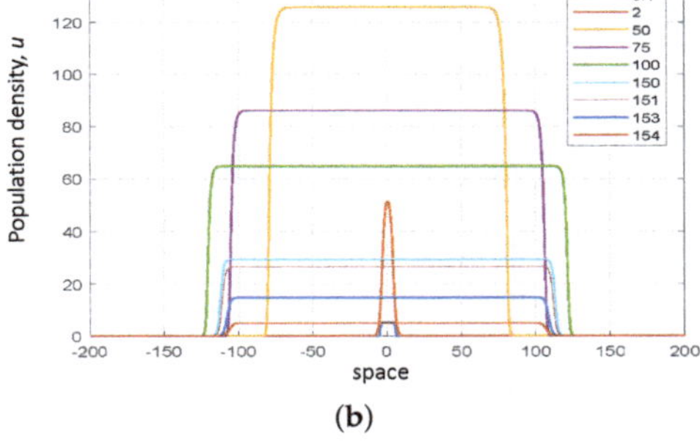

(b)

Figure 5. Solution of Model (8) shown at different moments of time (as is explained in the figure legend) obtained for $m_1 = 3$ and different values of the final withdrawal rate: (**a**) for $m_0 = 5.4$ and (**b**) for $m_0 = 5.8$. Other parameters are $\epsilon = 3$, $\epsilon_1 = 0$, $a = 150$, $h = 35$, $b = 0.01$, and $D = 1$.

We mention here that we are not aware of any well developed mathematical theory of travelling waves in non-autonomous reaction-diffusion systems, i.e., with reaction rates explicitly depending on time. It seems, however, possible to explain the numerical results in terms of autonomous systems. In the baseline case of constant m, the reaction term of Equation (8), that is:

$$\mathcal{F}(u) = \epsilon_0 + \epsilon u + \frac{au^2}{h^2 + u^2} - mu, \tag{10}$$

is a sigmoidal curve, which, in the general case $m_* < m < m^*$ (cf. Figure 1), has three roots, $0 \le u_{lower} < u_{int} < u_{higher}$ (where $u_{lower} = 0$ if $\epsilon_0 = 0$). The direction of the front propagation is known to depend on the value of the following integral [38,39]:

$$M = \int_{u_{lower}}^{u_{higher}} \mathcal{F}(u)du, \tag{11}$$

so that the front propagates towards the low density area (and hence away from the place of the compact initial condition) if $M > 0$ and towards the high density area if $M < 0$. In the former case, the protests invade, and in the latter case, the protests retreat.

In our case, m depends on time, and strictly speaking, the above results are not applicable. However, one can expect that they may still work approximately. Indeed, it is readily seen that, for the parameters chosen as in Figure 5, $M > 0$ at an early time, but becomes negative at a large time, changing its sign at $t \approx 123$ (see Figure 6). This explains the reversing of the front.

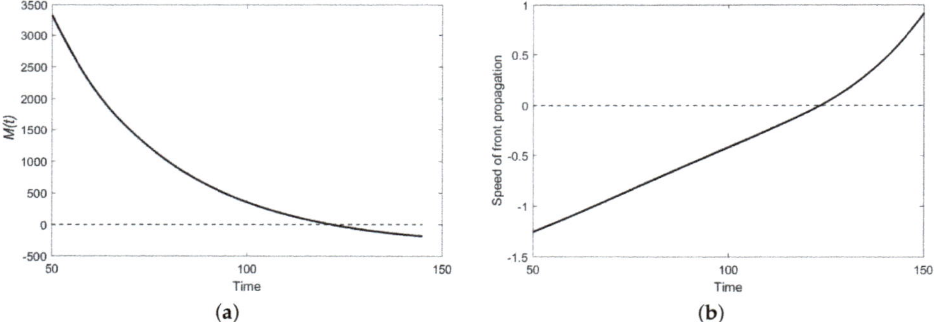

Figure 6. Dependence on time of (**a**) M and (**b**) the speed of the front calculated in the course of the travelling front propagation. The parameters are the same as in Figure 5a. It is readily seen that the moment ($t \approx 123$) when the front changes the direction of its propagation (invasion changes to retreat) coincides with the moment when M changes its sign.

We therefore conclude that the spatial dynamics exhibits some new properties that did not exist in its non-spatial counterpart, in particular by providing a new scenario of the termination of the social unrest. Recall that, in the non-spatial system, the unrest terminates only after the increase in m results in the disappearance of the upper positive steady state (see Figure 1). In the spatial system, the unrest termination can happen even when the positive steady state exists: for the termination to occur, it is sufficient that m becomes sufficiently large (but not necessarily supercritical) to reverse the front propagation.

The dependence of the total number of protesters on time, i.e., $N_{tot}(t) = \int_{[-L,L]} u(x,t)dx$, appears to be rather different in the spatial system compared to the corresponding non-spatial system; see Figure 7. Instead of the sharp peak, after the fast increase at the beginning, $N(t)$ then exhibits a much slower decrease than in the non-spatial system (cf. Figure 2), eventually accelerating to a fast decay to zero.

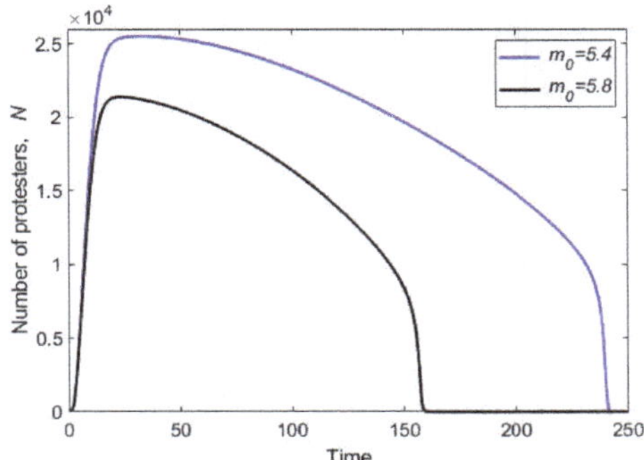

Figure 7. Total number of protesters vs. time as given by the solution of Model (8) obtained for $m_0 = 5.4$ (blue curve) and $m_0 = 5.8$ (black curve). Other parameters are the same as in Figure 5.

3.2. Two Component Model

We now consider the spatially explicit version of Model (7), which is given by the following equations:

$$\frac{\partial u(x,t)}{\partial t} = \epsilon_0 + (\epsilon - m)u + \frac{au^2}{h^2 + u^2} + D\frac{\partial^2 u}{\partial x^2}, \quad (12)$$

$$\frac{\partial m(x,t)}{\partial t} = \beta u + D_m \frac{\partial^2 m}{\partial x^2}. \quad (13)$$

For the sake of generality, we assume that the spatial dynamics of the withdrawal rate may occur with a different diffusion coefficient D_m.

For the initial conditions, similarly to the above, we consider the piece-wise constant distributions:

$$u(x,0) = u_0 \text{ for } -\Delta \le x \le \Delta, \quad \text{otherwise} \quad u(x,0) = 0, \quad (14)$$

$$m(x,0) = m_0 \text{ for } -\Delta_m \le x \le \Delta_m, \quad \text{otherwise} \quad m(x,0) = 0, \quad (15)$$

where $\Delta_m \le \Delta$ to reflect the assumption that there is no damage without protests. At the ends of the domain, we use the Dirichlet boundary condition, $u(-L,t) = u(L,t) = 0$ and $m(-L,t) = m(L,t) = 0$.

Mathematical Problem (12)–(15) was solved numerically by finite differences using the forward scheme. The mesh steps were chosen sufficiently small to avoid numerical artefacts, and it was checked that a further decrease of the steps value had no effect on the numerical solution.

Some typical results are shown in Figure 8. We readily observe that, as well as in the previous case, the evolution of the initial conditions results in the formation of two travelling waves propagating away from the location of the initial distribution. However, the properties of the wave are essentially different from those in the single species model. Instead of a smooth, monotonous front, we now have a travelling peak with a very high population density at its maximum. In the wake of the peak, the population density promptly decays to a much smaller value. The corresponding spatial distribution of the withdrawal rate is shown in Figure 9. After the peak reaches the domain boundary, it promptly disappears, and the total number of the protesters decays to zero.

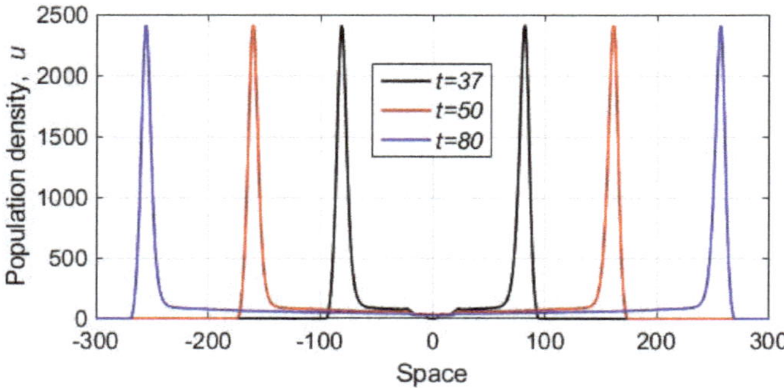

Figure 8. Population density vs. space obtained as a solution of Model (12) and (13) shown at different moments of time. Parameters are $\epsilon_0 = 0$, $\epsilon = 1$, $a = 115$, $h = 20$, $\beta = 0.00025$, $D = 1$, and $D_m = 0.1$.

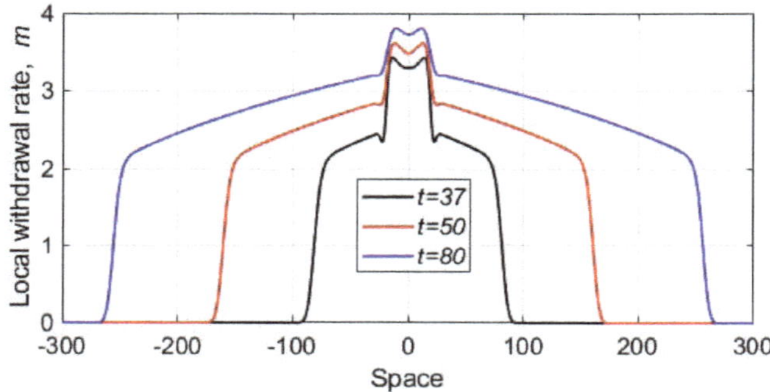

Figure 9. Distribution of the withdrawal rate over space as given by Model (12) and (13) shown at different moments of time. Parameters are the same as in Figure 8.

As we observed in numerous numerical simulations, the main properties of the solution, such as the formation of the population peaks propagating away from the location of the initial distribution, were generic and robust to parameter changes. In particular, we did not observe any significant changes in the system dynamics for different values of the diffusivity ratio D_m/D (having checked several values of the ratio in the range between one and 0.01). However, some quantitative features of the dynamics appeared to be somewhat sensitive to other parameter values. For instance, the height of the peak depended on ϵ rather strongly. An example shown in Figure 10 is obtained for $\epsilon = 0.1$. We readily observe that, whilst the dynamics is qualitatively the same as in the case $\epsilon = 1$ (see Figure 8), the height of the peak is almost an order of magnitude smaller. The speed of the peak propagation is slightly less as well. The distribution of damage, however, shows features similar to what was obtained for $\epsilon = 1$ (see Figure 9); we do not show it here for the sake of brevity.

Figure 11 shows the total number of protesters vs. time as is given by $N_{tot}(t) = \int_{[-L,L]} u(x,t)dx$. It is readily seen that the dynamics of the protests in the spatially explicit Model (12) and (13) is considerably different from its non-spatial counterpart (7). Recall that in the non-spatial system, the number of protesters reaches its maximum at early time and then decays monotonously, first very fast and later at a decreasing rate (see Figure 4a). In the spatial system, however, the fast increase at the beginning changes to a slower growth that goes on for a considerable length of time. The phase of slow growth corresponds to the peak travelling over space where the shape of the peak remains approximately the same, but the length of its tail grows steadily. The duration of this phase depends on the extent of the available space; the larger the spatial domain is, the longer the total number of protesters keeps growing slowly, but steadily. After the peak hits the boundary of the domain, the effect of the boundary conditions bring it down, resulting in the sharp decrease in the protesters' number.

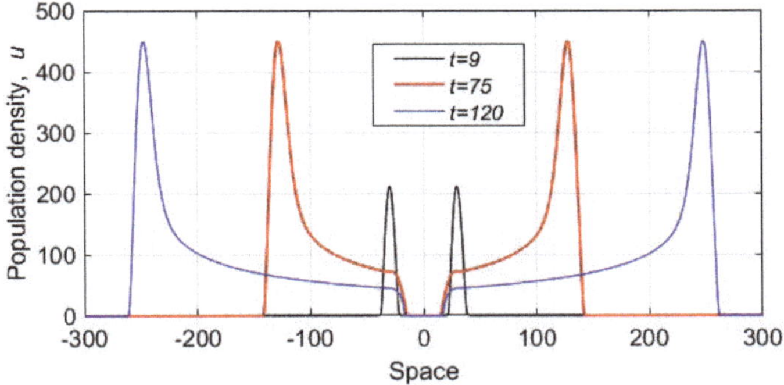

Figure 10. Population density vs. space obtained as a solution of Model (12) for $\epsilon = 0.1$ shown at different moments of time. Other parameters the same as in Figure 8.

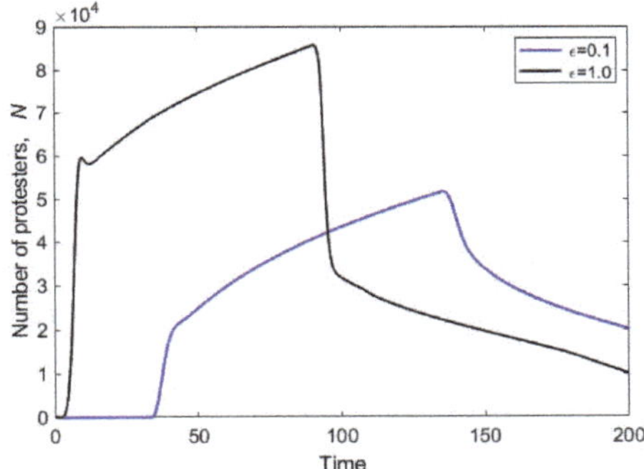

Figure 11. Total number of protesters vs. time as given by the solution of Model (12) and (13) with the initial conditions (14) and (15) obtained for $\epsilon = 1$ (black curve) and $\epsilon = 0.1$ (blue curve). Other parameters are the same as in Figure 8.

Note that the collapse of the travelling wave as a result of the collision with the boundary is a common phenomenon in reaction-diffusion systems of various origins [39–41]. Such a collapse is not very sensitive to the type of the boundary condition; in particular, in our model, it was observed both for the Dirichlet and Neumann types (for the sake of brevity, we do not show here the corresponding figures for the zero-flux boundary conditions). We also mention that the collapse of protests observed in our model when the wave reached the domain boundary could be insightful for understanding the dynamics of real events of social unrest. For example, in the 2005 French riots, it was reported that the street protests were quickly spreading across suburbs before they reached some quieter areas, after which the whole unrest went extinct [13]. Although the collapse of the 2005 unrest in Paris can be explained by a number of reasons, e.g., overall decrease of the protesters' motivation, increasing efficiency of the police, etc., reaching the edge of the area where the social or geographical conditions changes significantly is arguably a major factor.

Apparently, the rate of increase in the total number of protesters depends on the speed of the travelling peak. That raises an interesting mathematical question as to whether the speed can be estimated. We recall here that, in reaction-diffusion systems, the travelling wave usually arises as a heteroclinic connection between two equilibria (homoclinic in the case of a travelling peak) in the phase space of the system, the speed of the travelling wave being an eigenvalue of the corresponding nonlinear problem [39,42,43]. Our system (12) and (13) does not quite fit into this framework, as a positive equilibrium does not exist for $\epsilon_0 = 0$. This indicates that, in a strict mathematical sense, the travelling wave does not exist. However, arguably, the part of the solution that describes the peak still behaves as a travelling wave, as indeed is readily seen from the properties of the numerical solution, e.g., see Figures 8 and 10. Hence, we hypothesize that some of the tools normally used to estimate the speed of the wave may still be applicable.

Such tools must be applied with due care though. We first observe that the solution behaves as a travelling wave at the "leading edge" of the travelling peak, i.e., where both components u and m are very small. Equation (12) can therefore be linearized taking the following very simple form:

$$\frac{\partial u(x,t)}{\partial t} = \epsilon u + D\frac{\partial^2 u}{\partial x^2}, \tag{16}$$

where we consider $\epsilon_0 = 0$ here and below for the sake of simplicity. Note that the linearized equation does not contain m any more. For the linear reaction-diffusion equation with finite or semi-finite initial condition, it is well known that, although the travelling wave solution does not exist globally, the leading edge of the solution propagates as a travelling wave with the speed given by:

$$c = 2\sqrt{\epsilon D}, \tag{17}$$

e.g., see [37,38]. Estimate (17) obtained by linearizing the system (12) and (13) therefore coincides with the speed of the travelling front in a scalar reaction-diffusion equation with the generalized logistic growth [42,43].

However, the expression (17) appears to be in bad disagreement with numerical results predicting a value considerably less than was obtained in the simulations, e.g., two instead of 3.1 for the parameters of Figure 8. Even worse, it does not in any way predict the dependence of the speed on parameter values: for instance, for $\epsilon = 0.1$ Expression (17) gives $c \approx 0.63$, whilst the actual value is about 2.6. A more careful look readily reveals the reason for this disagreement: as a matter of fact, Expression (17) is hardly relevant at all as its validity is limited to the case where the growth function is convex [42,43], whilst in our case, it is not (see Figure 12).

We therefore have to apply a different approach, namely the approach based on the comparison principle for PDEs [44]. We first notice that Equation (12) can be written as the following differential inequality:

$$\frac{\partial u(x,t)}{\partial t} \leq \epsilon u + \frac{au^2}{h^2+u^2} + D\frac{\partial^2 u}{\partial x^2}. \tag{18}$$

We now build an auxiliary function $\tilde{\mathcal{F}}(u)$ as follows:

$$\tilde{\mathcal{F}}(u) = \alpha u \quad \text{for} \quad 0 \leq u \leq u_A, \tag{19}$$

$$\tilde{\mathcal{F}}(u) = \epsilon u + \frac{au^2}{h^2+u^2} \quad \text{for} \quad u > u_A, \tag{20}$$

where α is the slope of the tangent line to the graph of the reaction term and u_A is the abscissa of the conjunction point (Figure 12). After some simple algebra, we obtain that:

$$\alpha = \frac{a}{2h} + \epsilon. \tag{21}$$

Obviously, $\tilde{\mathcal{F}}(u)$ is an upper bound for the reaction term in the right hand side of (18), so that the following inequality holds:

$$\frac{\partial u(x,t)}{\partial t} \leq \tilde{\mathcal{F}}(u) + D\frac{\partial^2 u}{\partial x^2}. \tag{22}$$

By the virtue of the comparison principle, the solution of the following equation:

$$\frac{\partial \tilde{u}(x,t)}{\partial t} = \tilde{\mathcal{F}}(\tilde{u}) + D\frac{\partial^2 \tilde{u}}{\partial x^2}. \tag{23}$$

gives an upper bound for the solution $u(x,t)$ of Equation (12) provided the initial and boundary conditions are the same.

Function $\tilde{\mathcal{F}}(u)$ is convex; therefore, the linearization at the leading edge can be applied resulting, taking (21) into account, in the following upper bound for the speed of the wave:

$$c = 2\sqrt{\alpha D} = 2\sqrt{\left(\frac{a}{2h} + \epsilon\right) D}. \tag{24}$$

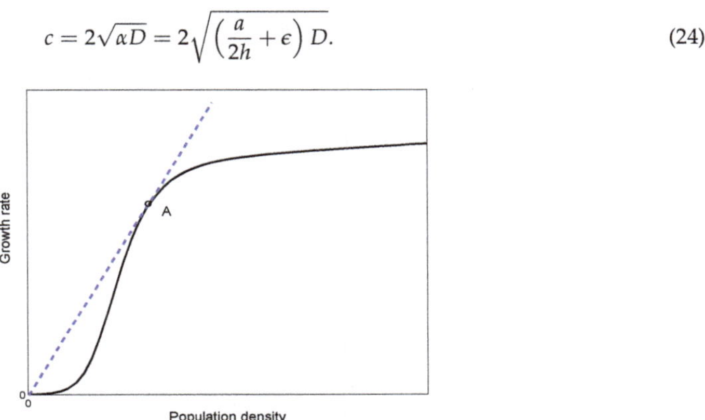

Figure 12. A sketch illustrating the idea of building an upper bound for the reaction term. The solid (black) curve shows the reaction term in the right hand side of (18) (or Equation (12) with $m \equiv 0$), and the dashed (blue) line shows the tangent line; see more details in the text.

Figure 13 shows the values of the speed calculated in simulations along with the analytical estimate (24). It is readily seen that, while the expression (24) somewhat overestimates the actual value, it manages to describes reasonably well the dependence of the speed on the parameter, i.e., the approximately linear increase along with an increase in ϵ.

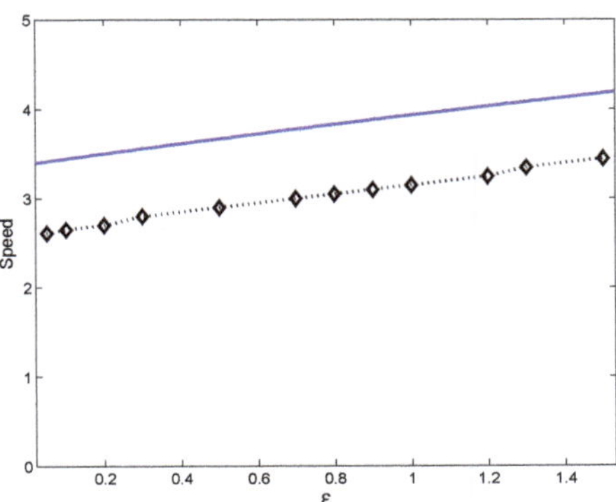

Figure 13. Dependence of the speed of the travelling peak on ϵ; diamonds show the numerical results, and the blue line shows the analytical estimate (24).

4. Discussion and Concluding Remarks

Social protests are a complex phenomenon that has a variety of important implications for the functioning of different social groups and organizations, as well as the society as a whole. Hence, a good understanding of the protests' dynamics is needed, in particular to identify the factors controlling their intensity and duration and to reveal the generic patterns of the protests' evolution in time and space. Whilst this is apparently a problem in the remit of social sciences, mathematical modelling has long been regarded as a powerful research tool that helps to achieve these goals [3,18].

In this paper, we considered a few generic, conceptual models of the social protests dynamics. We disregarded the social structure of the crowd and described the dynamics by a single dynamic variable, i.e., the number of protesters. Our study was partially motivated by the observation that there can be two apparently different scenarios of the protests' development in time: the one where the peak in the total number of protesters is followed by a fast decay and the other one where the decay in the wake of the main peak is very slow, thus creating a long tail. Correspondingly, our first goal was to understand whether these two apparently different cases can be embraced in the framework of the same model. We found that the two cases differed quantitatively rather than qualitatively, i.e., the different decay rate (and hence, the total duration of the protests) was determined by the model parameters. Interestingly, we obtained that the duration of the protests could, in principle, be indefinitely long, although the corresponding parameter range was small.

There have been several studies previously where riots and social unrest were treated in terms of nonlinear dynamical systems [2,11–13,17]. Adopting this modelling paradigm, the slow decay in the protests intensity could be linked to the so-called long transient dynamics (cf. [17]): a phenomenon inherent in many dynamical systems of different origin [30,31,45,46]. In a nutshell, a long transient means that the dynamical system mimics its stable asymptotical dynamics (e.g., steady state, periodical, or chaotic), being in fact in a transient regime. The quasi-asymptotical dynamics can persist for an indefinitely long time before the system experiences a fast transition to its true asymptotics. In a non-spatial system, the main mechanisms resulting in long transient are ghost attractors, crawl-bys, and slow-fast dynamics [30,31], and this is precisely what determines that long tail predicted by our models. The long tail obtained in the single species model (cf. Curves 1–3 in Figure 2b) was the effect of the ghost attractor that emerged after the upper stable steady state of the system disappeared in a saddle-node bifurcation when the final value of the withdrawal rate exceeded the critical value m^* (see Figure 1b). The long tail predicted by the two component model (e.g., see the green curve in Figure 4a) was a manifestation of the slow-fast dynamics that arose for small values of β.

We mention here that linking our findings to some general properties of the dynamical systems (such as the existence of long transient dynamics) is not a futile mathematical exercise, but brings insights that can have immediate practical value. In particular, the generic property of long transients that the slow decay in the protesters number is followed by a fast decay to zero may be used as a basis for a response strategy: if the authorities have enough patience to wait (or other reasons not to use excessive force in response to protests), it is likely that the protests eventually disappear due to their inherent dynamics (perhaps this is what is happening with the 2019 riots in Hong Kong where the authorities have been surprisingly mild in their response, probably just waiting until the riots burn out by themselves).

The second goal of this study was to highlight the role of the spatial dimension in the dynamics of social protests. We used mathematical modelling to reveal essential differences between the properties of the spatial systems and their non-spatial counterparts. The importance of space is well recognized in natural systems, in particular in ecology [22], but its effect on the social processes remains somewhat obscure. We obtained that both models predicted the emergence of a travelling wave of the protests' activity (albeit with rather different properties). This was in agreement with other recent modelling studies [13,14]; also, the spatial dynamics reminiscent of a spreading wave was observed in some data [13,47].

Our results suggested two essentially different scenarios of the protests' development in space, say, Type I and Type II. In the Type I scenario, the protesters are distributed over a certain area with approximately uniform density, this area initially growing with time. The growth of the area "invaded" by the protests is combined with a slow decrease in time (apart from a very early stage of the dynamics) of the protesters' population density uniformly over the whole invaded space, eventually resulting in the protests shrinking back to the place of their origin and collapsing (see Figure 5). This is a result of the global coupling where the withdrawal rate is the same everywhere in space: m depends on time, but not on space. In the Type II scenario, the distribution of protesters over space is highly heterogeneous. The large density of protesters is only observed inside a narrow range or small area, which in the course of time moves away from the place where the protests started (see Figure 8). The collapse of the protests occurs as a result of a specific "boundary forcing", i.e., when the peaks are pushed close to the domain boundary. This is a result of the interplay between the local coupling (where, generally speaking, the withdrawal rate is different at different locations) and the feedback mechanism through which the withdrawal rate depends on the protesters' density. One can further speculate as to how the difference between the two scenarios (and between the models, respectively) can be projected onto real life. Since the explicit dependence of the withdrawal rate on space may indicate the heterogeneity in the peoples' responses, one can suggest that the Type I scenario is more likely to occur in a well connected (hence global coupling), socially uniform society, while the Type II scenario is more likely to be seen in a poorly connected society with a broader range of social norms.

Note that both non-spatial models that we considered in Section 2 describe a very similar pattern in time, i.e., the main peak was followed by a tail, the rate of decay at the tail depending on the model parameters. Since different models were built to account for the effect of different factors, that evokes the old problem of the relation between the pattern and the process: if two different models predict the same pattern, how do we "let the right one in", i.e., what are the processes that are actually important? However, when the models are extended to include space, the dynamics become qualitatively different. Therefore, here, we showed that taking into account the spatial dimension of the system could be one possible way to solve this problem.

Our study leaves a few open questions. Firstly, in our models, we assumed that the "population" of protesters consisted of essentially identical individuals. In reality, people are different in terms of beliefs, responses, social status, etc. [15,18,27,48,49]. How the heterogeneity of the population can modify our findings remains unclear. In particular, the recruitment of protesters can be facilitated by activists, as was the case in the Russian revolutions in 1905–1907 and later in 1917 [1,19,20]. Currently, such activists are active Internet users and social engineers and may have even a stronger influence. Secondly, with regard to the spatial extension of the models, we restricted our analysis to the hypothetical 1D case. How the patterns of spread may change in the more realistic 2D system remains to be seen. Heterogeneity of space can also be an important factor affecting the dynamics of protests, as was shown in a number of works [9,12,13], and that may require a mathematical framework different from the standard Fickian diffusion, e.g., networks. Consideration of these factors should become a focus of future research.

Author Contributions: The authors contributed equally to this work. All authors have read and agreed to the published version of the manuscript.

Funding: The publication has been prepared with the support of the "RUDN University Program 5-100" (to S.P.).

Conflicts of Interest: The authors declare no conflict of interest.

References

1. Andreev, A.; Borodkin, L.; Levandovskii, M. Using methods of non-linear dynamics in historical social research: Application of chaos theory in the analysis of the worker's movement in pre-revolutionary Russia. *Hist. Soc. Res./Historische Sozialforschung* **1997**, 22, 64–83.

2. Epstein, J.M. *Nonlinear Dynamics, Mathematical Biology, and Social Science*; Addison-Wesley: Reading, MA, USA, 1997.
3. Epstein, J.M. Modeling civil violence: An agent-based computational approach. *Proc. Natl. Acad. Sci. USA* **2002**, *99* (Suppl. 3), 7243–7250. [CrossRef] [PubMed]
4. Turchin, P. *Historical Dynamics: Why States Rise and Fall*; Princeton University Press: Princeton, NJ, USA, 2003.
5. Eguiluz, V.M.; Zimmermann, M.G.; Cela-Conde, C.J.; San Miguel, M. Cooperation and emergence of role differentiation in the dynamics of social networks. *arXiv* **2006**, arXiv:physics/0602053v1.
6. Brantingham, P.J.; Tita, G.E.; Short, M.B.; Reid, S.E. The ecology of gang territorial boundaries. *Criminology* **2012**, *50*, 851–885. [CrossRef]
7. Fonoberova, M.; Fonoberov, V.A.; Mezic, I.; Mezic, J.; Brantingham, P.J. Nonlinear dynamics of crime and violence in urban settings. *J. Artif. Soc. Soc. Simul.* **2012**, *15*, 2. [CrossRef]
8. Smith, L.M.; Bertozzi, A.L.; Brantingham, P.J.; Tita, G.E.; Valasik, M. Adaptation of an ecological territorial model to street gang spatial patterns in Los Angeles. *Discret. Contin. Dyn. Syst.* **2012**, *32*, 3223–3244. [CrossRef]
9. Davies, T.P.; Fry, H.M.; Wilson, A.G.; Bishop, S.R. A mathematical model of the London riots and their policing. *Sci. Rep.* **2013**, *3*, 1303. [CrossRef]
10. Turalska MWest, B.J.; Grigolini, P. Role of committed minorities in times of crisis. *Sci. Rep.* **2013**, *3*, 1371. [CrossRef]
11. Khosaeva, Z.H. The mathematics model of protests. *Comp. Res. Model.* **2015**, *7*, 1331–1341. (In Russian) [CrossRef]
12. Berestycki, H.; Nadal, J.-P.; Rodíguez, N. A model of riots dynamics: Shocks, diffusion and thresholds. *Netw. Heterog. Media* **2015**, *103*, 443–475. [CrossRef]
13. Bonnasse-Gahot, L.; Berestycki, H.; Marie-Aude Depuiset, M.A. Epidemiological modelling of the 2005 French riots: A spreading wave and the role of contagion. *Sci. Rep.* **2018**, *8*, 107. [CrossRef] [PubMed]
14. Yang, C.; Rodriguez, N. A numerical perspective on traveling wave solutions in a system for rioting activity. *Appl. Math. Comput.* **2020**, *364*, 124646. [CrossRef]
15. Gavrilets, S. Collective action problem in heterogeneous groups. *Philos. Trans. R. Soc. Lond. B* **2015**, *370*, 20150016. [CrossRef] [PubMed]
16. Lang, J.C.; De Sterck, H. The Arab Spring: A simple compartmental model for the dynamics of a revolution. *Math. Soc. Sci.* **2014**, *69*, 12–21. [CrossRef]
17. Morozov, A.; Petrovskii, S.V.; Gavrilets, S. Dynamics of Social Protests: Case study of the Yellow Vest Movement. *SocArXiv* **2019**. [CrossRef]
18. Granovetter, M. Threshold models of collective behavior. *Am. J. Sociol.* **1978**, *83*, 1420–1443. [CrossRef]
19. Haynes, M.J. Patterns of conflict in the 1905 revolution. In *The Russian Revolution of 1905: Change through Struggle*; Glatter, P., Ed.; Porcupine Press: London, UK, 2005; pp. 215–233.
20. Perrie, M. The Russian peasant movement of 1905–1907: Its social composition and revolutionary significance. *Past Present* **1972**, *57*, 123–155. [CrossRef]
21. Gelbwestenbewegung. Wikipedia. Available online: https://de.wikipedia.org/wiki/Gelbwestenbewegung (accessed on 1 October 2019). (In German)
22. Malchow, H.; Petrovskii, S.V.; Venturino, E. *Spatiotemporal Patterns in Ecology and Epidemiology: Theory, Models, Simulations*; Chapman & Hall/CRC Press: Boca Raton, FL, USA, 2008.
23. Gnedenko, B.V.; Kolmogorov, A.N. *Limit Distributions for Sums of Independent Random Variables*; Addison-Wesley Mathematics Series; Addison-Wesley: Cambridge, MA, USA, 1954.
24. Altbach, P.G.; Klemencic, M. Student activism remains a potent force worldwide. *Int. High. Educ.* **2014**, *76*, 2–3. [CrossRef]
25. Mohamed, E.; Gerber, H.R.; Aboulkacem, S. (Eds.) *Education and the Arab Spring*; Sense Publishers: Rotterdam, The Netherlands; Boston, MA, USA; Taipei, Taiwan, 2016.
26. Biggs, M. Size matters: Quantifying protest by counting participants. *Sociol. Methods Res.* **2018**, *47*, 351–383. [CrossRef]
27. Raafat, R.M.; Chater, N.; Frith, C. Herding in humans. *Trends Cognit. Sci.* **2009**, *13*, 420–428. [CrossRef]
28. Courchamp, F.; Berek, L.; Gascoigne, J. *Allee Effects in Ecology and Conservation*; Oxford University Press: Oxford, UK, 2008.

29. Schussman, A.; Soule, S.A. Process and protest: Accounting for individual protest participation. *Soc. Forces* **2005**, *84*, 1083–1108. [CrossRef]
30. Hastings, A.; Abbott, K.C.; Cuddington, K.; Francis, T.; Gellner, G.; Lai, Y.C.; Morozov, A.; Petrovskii, S.; Scranton, K.; Zeeman, M.L. Transient phenomena in ecology. *Science* **2018**, *361*, eaat6412. [CrossRef] [PubMed]
31. Morozov, A.; Abbott, K.; Cuddington, K.; Francis, T.; Gellner, G.; Hastings, A.; Lai, Y.C.; Petrovskii, S.; Scranton, K.; Zeeman, M.L. Long transients in ecology: Theory and applications. *Phys. Life Rev.* **2019**, in press. [CrossRef]
32. Braha, D. Global civil unrest: Contagion, self-organization, and prediction. *PLoS ONE* **2012**, *7*, e48596. [CrossRef]
33. Brockmann, D.; Hufnagel, L.; Theo Geisel, T. The scaling laws of human travel. *Nature* **2006**, *439*, 462–465. [CrossRef]
34. Kendall, D.G. Discussion of "Measles periodicity and community size" by M.S. Bartlett. *J. R. Stat. Soc. Ser. A* **1957**, *120*, 64–67.
35. Mendez, V.; Fedotov, S.; Horsthemke, W. *Reaction-Transport Systems: Mesoscopic Foundations, Fronts, and Spatial Instabilities*; Springer: Berlin, Germany, 2010.
36. Mendez, V.; Campos, D.; Bartumeus, F. *Stochastic Foundations in Movement Ecology: Anomalous Diffusion, Front Propagation and Random Searches*; Springer: Berlin, Germany, 2014.
37. Lewis, M.A.; Petrovskii, S.V.; Potts, J. *The Mathematics behind Biological Invasions*; Interdisciplinary Applied Mathematics; Springer: Berlin, Germany, 2016; Volume 44.
38. Murray, J.D. *Mathematical Biology*; Springer: Berlin, Germany, 1989.
39. Volpert, A.; Volpert Vit Volpert, V.L. *Traveling Wave Solutions of Parabolic Systems*; AMS: Providence, RI, USA, 1994.
40. Fife, P.C. *Mathematical Aspects of Reacting and Diffusing Systems*; Lecture Notes in Biomathematics; Springer: Berlin, Germany, 1979; Volume 28.
41. Zeldovich, Y.B.; Barenblatt, G.I.; Librovich, V.B.; Makhviladze, G.M. *The Mathematical Theory of Combustion and Explosions*; Consultants Bureau: New York, NY, USA, 1985.
42. Kolmogorov, A.N.; Petrovskii, I.G.; Piskunov, N.S. A study of the diffusion equation with increase in the quantity of matter, and its application to a biological problem. *Bull. Moscow Univ. Math. Ser. A* **1937**, *1*, 1–25.
43. Kolmogorov, A.N.; Petrovskii, I.G.; Piskunov, N.S. A study of the diffusion equation with increase in the amount of substance, and its application to a biological problem. In *Selected Works of A. N. Kolmogorov*; Tikhomirov, V.M., Ed.; Kluwer Academic: Dordrecht, The Netherlands, 1991; pp. 242–270.
44. Protter, M.H.; Weinberger, H.F. *Maximum Principles in Differential Equations*; Springer: New York, NY, USA, 1984.
45. Hastings, A.; Higgins, K. Persistence of transients in spatially structured ecological models. *Science* **1994**, *263*, 1133–1136. [CrossRef]
46. Lai, Y.C.; Tél, T. *Transient Chaos—Complex Dynamics on Finite-Time Scales*; Springer: New York, NY, USA, 2011.
47. Map of White Supremacy Mob Violence. Available online: http://www.monroeworktoday.org/explore/ (accessed on 1 October 2019).
48. Homer-Dixona, T.; Maynardb, J.L.; Mildenbergerc, M.; Milkoreita, M.; Mocka, S.J.; Quilleyd, S.; Schröder, T.; Thagarde, P. A complex systems approach to the study of ideology: Cognitive-affective structures and the dynamics of belief systems. *J. Soc. Political Psychol.* **2013**, *1*, 337–363.
49. Mistry, D.; Zhang, Q.; Perra, N.; Baronchelli, A. Committed activists and the reshaping of the status-quo social consensus. *Phys. Rev. E* **2015**, *92*, 042805. [CrossRef]

MDPI
St. Alban-Anlage 66
4052 Basel
Switzerland
Tel. +41 61 683 77 34
Fax +41 61 302 89 18
www.mdpi.com

Mathematics Editorial Office
E-mail: mathematics@mdpi.com
www.mdpi.com/journal/mathematics

www.ingramcontent.com/pod-product-compliance
Lightning Source LLC
LaVergne TN
LVHW070057170726
843515LV00007B/1558